FIELD PRACTICE

DATA BOOK for CIVIL ENGINEERS

By ELWYN E. SEELYE

VOLUME ONE—DESIGN—
Second Edition

VOLUME TWO — SPECIFICATIONS AND
COSTS—Second Edition

VOLUME THREE — FIELD PRACTICE —
Second Edition

DATA BOOK FOR CIVIL ENGINEERS

FIELD PRACTICE

SECOND EDITION

The Late ELWYN E. SEELYE

Song of the Book

I travel with jeeps and soaring wings.
I'm sandwiched 'tween the augers and the sieves.

Apologies to Kipling

JOHN WILEY & SONS, Inc.

New York • Chichester • Brisbane • Toronto

SECOND EDITION

16

ISBN 0 471 77352 2
PRINTED IN THE UNITED STATES OF AMERICA

PREFACE

Field practice embraces the inspection and sometimes supervision of construction of engineering works by a field man who may have the background of an inspector, a designer, a clerk-of-the-works, a contractor's superintendent, or a surveyor. If the inspection and supervision are performed in accordance with modern practice, the field man merits the dignity that is implied by the title of engineer.

Modern practice for field engineers comprises extensive technological advances, many of them made within the past decade. The purpose of this volume is to enable the inspector or field engineer to brief himself as to the essentials in the inspection and supervision of the work which he is to undertake. Its purpose is also to enable him to bring to the field the basic data which he will require.

For example, sampling of material for laboratory tests should be done in accordance with certain rules. The method of taking a concrete sample for a compression or flexure test is rigidly prescribed. Any deviation from the rules will detract from the validity of the test. Hence "Rules for Sampling" are included in this book.

Certain field tests, such as the concrete slump test, the penetration of asphalt test, and soil tests, are required to control the quality of construction. These tests should be performed according to certain rules; hence, "Instructions for Field Tests."

Field engineering includes the checking of material so that size, quality, and other properties are in accordance with plans and specifications. Therefore, tables such as the detailed dimensions of steel beams and of culvert pipe are included herein to enable an inspector to identify the exact size of a steel beam or the classification of a reinforced-concrete pipe.

A whole series of special tests have been developed in connection with the science of soil mechanics. A field engineer may be required to make these tests and to furnish information concerning them. In order that he may do so, detailed information is given to determine density, grain size, Atterberg limits, optimum moisture, field shear tests, C.B.R. values, and related data.

What items should be checked by an inspector? A check list for inspectors is included for such work as concrete, bituminous paving, steel, welding, and timber. Complete information for inspecting pile driving is also given. In addition, report forms are presented so arranged that the report becomes not only a progress report but also an inspector's checking list. This is illustrated by the steel inspector's reports, of which Part I is a list of items to be checked off by the inspector and Part II is a progress report.

The importance of surveying to field engineering has been recognized, and a section of this volume provides the data a construction surveyor requires. Under "Surveying" are stadia reduction tables, stakeout problems, curve data, railroad turnout data, earthwork tables, transit and level problems, azimuth determination, isogonic chart, instrument adjustments, tape data, plotting problems, mapping symbols, and tables of measure, trigonometric formulas, and trigonometric functions.

The identification of common building stone and timber is assisted by photographs of different types or species placed in juxtaposition to emphasize points of difference.

A few words on job power together with cuts of construction machinery are given to assist the field engineer in talking to the contractor in his own language.

In this second edition, all tables and other data have been brought up to date and new material has been added. The section on air-entrainment has been greatly extended and enough theory of the subject included to enable the engineer or inspector to use it intelligently.

The principal types of piles are illustrated, the uses and characteristics of each explained, and Red Lights are pointed out for the inspector's guidance.

Additional new material includes the following:

> *Additional heavy equipment.*
> *Practical methods of erection.*
> *Additional concrete data.*
>> Correct and incorrect methods of placing concrete and storing materials.
>> Tests for organic impurities.
>> Table for removal of forms.
>> Gradation limits for aggregates.
>> New tables—concrete mixes and characteristics.
> *Masonry and brickwork.*
>> Common types of cracking in walls.
>> Salmon brick.
>> Bonds and types of brick and stonework.
>> Predeflection of new lintel in old walls.
> *Metals.*
>> Corrosion.
>> Identification by fractures and spark tests.
> *Timber.*
>> Identification of additional wood species.
>> Prevention of shrinkage cracks.

Soil.

    Soil mechanics—a brief course.

        Soil tests.

        Use and treatment P.R.A. groups.

        Casagrande classification.

        Constants for pavement design.

        Consolidation by sand drains.

        Boring inspection check list.

Safety provisions for the inspector.

Airports—list of tests.

                                  ELWYN E. SEELYE

ACKNOWLEDGMENTS—FIRST EDITION

In addition to the sources listed in the text, the author wishes to thank the following for their aid.

Dr. Arthur S. Tuttle, for his help and advice and for making available to the author his wide contacts for the collection of information.

Dean S. C. Hollister, of the College of Engineering, Cornell University, who has given the author advice and encouragement, particularly in regard to the general scope of the book.

Mr. Frank H. Alcott, of National Lumber Manufacturers Association.

Mr. L. E. Andrews, Regional Highway Engineer, Portland Cement Association.

Mr. Clinton L. Bogert.

Mr. Gilbert D. Fish.

Mr. E. M. Fleming, District Manager, Portland Cement Association.

Mr. Bedrich Fruhauf.

Mr. Elliott Haller, of Haller Testing Laboratories, Inc.

Mr. John Hoffhine.

Mr. Prevost Hubbard and Mr. Bernard E. Gray, of The Asphalt Institute.

Mr. Elson T. Killam.

Mr. Ralph H. Mann, of American Wood-Preservers Association.

Mr. Edward A. Robinson, of Haller Testing Laboratories, Inc.

Mr. John M. Stratton.

Mr. Russell Wise, C.E.

The author wishes to make special mention of the assistance received from the following members of his organization.

Mr. J. U. Wiesendanger for his work on general editing.

Colonel Burnside R. Value and Mr. A. L. Stevenson for general advice and counsel.

Messrs. E. G. Whitten, Lt. Comdr. W. D. Bailey, V. J. Pacello, C. M. Throop, Patrick H. Murphy, A. H. Jorgensen, Laurence Vought, and Lt. S. D. Teetor.

Miss Jean M. Luckett for her assistance in editing.

ACKNOWLEDGMENTS—SECOND EDITION

In addition to the sources listed in the text, the author wishes to thank the following for their aid.

Mr. L. E. Andrews, Regional Highway Engineer, Portland Cement Association.

Mr. Donald V. Buttenheim and Mr. Albert C. Smith, of the Buttenheim-Dix Publishing Company.

Mr. S. A. Greenberg, Technical Secretary, American Welding Society.

Mr. D. G. Joy, American Steel and Wire Company.

Mr. Leo Kamp, Chemist-in-Charge of Laboratory, Office of Borough President of Queens, City of New York.

Mr. Martin J. O'Reilly, Director of Engineering Services, Department of Public Works, City of New York.

Mr. John A. Procaccino, New York City Housing Authority.

Mr. Allan S. Platt, Director, Structural Clay Products Institute.

Mr. J. M. Sprague, Clay Sewer Pipe Association.

Mr. Carl J. Stenz, New York Trap Rock Corporation.

Mr. L. N. Whitcraft, Wait Associates.

Mr. U. Sena Zanetti, Poirier & McLane Corporation.

The author wishes to make special mention of the assistance received from the following members of his organization:

Mr. C. C. Hurlbut, Mr. F. C. Zeigler, and Mr. P. P. Page, Jr., for work on general editing.

Messrs. G. D. Fish, A. H. Jorgensen, E. A. Robinson, and M. H. Whitnall.

Miss Jean M. Luckett for her assistance in editing.

Mrs. Lucille T. Burnham, Mrs. Grace M. Carlock, and Miss Rita Larsen for secretarial assistance.

CONTENTS

PART I

INSPECTION

PAGE

PART I

INSPECTION

TYPICAL HEAVY CONSTRUCTION EQUIPMENT

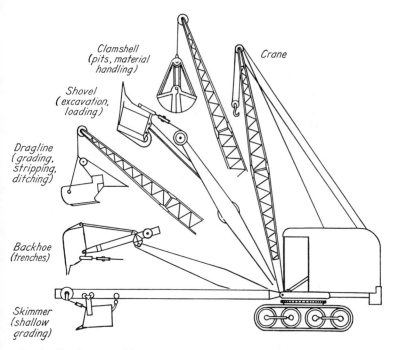

Clamshell
(pits, material
handling)

Crane

Shovel
(excavation,
loading)

Dragline
(grading,
stripping,
ditching)

Backhoe
(trenches)

Skimmer
(shallow
grading)

FIG. 1.   Lorain crane with attachments.   *Courtesy of the Thew Shovel Company.*

FIG. 2.   Four-wheel scraper.   (Earth-moving, grading, excavation.)   *Courtesy of Bucyrus-Erie Company.*

3

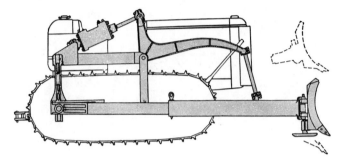

Fɪɢ. 3. Bulldozer. (Clearing, stripping, grading, earth-moving.) *Courtesy of Bucyrus-Erie Company.*

Fɪɢ. 4. Motor grader "motor patrol." (Shaping subgrades and surfaces, soil mixing.) *Courtesy of the Galion Iron Works and Manufacturing Company.*

FIG. 5. Tamping roller "sheepsfoot." (Compacting fills.) *Courtesy of the Baker Manufacturing Company.*

Fig. 6. Eight-ton three-wheel roller. *Courtesy of Huber Manufacturing Company.*

FIG. 7. Five- to eight-ton tandem roller. *Courtesy of Huber Manufacturing Company.*

FIG. 8. Pulvi-Mix. (Mixing earth and stabilizing agents—pulverizing.) *Courtesy of Seaman Motors.*

Fig. 9. Trencher. (Trench excavation in earth.) *Courtesy of the Parsons Company.*

Fig. 10. Concrete paver. *Courtesy of Ransome Machinery Company.*

Fig. 11. Rex transit-mix truck. *Courtesy of Chain Belt Company.*

Fig. 12. Aggregate batching plant. Bin division must be high enough to prevent mixing of materials. *Courtesy of Blaw-Knox Company.*

Fɪɢ. 13. Finishing machine for roads and airports. *Courtesy of Blaw-Knox Company.*

Fɪɢ. 14. Compressor. *Courtesy of the Jaeger Machine Company.*

Fig. 15. Grading, concreting, finishing operation. (Concrete paving.) *Courtesy of Blaw-Knox Company.*

Fig. 16. Barber-Greene Tamping leveling and finishing machine. (Bituminous pavements.)

Fig. 17. Bros Straight Wheel giant model roller and Allis-Chalmers tractor. (Compaction of fills and grades.) *Courtesy Contractors and Engineers Monthly.*

TYPICAL SMALL CONSTRUCTION EQUIPMENT

FIG. 18.   Clipper concrete saw.   (Cutting contraction joints in pavement.)   *Courtesy Ray Day Photo.*

Fɪɢ. 19.   Triplex pneumatic backfill tamper.   *Courtesy Chicago Pneumatic Tool Co.*

Fig. 20.   Concrete vibrator.   *Courtesy of Syntron Power Tool Co.*

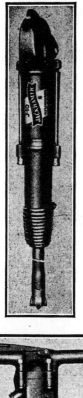

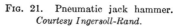

Fig. 21. Pneumatic jack hammer.
Courtesy Ingersoll-Rand.

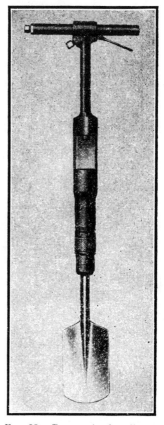

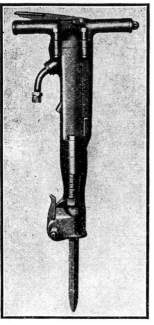

Fig. 22. Pneumatic paving
breaker. *Courtesy Ingersoll-Rand.*

Fig. 23. Pneumatic clay digger.
Courtesy Ingersoll-Rand.

Fig. 24.   Conversion head for pneumatic sheeting hammers.   *Courtesy Ingersoll-Rand.*

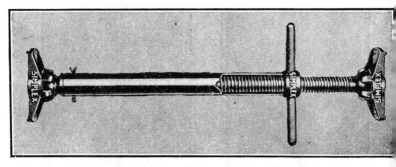

Fig. 25.   Simplex trench brace.   (Bracing trench walls.)   *Courtesy Templeton Kenly & Co.*

FIG. 26. Salamander with firing gun. (To prevent freezing of concrete.) *Courtesy Hauck Manufacturing Co.*

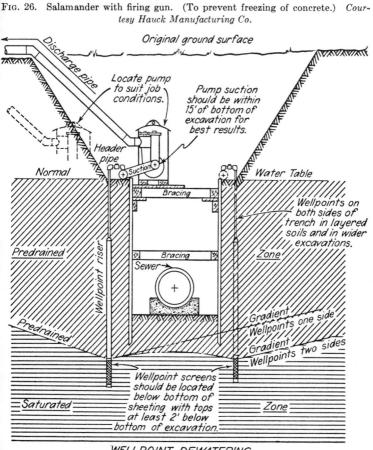

WELLPOINT DEWATERING
SHEETED EXCAVATION

FIG. 27. Wellpoint dewatering sheeted excavation. *Courtesy Moretrench Corporation.*

ERECTION METHODS

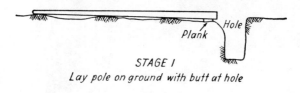

STAGE 1
Lay pole on ground with butt at hole

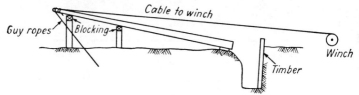

STAGE 2
Raise end of pole with jacks, derrick or other hoist and block up.
Prevent side sway with guy ropes

STAGE 3
Draw pole up with cable until it slides into hole against timber
Draw up vertical with cable and guy ropes

A guyed gin pole for use as a derrick can be erected in the
same way, but seated on an anchored timber base instead
of a hole in the ground

Fig. 28.   Method of erecting a heavy pole.

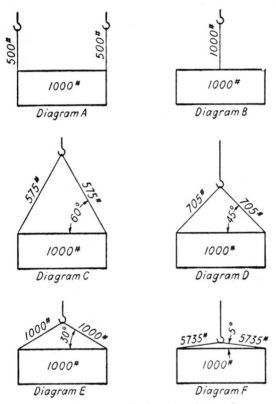

FIG. 29. Stresses on sling legs at various angles.

Note: Figures 29 to 36 inclusive are from *Technical Manual 5-225*, by courtesy of the War Department.

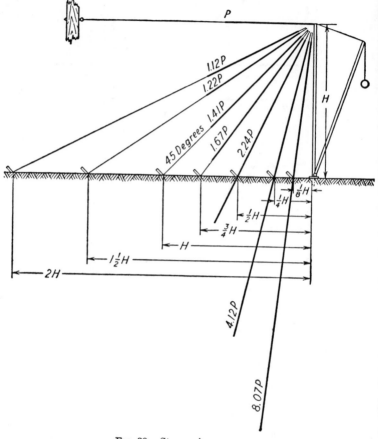

FIG. 30.   Stresses in guy ropes.

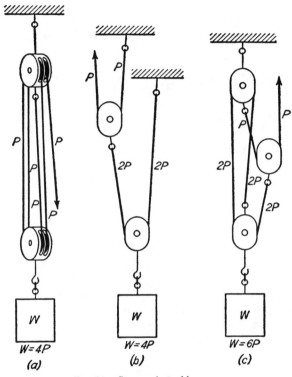

Fig. 31. Stresses in tackle ropes.

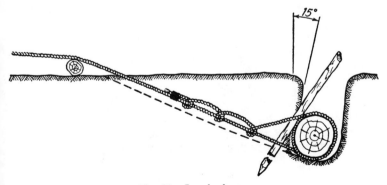

Fig. 32. Log deadman.

Fig. 33.   Standard hold-fast (steel pickets).

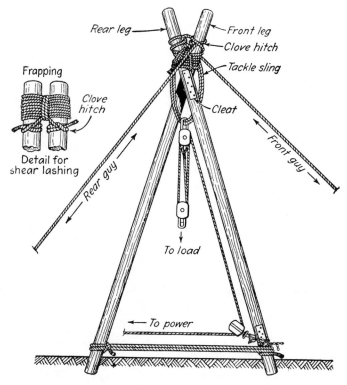

Fig. 34.   Shears.

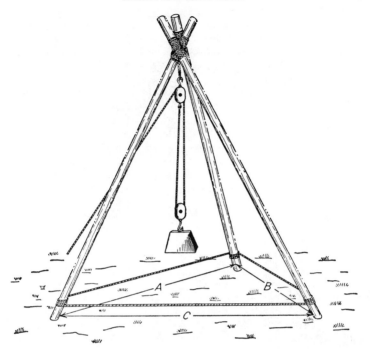

Fig. 35. Tripod. Spread *A*, *B*, and *C* are equidistant, not less than one-half nor more than two-thirds height of cross.

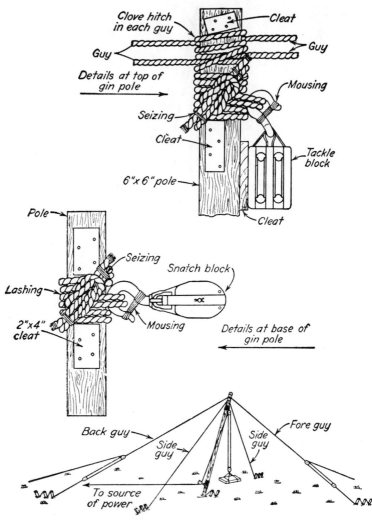

FIG. 36. Gin pole.

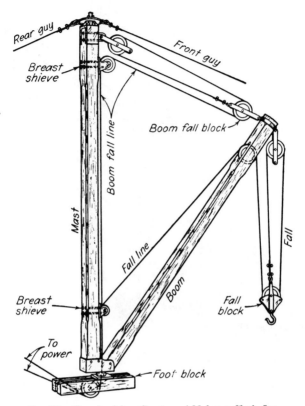

FIG. 37. Guy derrick. *Courtesy of Mahoney-Clark, Inc.*

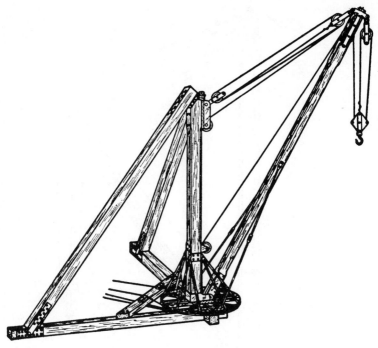

FIG. 38.   Stiff-leg derrick.   *Courtesy of Mahoney-Clark, Inc.*

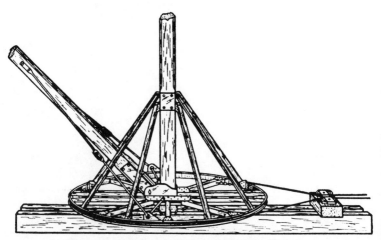

FIG. 39.   Bullwheel.   *Courtesy of Mahoney-Clark, Inc.*

MATERIAL HANDLING

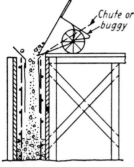

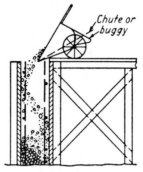

CORRECT

Separation is avoided by discharging concrete into hopper
feeding into drop chute. This
arrangement also keeps forms
and steel clean until concrete
covers them.

INCORRECT

Permitting concrete from chute
or buggy to strike against form
and ricochet on bars and form
faces causes separation and
honeycomb at the bottom.

PLACING IN TOP OF NARROW FORM

CORRECT

Start placing at bottom of
slope so that compaction is
increased by weight of newly
added concrete. Vibration
consolidates the concrete.

INCORRECT

When placing is begun at top
of slope the upper concrete
tends to pull apart especially
when vibrated below as this
starts flow and removes
support from concrete above.

*WHEN CONCRETE MUST BE
PLACED IN A SLOPING LIFT*

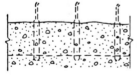

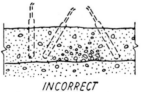

CORRECT

Vertical penetration of vibrator
a few inches into previous lift
(which should not yet be rigid)
at systematic regular intervals
will give adequate consolidation.

INCORRECT

Haphazard random penetration
of the vibrator at all angles and
spacings without sufficient depth
will not assure intimate combination
of the two layers.

SYSTEMATIC VIBRATION OF EACH NEW LIFT

FIG. 40.   Placing concrete.   *Adapted from American Concrete Institute*
(ACI 614-42).

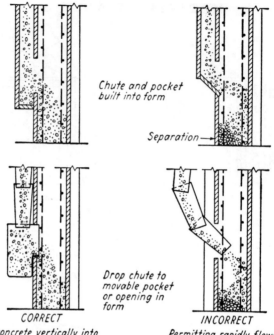

Chute and pocket built into form

Separation →

Drop chute to movable pocket or opening in form

CORRECT

Drop concrete vertically into outside pocket under each form opening so as to let concrete stop and flow easily over into form without separation.

INCORRECT

Permitting rapidly flowing. concrete to enter forms on an angle invariably results in separation.

PLACING IN DEEP NARROW WALL THROUGH PORT IN FORM

CORRECT
Concrete should be dumped into face of previously placed concrete.

INCORRECT
Dumping concrete away from previously placed concrete causes separation.

PLACING SLAB CONCRETE FROM BUGGIES

Fig. 41.   Placing concrete.   *Adapted from ACI 614-42.*

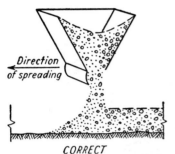

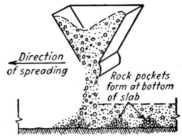

CORRECT
Bucket should be turned so that separated rock falls on concrete where it may be readily worked into mass.

INCORRECT
Dumping so that free rock falls out on forms or subgrade results in rock pockets.

IF SEPARATION HAS NOT BEEN ELIMINATED IN FILLING PLACING BUCKETS
(A temporary expedient until correction has been made)

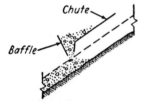

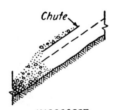

CORRECT
A baffle and drop at end of chute will avoid separation and concrete remains on slope.

INCORRECT
Discharging concrete from free end chute onto a slope causes separation of rock which goes to bottom of slope. Velocity tends to carry concrete down the slope.

PLACING CONCRETE ON A SLOPING SURFACE
From design and control of concrete mixtures by Portland Cement Association

Fig. 42.   Placing concrete.   *Adapted from ACI 614-42.*

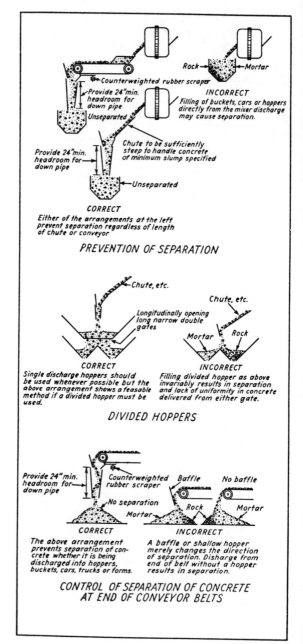

Rock — Mortar

Counterweighted rubber scraper

*Provide 24″min.
headroom for
down pipe*

Unseparated

INCORRECT
*Filling of buckets, cars or hoppers
directly from the mixer discharge
may cause separation.*

*Chute to be sufficiently
steep to handle concrete
of minimum slump specified*

*Provide 24″min.
headroom for
down pipe*

Unseparated

CORRECT
*Either of the arrangements at the left
prevent separation regardless of length
of chute or conveyor*

PREVENTION OF SEPARATION

Chute, etc.

*Longitudinally opening
long narrow double
gates*

Chute, etc.

Mortar — Rock

CORRECT
*Single discharge hoppers should
be used whenever possible but the
above arrangement shows a feasible
method if a divided hopper must be
used.*

INCORRECT
*Filling divided hopper as above
invariably results in separation
and lack of uniformity in concrete
delivered from either gate.*

DIVIDED HOPPERS

*Provide 24″min.
headroom for
down pipe*

*Counterweighted
rubber scraper*

Baffle — No baffle

No separation
Mortar — Rock — Mortar

CORRECT
*The above arrangement
prevents separation of concrete whether it is being
discharged into hoppers,
buckets, cars, trucks or forms.*

INCORRECT
*A baffle or shallow hopper
merely changes the direction
of separation. Discharge from
end of belt without a hopper
results in separation.*

CONTROL OF SEPARATION OF CONCRETE
AT END OF CONVEYOR BELTS

FIG. 43. Placing concrete. *From*

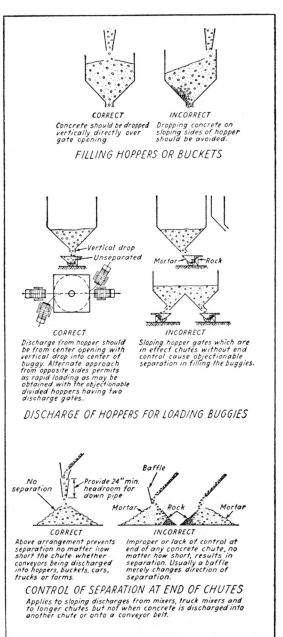

CORRECT

Concrete should be dropped vertically directly over gate opening.

INCORRECT

Dropping concrete on sloping sides of hopper should be avoided.

FILLING HOPPERS OR BUCKETS

Vertical drop
Unseparated

Mortar — *Rock*

CORRECT

Discharge from hopper should be from center opening with vertical drop into center of buggy. Alternate approach from opposite sides permits as rapid loading as may be obtained with the objectionable divided hoppers having two discharge gates.

INCORRECT

Sloping hopper gates which are in effect chutes without end control cause objectionable separation in filling the buggies.

DISCHARGE OF HOPPERS FOR LOADING BUGGIES

No separation

Provide 24" min. headroom for down pipe

Baffle

Mortar — *Rock* — *Mortar*

CORRECT

Above arrangement prevents separation no matter how short the chute whether conveyors being discharged into hoppers, buckets, cars, trucks or forms.

INCORRECT

Improper or lack of control at end of any concrete chute, no matter how short, results in separation. Usually a baffle merely changes direction of separation.

CONTROL OF SEPARATION AT END OF CHUTES

Applies to sloping discharges from mixers, truck mixers and to longer chutes but not when concrete is discharged into another chute or onto a conveyor belt.

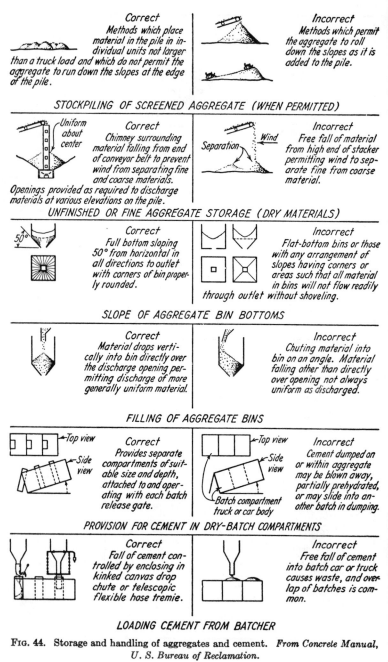

Correct
Methods which place material in the pile in individual units not larger than a truck load and which do not permit the aggregate to run down the slopes at the edge of the pile.

Incorrect
Methods which permit the aggregate to roll down the slopes as it is added to the pile.

STOCKPILING OF SCREENED AGGREGATE (WHEN PERMITTED)

Correct
Chimney surrounding material falling from end of conveyor belt to prevent wind from separating fine and coarse materials. Openings provided as required to discharge materials at various elevations on the pile.

Uniform about center

Incorrect
Free fall of material from high end of stacker permitting wind to separate fine from coarse material.

UNFINISHED OR FINE AGGREGATE STORAGE (DRY MATERIALS)

Correct
Full bottom sloping 50° from horizontal in all directions to outlet with corners of bin properly rounded.

Incorrect
Flat-bottom bins or those with any arrangement of slopes having corners or areas such that all material in bins will not flow readily through outlet without shoveling.

SLOPE OF AGGREGATE BIN BOTTOMS

Correct
Material drops vertically into bin directly over the discharge opening permitting discharge of more generally uniform material.

Incorrect
Chuting material into bin on an angle. Material falling other than directly over opening not always uniform as discharged.

FILLING OF AGGREGATE BINS

Correct
Provides separate compartments of suitable size and depth, attached to and operating with each batch release gate.

Top view
Side view

Incorrect
Cement dumped on or within aggregate may be blown away, partially prehydrated, or may slide into another batch in dumping.

Top view
Side view
Batch compartment truck or car body

PROVISION FOR CEMENT IN DRY-BATCH COMPARTMENTS

Correct
Fall of cement controlled by enclosing in kinked canvas drop chute or telescopic flexible hose tremie.

Incorrect
Free fall of cement into batch car or truck causes waste, and overlap of batches is common.

LOADING CEMENT FROM BATCHER

FIG. 44. Storage and handling of aggregates and cement. *From Concrete Manual, U. S. Bureau of Reclamation.*

SAFETY PROVISIONS

To be looked out for by the inspector

The inspector should insist that the contractor provide adequate and accessible scaffolding when riveted or welded work is to be inspected, and safe and stable means of access between stories or different levels.

The safety of the men or work is primarily the contractor's responsibility, but the inspector should see that the contractor follows safe practices, including: not overloading steel or concrete framing with construction loads; allowing sufficient time before stripping forms; adequate bracing for sheet piling; safe scaffoldings and runways; barricades around openings; adequate guys or braces for wind during construction; shoring of rock faces to prevent shearing failures or caves.

In underpinning work, a stage plan showing the sequence of operations should be provided, which the inspector should check and see carried out.

The inspector should be on the lookout for operations tending to undermine or threaten adjacent foundations, or permitting pumping water or drainage to draw sand from adjacent footings.

In making load tests on floor slabs or beams, the inspector should see that shores are provided or other provisions made, so that a failure of the slab being tested will not involve a failure of the slab below it or of other parts of the structure. *All load tests should be conducted on the assumption that, if the test area should fail, no one would be injured.*

The inspector should take no chances with ice in freezing weather.

Engineers should carry public liability and property damage insurance against injury caused by their inspectors or representatives.

CONCRETE—GENERAL

CHECK LIST FOR INSPECTORS

Inspectors' Equipment

Complete set of plans and specifications and approved set of reinforced-concrete working drawings.

Supply of required forms, sample tags, bags and boxes for samples.

Balance, capacity 2 kg., sensitive to 0.1 gram.

Set of square-mesh sieves of specified aggregate sizes and cleaning brush.

Fruit jar pycnometer or hot plate and pan for moisture content of aggregates.

12-oz. graduate bottle and 1 lb. of sodium hydroxide (caustic soda) for colorimetric test.

Pint milk bottle for silt and clay test.

6 in. by 12 in. metal or paraffined cardboard molds for concrete test cylinders, shipping boxes for same, and scoop for filling.

Slump cone, ⅝ in. by 24 in. tamping rod, and mason's trowel.

⅓ or ½ cu. ft. calibrated bucket and scale for unit weight tests, when specified.

Thermometer similar to Weston All-Metal type, 0 to 180° F. for cold-weather concreting.

6-ft. rule and 50-ft. steel tape.

Plumb bob and marking keel.

Field book and pencils for records and diary.

Pressure meter for measuring air.

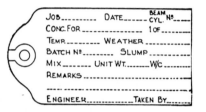

Job.......... Date...... BEAM N°....
                         CYL.
Conc. For .... 1 of........
Temp.......... Weather..........
Batch N°.......... Slump..........
Mix........ Unit Wt........ W/c........
Remarks...........................
----...............................
Engineer..............Taken By......

Fig. 45. Cloth tag for attaching to concrete test beams or cylinders.

Procedure in Inspection

Tested and Approved Materials. Cement, aggregates, reinforcing steel, and water tested and source approved before use.

Schedule of required field tests adhered to.

Prompt shipment of samples of materials delivered at site.

Prompt reporting of field tests.

Accurate and complete daily reports and records.

Removal of rejected materials from site of work.

Storage and Handling of Materials. Aggregates stockpiled in 2-ft. to 4-ft. layers on mats or planking.

Aggregate segregation avoided; see Fig. 44.

Cement protected from moisture and weather.

Cement handled to avoid loss by blowing or leakage, see Fig. 44.

Reinforcing steel protected from rusting, bending, or distortion and kept free from oil or grease.

Batch Plant Inspection

Batching Plant. Inspected and approved before use.

Daily check of weighing scales, accurate to tolerance of 0.004.

Use ten 50-lb. weights, check in 500-lb. increments to greatest batch weight *or* have scales checked and sealed by certified scale master.

Adequate visibility of weighing and batching.

Telltale dial or balance indicator for correct quantities in hoppers.

Positive shut-off for bulk cement.

Prompt removal of excess material in hoppers.

Protection for weighing equipment from dust or damage.

Oscillating beams normally horizontal with equal play.

Beam scale for each aggregate usually required.

Control of Concrete. Determine percentage of surface moisture in aggregates; also gradation.

Check at least 3 times daily, or more often when slump of concrete or condition of aggregate changes. Reject any segregated concrete.

Translate the design into batch weights, see pp. 57–61.

Run trial batch to check on slump and unit weight of mixture.

Check on cement factor during operations to detect bulking due to voids, air entrainment, or batching inaccuracies.

Adjust batch weights to produce required cement content per cubic yard and yield of concrete per batch.

Check actual amount of cement used to concrete laid each day as check on dimensions of concrete and accuracy of batching.

Note. The inspector should not vary the approved design mix without authority from the project or resident engineer.

Transporting Materials. Record of batch weights and number of batches dispatched; check with mixer inspector daily.

Tight truck partitions high enough to prevent intermingling of aggregates and loss of cement. Separate cement partitions, when specified.

Required amount of cement placed in batch partitions.

Covers for batch trucks provided.

Cement carried in sacks if specified.

Field Inspection. Drawings approved and up to date.

Forms. Check surfaces of forms against specifications. Where surfaces are in view or exposed to weather, form ties shall be no closer to the surface than 1 in., and no wire or snap ties shall be allowed because of eventual rust stain. Check correct alignment and elevation.

Centering true and rigid with horizontal and diagonal bracing.

Tight enough to prevent mortar leakage.

Columns plumb, true, and cross braced.

Floor and beam centering crowned ¼ in. per 16 ft. of span.

Beveled chamfer strips at angles and corners when specified.

Inside of forms oiled or wetted. Oil applied before placing of reinforcing.

Check installation of bolts, sleeves, inserts, and embedded items against plan details. These items must not infringe on structural items.

Check cleaning and removal of debris through temporary openings.

Check slab depths, beam and column sizes.

Provide temporary relief of hydrostatic pressure on walls and slabs by relief holes, drainage, or other means, and maintain relief until engineer in charge approves otherwise.

Removal of Forms and Shoring. Record of date forms poured and date forms removed. See Table 1.

Forms not removed until concrete is set, should ring under a hammer blow; follow job specifications.

Reshores placed after forms removed—generally reshore for 2 floors. Use 4 x 4 braced at middle or 6 x 6 unbraced for conventional heights. If in doubt, check with superiors.

Inspect surface at once after form removal. Notify superior of serious defects. Follow specifications for removal of defective concrete.

Removal of Forms.* No safe rules can be given for the time of removal of forms unless cylinders are taken and tested. The suggestions in Table 1 are subject to the approval of the engineer in each case.

TABLE 1

Temperature (Fahrenheit)

	Over 95° †	70°–95°	60°–70°	50°–60°	Below 50°
Walls	5 days	1½ days	2 days	3 days	Do not remove until
Columns	7 days	2 days	3 days	4 days	site-cured test cyl-
Beams	10 days	4 days	5 days	6 days	inder develops
Slabs	10 days	5 days	6 days	7 days	50% of 28-day
					strength

Remove forms carefully and avoid damage to green concrete.

Repair surface defects immediately upon removal of forms. Avoid feather edges in making patches. Fill bolt holes completely. Cut back tie wires where permitted an inch from surface and patch.

Patching. All tie holes, voids, stone pockets, or other minor defective areas shall be patched on removal of forms. Defective areas shall be chipped away with all edges perpendicular to the surface. Patched and adjoining areas shall be kept moist for several hours before patching. Patches on exposed work shall be carefully matched to adjoining work by the addition of white cement if necessary. All patches shall be cured in the same manner as adjoining surfaces.

Reinforcing Steel. Clean and free of scale, oil, and defects. Can be rubbed down with burlap sacks or wire brushes.

---

* New York City Housing Authority.

† Where exposed surfaces of concrete can be effectively sealed to prevent loss of water, these times may be reduced to the 70°–95° times.

Accurately fabricated to plan dimensions.

Supports rigid, metal preferable; do not allow use of rocks, brickbats, old concrete fragments, etc., to support steel.

Check minimum clear spacing between bars; $1\frac{1}{2}$ diameters for round bars and 2 times side dimension for square bars.

2-in. cover for steel in exposed exterior surfaces or as specified or detailed.

Check, from approved working drawings, the quantity, size, placing, bending, splicing, and location of reinforcing.

Check prebent steel against bending schedule upon delivery.

Mixing Concrete. Mixer in good condition and kept clean of hardened concrete.

Mixer blades not worn, and drum watertight.

Check drum speed, usually 200 to 225 peripheral feet per minute.

Check mixing time frequently; should be $1\frac{1}{2}$ minutes minimum.

No retempering of concrete. Mixer completely emptied before starting new batch.

Adherence to specified water content. Amount of mix water based on moisture content of aggregates obtained from batch plant inspector and correct amount added at mixer.

Check consistency; make slump test at least 2 or 3 times daily.

Check for full cement content in each batch if cement is batched at mixer.

Ready-Mixed Concrete, Transit Mixers. Strict adherence to job specifications.

Calibration of water-discharge mechanism plainly marked.

Error in water measurement should not exceed 1%.

No leakage in valves; should be tight when closed.

Drums should be watertight. Check specified revolutions, usually 50 to 150 allowed for mixing.

Number, arrangement, and dimensions of mixer blades checked against manufacturer's statement. Blades not worn more than 15% of stated width.

Main water tank provided against loss by leakage or surging. To discharge full volume for mixing in not more than 5 minutes.

Volume of concrete mixed not more than 58% gross volume of drum. (If concrete is central mixed and only transported in truck mixers, 80% of volume is usually allowed.)

All truck mixers inspected and approved.

Complete removal of wash water or remaining concrete after each mixer discharge.

Wash water transported in auxiliary tank with gage and watertight valve.

Adherence to specified mixing time and any restrictions on mixing en route.

Drum to be revolved during transfer of water into drum.

Adherence to correct amount of water. Inspector should approve adding additional water. If necessary to add water to discharge, dry cement should be added at required W/C ratio.

Concrete containing air-entraining agent not to be mixed en route.

For transit trucks, the concrete shall be mixed not less than 50 revolutions of the drum at the manufacturer's specified speed.

Placing of Concrete. Forms inspected and approved before concreting. Steel reinforcing in place and inspected.

Earth under footings to be undisturbed, original soil.

Rock or ledge should be well cleaned off and washed; dirt and loose rock fragments should be removed. Sloping rock stepped or benched for level bearing.

Concrete shall not be placed in running water. Accumulated water to be removed where possible; where impossible, concrete to be placed by tremie as directed by engineer.

Segregation, rehandling, or flowing to be avoided.

Check concrete especially when placing in subsurface structures to avoid segregation of aggregates, excessive water, accumulation of laitance, batches over or under sanded. (Sometimes caused by sand spilling over the bin barrier into the stone, or vice versa.)

Each unit to be placed continuously between construction joints as approved by engineer.

Spading and vibrating to maximum subsidence without segregation and next to forms and joints. Avoid excessive internal vibration. Internal vibrators should operate at a speed of not less than 5000 vibrations per minute.

Reinforcing bars shaken to insure bond ,with concrete, but excessive vibration and manipulation to be avoided.

In thin high sections avoid having concrete stick and harden on steel and forms above placing level.

Mold required number of test cylinders each day.

See that wood form spreaders are removed and not buried as concrete is placed.

Concrete placed as close to final position as possible in continuous horizontal layers.

Embedments. Pipe, conduit, and castings in concrete shall be embedded at least 3 in., except, in thin floor slabs, place on top of lower reinforcement. See that inserts and other fixtures are set as called for and that pipe sleeves on different floors are so aligned as not to require pipe offsets. Excessive size or number of embedments should be reported.

Construction Joints. Avoid if possible, except as detailed on plans.

If necessary at end of day's pour, install plumb, at right angles to plane of stress and in area of minimum shear.

When concreting is resumed at a construction joint, hardened concrete shall be cleaned and wetted and a thin layer of mortar spread over it.

Check on placing of dowels, keys, waterstops, and other details as shown on plans.

Floors. Check and remove laitance when concrete reaches required level. If excessive, cut down on mix water or overworking of concrete.

Finish floor as specified.

Pumping and Conveying. Only if approved or specified.

Equipment cleaned before and after pouring.

Continuous flow of concrete; no segregation.

Exposed Surfaces. Retain original surface film and form marks; do not rub, except as specified.

Fins and projections removed.

Small voids filled with 1:2 mortar.

Construction joints only as detailed on plans.

To match color, use trial pats with white cement added to mix.

Metal ties, chairs and spacers covered with $1\frac{1}{2}$ in. of concrete.

Curing Concrete. Kept moist for 1 week minimum or sprayed with approved colorless liquid sealing compound.

Continuous saturation by sprays or wet fabric is preferred to intermittent sprinkling by hand. On vertical surfaces see that wet fabric is kept in contact with concrete.

Prompt application of curing materials as soon as possible after finishing concrete. Check that they are on hand ready for use.

Cold-Weather Concreting. Do not heat cement. Aggregates and water heated to not over 165° F. No snow or frozen lumps in aggregate.

Check temperature of concrete as placed, not less than 60° F. nor more than 100° F. Use immersion thermometer inserted in concrete near forms or surface.

Remove ice and snow from forms or other place of deposit and from reinforcement before placing concrete.

Do not allow placing of concrete on frozen ground.

Frost Protection. Provided by full enclosure of concrete and temperature of not less than 60° F. maintained for 7 days or as specified. Keep humidity high in enclosure.

Check that heating units will not harm concrete or damage surface for applied finishes.

Or, by consent of engineer, frost protection, provided by protecting surface with straw, hay, or fabric for 7 days. In buildings enclose story below and heat to 50° F. for 7 days.

Temperature protection gradually removed to prevent sudden temperature shock to concrete; 15° in 24 hours recommended.

Accelerating Admixtures (Calcium Chloride).   Use only if specified. Tested before use.

Delivered in moisture-proof bags or airtight drums.

Quantity used not over 2 lb. per sack of cement.

Dissolve 1 lb. per quart of water, and add not more than 2 qt. per sack of cement to mixing water.   Subtract amount of solution from normal quantity of mixing water.

Dry calcium chloride not to be added to aggregate in mixer skip or placed in contact with dry cement.

For cold-weather placing and curing, provide same precautions as for plain cement.

High-Early-Strength Cement.   Use only if specified.   Mixing and placing same as standard cement.

Prompt finishing (delay will ruin finish).

Curing temperature maintained as specified (usually 70° F. for 2 days or 60° F. for 3 days).   Must be kept moist for at least 3 days.

Load Tests.   May be required for faulty workmanship, violation of specification, or concrete suspected of having been frozen.

Notify superiors if necessary.

Pay Items

Accurate record kept of all pay items in contract, such as:

Volume of concrete placed and batches wasted.

Volume of openings or embedded structures if payment for such is not made.

Amount of reinforcing steel in pounds or tons actually placed.

Number and length of extra dowels and dowel holes drilled.

Embedded items or structures.

Any other contract pay items.

CHECK LIST FOR INSPECTORS

CONCRETE—PAVING

Procedure in Inspection

It is assumed that batching has been performed and inspected; see p. 34. For transit-mix concrete, see p. 37.

Field Inspection

Subgrade. Drainage, stability, compaction. Wet down ahead of placing. Moist, not muddy.

Grade and cross section. Full depth of pavement at all points.

Check ordinates to subgrade templates and scratch boards.

Forms. Approved type with true face, top, and base.

Connections rigid and true.

Alignment and grade.

Staked solidly with adequate base support.

Cleaned and oiled each time used.

Reinforcing and Joint Assemblies. Tested and approved reinforcing steel placed to secure final position shown on drawings.

Transverse joint assemblies at correct locations staked solidly. Accurate to line and perpendicular to subgrade. Joint material tight against forms or adjacent joint.

Approved dowels, painted and greased, held rigidly parallel to surface and axis of pavement. Correctly spaced. Approved expansion caps in place.

Correctly aligned longitudinal joints with correctly spaced tie bars held securely in place, normal to joint and parallel to surface.

Mixing and Placing Concrete. Full cement content of batch. Empty bags and count at end of each run to check cement factor. Provide against loss of bulk cement by blowing away.

Approved mixer with accurate timing and bell. Provision to lock discharge lever until mixing time is complete. Mixer drum not loaded more than 10% above rated capacity (29.7 cu. ft. for 27-E paver).*

Full mixing time for each batch after all ingredients are in drum. Check time frequently. Allow 1½ minute minimum unless otherwise specified. Check specified revolutions of drum, usually 14 to 20 r.p.m., and peripheral speed.*

* Does not apply to transit-mix concrete.

Slump not to exceed 3 in. in any case, preferably 2 in.

Thorough compaction of concrete. Spade or vibrate against forms and existing concrete. Do not vibrate or manipulate too much.

Daily check of cement content, yield, water cement ratio, adherence to design mix, aggregates, and cement used. Check of slump and unit weight, several tests daily.

Adequate protection at hand (burlap, cotton mats, tarpaulins, etc.), for sudden rain or drop in temperature. Assembled construction joint ready to install for stoppage over 30 minutes.

Uniform amount of concrete carried ahead of strike-off. Workmen to avoid walking on soft concrete or reinforcement assemblies. Deposit concrete in final position. Do not dump on joint assemblies.

Finishing and Curing. Surface finished at proper time with approved tools and appliances. Systematic checking with tested straightedge.

Ordinates checked to all screeds. For parabolic ordinates, see p. 315.

Overfinishing avoided, may produce scaling. High or low spots corrected.

Good workmanship on tooling of joints and edges; specified edge rounding radius and width of tooling.

Prompt application of approved curing agents. Curing for full period specified.

Care in removing forms and bending tie bars. Do not pry against green concrete.

Ample protection from traffic until cured.

Sealing Joints, Opening to Traffic. Careful cleaning and sealing of joints and cracks.

Final check for surface roughness, high joints, fractured slabs, flush sealing of joints. Correction and repair as directed.

Temporary shoulder for edge protection before traffic is allowed.

Adequate structural strength.

Air pressure in mixing drum of transit mixers. Leave discharge door partly opened and vent end of drum with four $\frac{5}{8}$-in.-diameter holes kept open at all times. Report excessive air content to engineer.

Cold-Weather Concrete

Concrete not placed on frozen subgrade.

Aggregate and water heated to produce temperature of concrete, at placing, of 70° F. minimum and 100° F. maximum or as specified.

Curing temperature of 50° to 100° F. maintained for specified period.

No admixtures or extra cement used unless specified.

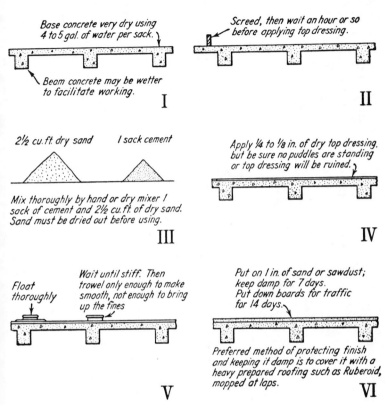

FIG. 46. Rules for construction of monolithic floor.

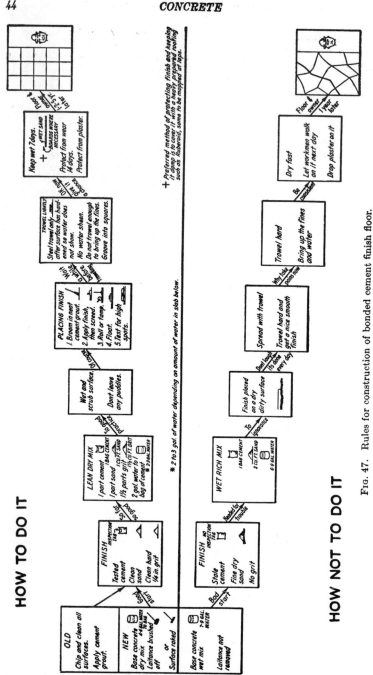

Fig. 47. Rules for construction of bonded cement finish floor.

AGGREGATES

FIELD TESTING

Specific Gravity and Surface Moisture

Use fruit jar (see Fig. 106) and 2-kilo. (5-lb.) balance accurate to $\frac{1}{10}$ gram.

Specific Gravity. Weigh jar full of water. Empty jar; place therein 700 grams surface-dry sample. Fill jar with water and weigh. Determine specific gravity from nomograph. See Fig. 106. Page 185

Surface Moisture. Same procedure except 700-gram sample is moist aggregate to be tested.

Precautions. Roll submerged sample to remove air. Jar must be dry outside when weighed. Use eye-dropper to insure completely filling with water. Remove foam.

Surface Moisture, Heat Method. Heat a weighed sample at 212° F., in open pan until surface water disappears (3 to 10 minutes). Weigh again. The difference between the original and the final weight is calculated as per cent of surface moisture.

Total Moisture Content. Heat weighed sample in open pan above 212° F. for 30 minutes or to constant weight. The difference between the original and the final weights is calculated as per cent of total moisture.

TABLE 2. APPROXIMATE SURFACE MOISTURE

(Use only when testing is impracticable)

CONDITION OF AGGREGATE	PER CENT BY WEIGHT
Very wet sand	6 to 8
Average stock pile sand, drained	$3\frac{1}{2}$ to 4
Moist sand	2
Moist gravel or crushed rock	2

Tests of Gradation. Sieve Analysis, A.S.T.M. C-136

Quarter sample until sufficient material remains to give a dry sample as follows: sand under No. 10, 100 grams (0.2 lb.); sand under No. 4, 500 grams (1.1 lb.); coarse sand, 1000 grams (2.2 lb.); coarse aggregate under 1 in. maximum, 10 kg. (22 lb.); 2 in. maximum, 20 kg. (44 lb.); 3 in. maximum, 30 kg. (66 lb.). Use square- or round-aperture sieves as specified and of the sizes specified. If not specified, use square-mesh sieves as follows: bituminous aggregates, Nos. 200, 80, 40, 10, 4, $\frac{3}{8}$ in., $\frac{1}{2}$ in., $\frac{3}{4}$ in., 1 in., $1\frac{1}{4}$ in., $1\frac{1}{2}$ in.; concrete aggregates, Nos. 100, 50, 30, 16, 8, 4, $\frac{3}{8}$ in., $\frac{3}{4}$ in., $1\frac{1}{2}$ in., 3 in. Use 8-in.-diameter sieves for samples of 5 kg. (11 lb.) or less and 16-in.-diameter sieves for larger samples. Use

CONCRETE

TABLE 3. FIELD SAMPLING

Material and Method	When Sampled	Size of Sample	Instructions
Cement, A.S.T.M. E-183	Each 2000 sacks or equivalent bulk	8 lb. min.	Sacked cement: compose sample from portions taken from 1 sack in 40. Bulk cement: sample from different locations with small scoop. Ship in container sealed airtight with paraffin.
Aggregates, A.S.T.M. D-75	Each source First shipment and if any change	Sand, 30 lb. Stone and slag, 100 lb. Gravel, 100 lb. over ½-in. size	Quarter aggregates by placing on canvas square or clean surface. Mix thoroughly. Form into conical pile. Flatten pile. Cut into 4 pie-shape parts. Discard 2 opposite quarters including dust. Remix remainder. Repeat till desired size, but not less than twice. Ship in strong, tight bag or box.
Steel Reinforcement, A.S.T.M. A-15 A-16 A-160	Each 10 tons Each lot or shipment	3 pieces of each size, 18 in. long min.	Wire pieces together and wrap in burlap.
Bar or rod mats, A.S.T.M. A-184	Each order or each 500 mats	2 ft. by 2 ft.	Cut sample from 2 mats in each order. Ship crated.
Wire fabric, A.S.T.M. A-185 and A-82	Each order or each 75,000 sq. ft.	2 ft. by 2 ft.	If heavy edge wire type include edge in square. Ship crated.
Expansion joint filler, A.S.T.M. D-545	Each 1000 sq. ft.	3 ft. long min. by full depth	Ship crated. Seal cork type in waterproof paper.
Joint sealer, A.A.S.H.O. T40	Each lot or shipment	1 qt. min.	Place in friction lid can. Ship crated or boxed.
Curing liquids, A.S.T.M. C-156	Each lot or shipment	1 qt. min.	Ship in small-mouth can with cork-lined screw top.

Concrete test cylinders, A.S.T.M. C-31 and C-172	As specified, or 4 for each 250 cu. yd. or 2000 sq. yd. of slabs	6 in. dia. by 12 in. high for aggregate 2 in. and under; 8 in. dia. by 16 in. high for aggregate over 2 in.	Use paraffined cardboard or metal mold. Place sample in mold in 3 equal layers, rodding each layer 25 strokes with ⅝ in. by 24 in. bullet pointed rod. Strike off top with rod. Cover and keep moist at 60°–80° F. Do not move for 24 hr., then remove molds and paint identification on cylinder. Cure laboratory control cylinder moist at 70° F. till tested. Cure field control cylinders same as corresponding concrete. Pack in damp sand or burlap, and ship in strong box.
Concrete test beams, A.S.T.M. C-31 and C-172	3 or 4 beams for every 2000 sq. yd. of pavement or slab	6 in. by 6 in. by 30 in. or 36 in.	Use rigid wood or metal form (6-in. channels) lightly oiled or paraffined. Place concrete in 2 equal layers, each layer rodded 50 times per sq. ft. Spade sides and edges with trowel, and strike off top. Finish with cork float. Cover at once with damp burlap. After 24 hr. remove forms and cure moist at 60° to 75° F. for laboratory control. Paint identifying marks or symbols. Cure field control beams same as corresponding concrete. Pack in damp sand or burlap, and ship in strong box.
Calcium chloride, A.S.T.M. D-345	Each lot or shipment	1 qt. min.	Ship in airtight container.
Water, A.A.S.H.O. T-26	Each source	2 qt.	Ship in crated glass jar with glass stopper.

MARKING SAMPLES—ALL MATERIALS

Place one tag inside container, and attach one tag firmly outside. Record all shipments and data in field book. Mark tags with name and address of laboratory; date; project; contractor; engineer; sampler; quantity represented; any special test desired if other than routine; vendor's or manufacturer's name and brand name if any; location or part of structure affected; sample number; address to send report; any other pertinent information. See Fig. 45 for sample tag.

Cement. Railroad car number; sacked or bulk; type; mill.

Aggregates. Kind; quantity in source; name of plant pit or quarry, and location.

Reinforcing. Lot number; markings on rods.

Test Cylinders and Beams. Date molded; station or location in structure; mix proportions; W/C ratio, gallons per sack; cement, sacks per cubic yard; slump; unit weight, pounds per cubic foot; cement brand, type, mill, and car number; type and source of aggregate, by whom made.

Note. Use envelope-style tags with name and address of laboratory and shipper on envelope and complete data on tag or card inside envelope tag.

balance or scale sensitive to 0.1% of sample weight. Set sieves in sequence with smallest size on bottom. Weighed sample is set on top sieve, and sieves are vibrated by lateral and vertical motion with jarring action. Weigh amount retained on each sieve and in pan, and compute percentage.

Fineness Modulus

Add cumulative per cent retained on each of U. S. Standard Sieves listed above for concrete. Divide sum by 100; result equals fineness modulus.

Material Finer than No. 200 Sieve—Silt and Clay in Fine Aggregate, A.S.T.M. C-117-37

Use two sieves, No. 200 and No. 16, and a vessel large enough to contain the sample covered with water, and permit agitation. Select a moist sample large enough to weigh 500 grams (1.1 lb.) when dry. The sample after being dried to constant weight is placed in the container and covered with water. The contents of the container are agitated vigorously and the wash water is poured over the nested sieves, the No. 16 being on top. The operation is repeated until the wash water is clear. The washed aggregate is dried to constant weight and weighed to nearest 0.02%.

% of minus No. 200 material

Fig. 48. Sieves.

$$= \frac{\text{original dry weight} - \text{dry weight after washing}}{\text{original dry weight}} \times 100$$

Approximate Amount of Silt and Clay

Place fine aggregate in a pint bottle to a height of 4 in.; then add water until the bottle is nearly full. Shake thoroughly, and allow to settle for 1 hr. or until the water is clear. Silt and clay will settle on top. The thickness of this layer should not be over $\frac{1}{8}$ in. Alternative: Place 5 oz. of sand in 12-oz. graduated bottle and add water until the mixture equals 10 oz. after shaking. Allow to settle as above. If silt and clay content is more than 3% or as specified, sand should be washed or additional laboratory tests made.

Organic Impurities in Fine Aggregate (Colorimetric Test), A.S.T.M. C-40

This method of test covers the procedure for an approximate determination of the presence of injurious organic compounds in natural sands which are to be used in cement mortar or concrete. The principal value of the test is to furnish a warning that further tests of the sands are necessary before they are approved for use.

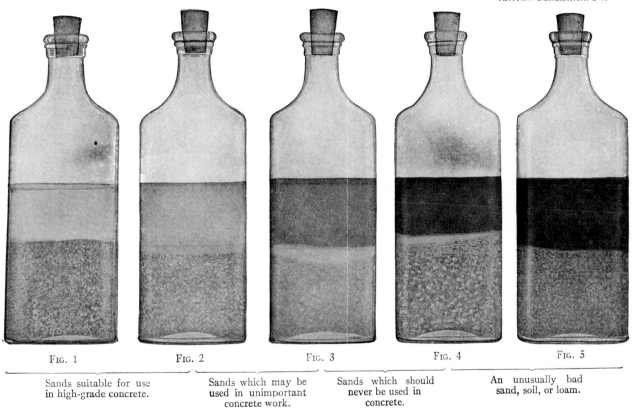

PLATE I
1946 BOOK OF A.S.T.M. STANDARDS, PART II

STANDARD METHOD OF TEST FOR
ORGANIC IMPURITIES IN SANDS FOR CONCRETE
A.S.T.M. DESIGNATION: C 40

FIG. 1   FIG. 2   FIG. 3   FIG. 4   FIG. 5

Sands suitable for use in high-grade concrete.

Sands which may be used in unimportant concrete work.

Sands which should never be used in concrete.

An unusually bad sand, soil, or loam.

FIG. 49. Colors of treated sands with suggested ranges of application.

A representative test sample of sand weighing about 1 lb. shall be obtained by quartering or by the use of a sampler.

Fill a 12-oz. graduated prescription bottle to the 4½-oz. mark with the sample to be tested. Add a 3% solution of caustic soda, known as sodium hydroxide, until the volume of sand and solution after shaking reaches the 7-oz. mark. Let the bottle stand for 24 hr., then observe the color of the liquid above the sand. If colorless or light amber color, the sand may be considered satisfactory. If it is light brown or darker, the sand should be sent to laboratory for additional tests.

Sulfate Test. The soundness of aggregates is tested by the use of sodium or magnesium sulfate. This test is of particular value in determining the presence of excessive amounts of clay or shale. It is described in full in A.S.T.M. Designation C-88.

Unit Weight of Aggregate, Dry Rodded Method, A.S.T.M. C-29

Use a calibrated bucket of minimum No. 11 gage metal, a ⅝-in. by 24-in. bullet-pointed tamping rod, and a scale accurate to 0.5%. The capacity of the bucket in cubic feet should be as follows: ½-in. maximum aggregate size use ⅒ cu. ft.; 2-in. maximum aggregate size use ⅓ or ½ cu. ft.; 4-in. maximum aggregate size use 1 cu. ft. Aggregate should be room dry and thoroughly mixed. Fill the measure in 3 equal layers, rodding each layer 25 times. Strike off top layer and determine net weight. Calculate weight per cubic foot (unit weight). *Note.* In rodding use only enough force to penetrate the layer being rodded. The rod should not strike the bottom of the bucket.

Voids in Aggregate, A.S.T.M. C-30

$$\% \text{ of voids} = \frac{(\text{specific gravity of aggregate} \times 62.4) - \text{weight}}{(\text{specific gravity of aggregate} \times 62.4)} \times 100$$

where weight equals the weight in pounds per cubic foot of the aggregate as determined by the unit weight test above (A.S.T.M. C-29). Specific gravity is determined by nomograph, Fig. 106, or by laboratory.

Absorption of Aggregates

The following table may be used as a guide for the field where A.S.T.M. Tests C-127 and C-128 are not practicable.

TABLE 4. APPROXIMATE ABSORPTION OF WATER BY AGGREGATES

	Per Cent by Weight
Average sand	1.0
Calcareous pebbles and crushed limestone	1.0
Trap rock and granite	0.5
Porous sandstone	7.0

TUTTLE, SEELYE, PLACE AND RAYMOND
ARCHITECT-ENGINEER
FORT DIX NEW JERSEY

Contract No. ________ Date of test ________
Contractor ________ Type construction ________
Source of material ________ Plant ________
Sampled at ________ Used at station or building ________
Specification ________ Material ________

REPORT ON AGGREGATES—SIEVE ANALYSIS

Screen or Sieve Size	Round or Square Shape	Weight Retained	Weight Passing	% Passing	% Spec. Reqmts.
					Min. Max.
3″					
2½″					
2¼″					
2″					
1½″					
1¼″					
1″					
¾″					
½″					
⅜″					
¼″					
No. 4					
No. 8					
No. 16					
No. 30					
No. 50					
No. 100					
No. 200					
Pan					

WEIGHTS OF SAMPLE

Total weight ________

Dry weight ________

Difference ____ % moisture

After washing ________

% gravel (over ¼″) ____

Clay, etc. ____ % ____

Material retained on ____

____ Sieve ____ ____ %

Fineness modulus = sum cumulative % retained on each of Nos. 100, 50, 30, 16, 8, and 4, ⅜-in., ¾-in., 1½-in., and 3-in. sizes ÷ 100 =

Remarks:

Remarks:

Tested by:

____
Inspector

Approved Disapproved

____
Engineer

FIELD TESTING

Slump Test for Consistency, A.S.T.M. C-143. Use a standard slump cone made of No. 16 gage galvanized metal in the form of a frustum of a cone with the base 8 in. in diameter, the top 4 in. in diameter, and the altitude 12 in. Provide mold with foot pieces and handles.

Take 5 samples of concrete, and thoroughly mix to form test specimen. Sample from discharge stream of mixer, starting at beginning of discharge and repeating until batch is discharged. For paving concrete, samples may be taken from the batch deposited on the subgrade. Before placing concrete, dampen the cone and place on a flat, moist, non-absorbent surface. In placing each scoopful of concrete move the scoop around the top edge of the cone as the concrete slides from it, in order to insure symmetrical distribution of concrete within the cone. Fill the mold in 3 equal layers, rodding each layer with 25 strokes of a $\frac{5}{8}$-in. ϕ rod 24 in. in length, bullet pointed at the lower end. Distribute the strokes in a uniform manner over the cross section of the cone and penetrate into the underlying layer. Rod the bottom layer throughout its depth. After the top layer has been rodded strike off the surface of the concrete with a trowel or board so that the cone is exactly filled. Immediately remove the cone from the concrete by raising it carefully in a vertical direction. Then measure the slump immediately by laying the 24-in. rod across the top of the cone and measuring down to the top of the sample. This is known as the slump, which is equal to 12 in. minus the height in inches, after subsidence, of the concrete specimen. The slump test should be made frequently, at least 3 or 4 times a day.

Unit Weight of Plastic Concrete, A.S.T.M. C-138. Use a calibrated bucket of minimum No. 11 gage metal, a $\frac{5}{8}$-in. by 24-in. bullet-pointed rod, and a scale accurate to 0.5% of total weight tested. Capacity of bucket should be $\frac{1}{10}$ cu. ft. for $\frac{1}{2}$-in. maximum aggregate; $\frac{1}{2}$ or $\frac{1}{3}$ cu. ft. for 2-in. maximum aggregate, and 1 cu. ft. for 4-in. maximum aggregate. Place a representative sample (selected as described for slump test above) in the bucket in 3 equal layers, rodding each layer 25 strokes as described for slump test. Vibrated concrete shall be compacted in the measure by vibration. Strike off surface, taking care that measure is just level full. Weigh to nearest 0.1 lb., subtract weight of bucket, and compute net weight of concrete in pounds per cubic foot.

Notes. 1. It is suggested that the inspector carefully sample about 1 cu. ft. or more of concrete and run slump test, unit weight test, and mold cylinders and beams in one sequence of operations. Complete data will then be obtained. 2. Concrete sample should not be taken from a chute, where partial segregation will be present.

APPROXIMATE DATA ON CONCRETE MIXES

TABLE 5. WATER-CEMENT RATIO (W/C) FOR VARIOUS STRENGTHS

Water Content		W/C Ratio		Strength of Concrete	
Gallons per Sack of Cement	W/C by Vol. Cu. Ft. per Sack	by Absolute Volume	W/C Ratio by Weight	at 28 Days	
				Compressive	Flexural
5 max.	0.668	1.38	0.444	5000 p.s.i.	750 p.s.i.
6 max.	0.802	1.66	0.533	4000 p.s.i.	600 p.s.i.
7 max.	0.936	1.93	0.621	3200 p.s.i.	500 p.s.i.
8 max.	1.069	2.21	0.710	2500 p.s.i.	450 p.s.i.

Note: Strengths should be determined by trial mixes (when practicable) based on fixed W/C. To allow for field conditions the strength values shown in table should be reduced by about 20%.

TABLE 6. EXPOSED CONCRETE—MAXIMUM WATER CONTENT IN GALLONS PER SACK

Type or Location of Concrete	Severe and Moderate Climate	Mild Climate
At waterline (intermittent saturation)		
Sea water	5½	5½
Fresh water	6	6
Not at waterline but frequent wetting		
Sea water	6	6½
Fresh water	6½	7
Ordinary exposed structures	6½	7
Completely submerged		
Sea water	6½	6½
Fresh water	7	7
Concrete deposited through water	5½	5½
Pavement slabs on ground		
Wearing slabs	5½	6
Base slabs	6½	7

Above water content includes free surface water in aggregates; see Table 8.

TABLE 7. RECOMMENDED PER CENT OF SAND TO TOTAL GRADED AGGREGATE FOR SMOOTH PLASTIC CONCRETE

Crushed stone, max. 1½-in. size	34 to 43
Crushed stone, max. ¾-in. size	35 to 43
Gravel, max. 1½-in. size	33 to 41
Gravel, max. ¾-in. size	35 to 42

Sand-Aggregate Ratio or percentage by weight or volume of sand to total aggregate in mix should be from 33 to 43%, with extreme limits of 28 and 49%. The most economical mix will be that with lowest sand-aggregate ratio producing the desired plasticity, workability, and consistency.

For air-entraining concrete, reduce weight of sand by approximately 8%. See Tables 12A, 12B, 12C.

TABLE 8. SUGGESTED CHARACTERISTICS OF CONCRETE FOR VARIOUS CLASSES OF WORK

Slump, in.	Size of Coarse Aggregate	28-Day, p.s.i.	Classes of Work
2–3	No. 4 to 2½ in. No. 4 to 2 in.	3000	Heavy sections such as dams, foundations, and walls
3–6	No. 4 to 1 in. No. 4 to ¾ in.	2500 to 3500	Unexposed reinforced concrete beams and slabs
2–3	No. 4 to 2 in.	3500	Submerged mass concrete
2–3	No. 4 to 2 in.	4000	Structures moderately exposed, subject to intermittent wetting
2–5	No. 4 to 1 in. No. 4 to 1½ in.	3000 4000	Foundation walls, building columns, and reinforced concrete in general
2–4	No. 4 to 1½ in.	4500	Severe exposure at water line
3–6	No. 4 to ¾ in.	4500	Very severe exposure of thin reinforced sections
3–6	No. 4 to 2 in.	5000	Concrete for depositing under water
2–3	No. 4 to ¾ in.	6000	Hand-cast internally vibrated concrete pipe
0–1	No. 8 to ⅜ in.	6000 to 7000	Machine-compacted concrete pipe
3–5	No. 8 to ½ in.		Electric ducts, conduits, fireproofing, and similar work

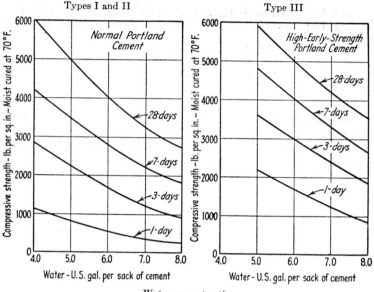

Fig. 50. Age-strength relation for normal and high-early-strength portland cements. The strengths indicated should be obtained on average construction projects where all materials, including the water, are controlled. On important work, tests should be made with the materials to be used on the project to establish job curves and fix design values.

Approximate Quantity of Surface Water Carried by Average Aggregates * †

Very wet sand	¾ to 1 gal. per cu. ft.
Moderately wet sand	about ½ gal. per cu. ft.
Moist sand	about ¼ gal. per cu. ft.
Moist gravel or crushed rock	about ¼ gal. per cu. ft.

Approximate Absorption of Aggregates *

Average sand	1.0% by weight
Pebbles and crushed limestone	1.0% by weight
Trap rock and granite	0.5% by weight
Porous sandstone	7.0% by weight
Very light and porous aggregate may be as high as	25% by weight

* *From Portland Cement Association.*

† The coarser the aggregate, the less free water it will carry.

TABLE 9. RECOMMENDED GRADATION LIMITS FOR GRADED AGGREGATES TO OBTAIN SMOOTH, DURABLE, AND DENSE CONCRETE (SQUARE MESH SIEVES) *

1. Fine aggregate.

For all purposes:

Sieve	⅜ in.	No. 4	No. 16	No. 50	No. 100
% passing	100	95–100	45–80	10–30	2–10

2. Coarse aggregate.

(a) *For general reinforced concrete:*

Sieve	1 in.	¾ in.	⅜ in.	No. 4	No. 8
% passing	100	90–100	20–55	0–10	0–5

(b) *For paving:*

Sieve	2 in.	1½ in.	¾ in.	⅜ in.	No. 4
% passing	100	95–100	35–70	10–30	0–5

Keep ¾ and smaller aggregate in separate bins from larger sizes, and recombine after weighing to charge the mixer.

(c) *For mass concrete:*

Sieve	3 in.	2½ in.	2 in.	1½ in.	¾ in.
% passing	100	90–100	35–70	0–15	0–5

Keep 2-in. and smaller aggregate in separate bins from larger sizes, and recombine after weighing to charge the mixer.

* Recommended by E. A. Robinson and based on A.S.T.M. C-33.

CONCRETE

MISCELLANEOUS DATA

TABLE 10. PERCENTAGE OF 70° MOIST-CURED COMPRESSIVE STRENGTH NORMAL PORTLAND CEMENT

Placed and Cured at	5 Gal. per Sack				6 Gal. per Sack			
	1 day	3 days	7 days	28 days	1 day	3 days	7 days	28 days
60° F.	77%	82%	86%	87%	69%	81%	83%	86%
50° F.	59%	66%	69%	74%	44%	60%	63%	68%
40° F.	32%	46%	55%	62%	25%	38%	45%	56%
30° F.	14%	32%	42%	42%	6%	26%	33%	45%

Placed and Cured at	7 Gal. per Sack				8 Gal. per Sack			
	1 day	3 days	7 days	28 days	1 day	3 days	7 days	28 days
60° F.	73%	75%	79%	85%	67%	75%	80%	82%
50° F.	36%	53%	58%	65%	33%	50%	58%	62%
40° F.	18%	37%	42%	51%	16%	33%	39%	48%
30° F.	9%	25%	31%	43%	16%	25%	36%	39%

TABLE 11. PERCENTAGE OF 70° MOIST-CURED COMPRESSIVE STRENGTH EARLY-STRENGTH PORTLAND CEMENT

Placed and Cured at	5 Gal. per Sack				6 Gal. per Sack			
	1 day	3 days	7 days	28 days	1 day	3 days	7 days	28 days
60° F.	84%	90%	92%	93%	82%	88%	90%	91%
50° F.	64%	76%	82%	86%	57%	72%	77%	82%
40° F.	43%	62%	69%	77%	36%	58%	65%	71%
30° F.	27%	44%	55%	63%	23%	38%	47%	58%

Placed and Cured at	7 Gal. per Sack				8 Gal. per Sack			
	1 day	3 days	7 days	28 days	1 day	3 days	7 days	28 days
60° F.	79%	85%	88%	89%	56%	86%	87%	88%
50° F.	51%	66%	72%	79%	36%	67%	74%	78%
40° F.	33%	53%	62%	69%	22%	55%	60%	66%
30° F.	18%	32%	43%	54%	11%	31%	40%	48%

Based on test data from Lone Star Cement Corp. Charts T-30 and T-31 prepared 11-11-50.

TABLE 12A. STRUCTURAL CONCRETE DESIGN MIXES

QUANTITIES OF INGREDIENTS TO PRODUCE GIVEN STRENGTHS AIR-DRY WEIGHTS FOR 27 CU. FT. OF CONCRETE NO. 4 TO ¾-IN. GRADED COARSE AGGREGATE

				Graded Natural Sand Fineness Modulus 2.70, Bulk Specific Gravity 2.65 Pounds of Sand			
				With Gravel		With Stone	
		Cement			Air-		Air-
28-Day Strength, p.s.i.	Slump, in.	Bags	Quantity, lb.	Regular Concrete	Entrained Concrete	Regular Concrete	Entrained Concrete
1	2	3	4	5	6	7	8
2000	3	4.6	432	1380	1280	1395	1325
	6	4.9	461	1310	1210	1325	1260
2500	3	5.0	470	1345	1245	1360	1295
	6	5.5	517	1260	1160	1280	1215
3000	3	5.5	517	1305	1205	1325	1260
	6	6.0	564	1215	1115	1240	1185
3500→3	→	6.0	→ 564	——1260	——1160	→1285 †	1230
	6	6.5	611	1180	1080	1200	1135
4000	3	6.5	611	1225	1125	1245	1180
	6	7.0	658	1145	1045	1160	1085
4500	3	7.0	658	1190	1090	1205	1130
	6	7.5	705	1105	1005	1120	1050
5000	3	7.5	705	1150	1050	1165	1095
Approximate total water, gal. per cu. yd.				36*–38	34*–36	40*–42	38*–40

No. 4 to ¾-In. Graded Coarse Aggregate

Gravel			Stone		
Dry Rodded Weight, lb. per cu. ft.	Bulk Specific Gravity	Pounds of Gravel	Dry Rodded Weight, lb. per cu. ft.	Bulk Specific Gravity	Pounds of Stone
9	10	11	12	13	14
103.8	2.62	1905	99.4	2.65	1825
104.2	2.63	1915	101.3	2.70	1860
104.6	2.64	1920	103.1	2.75	1895
105.0	2.65	1930	105.0→	2.80 →	1930 †
105.4	2.66	1935	106.9	2.85	1965
105.8	2.67	1945	108.8	2.90	2000
106.2	2.68	1950	110.6	2.95	2030

* For 3-in. and 6-in. slumps respectively.
† See typical problem, p. 59.

TABLE 12B. STRUCTURAL CONCRETE DESIGN MIXES

QUANTITIES OF INGREDIENTS TO PRODUCE GIVEN STRENGTHS AIR-DRY WEIGHTS FOR 27 CU. FT. OF CONCRETE NO. 4 TO 1½-IN. GRADED COARSE AGGREGATE

				Graded Natural Sand Fineness Modulus 2.70, Bulk Specific Gravity 2.65 Pounds of Sand			
		Cement		With Gravel		With Stone	
28-Day Strength, p.s.i.	Slump, in.	Bags	Quantity, lb.	Regular Concrete	Air-Entrained Concrete	Regular Concrete	Air-Entrained Concrete
1	2	3	4	5	6	7	8
2000	3	4.0	376	1400	1300	1445	1380
	6	4.3	404	1340	1240	1375	1310
2500	3	4.5	423	1360	1260	1405	1340
	6	5.0	470	1275	1175	1320	1255
3000	3	5.0	470	1320	1220	1365	1300
	6	5.5	517	1246	1140	1285	1210
3500	3	5.5	517	1285	1185	1330	1260
	6	6.1	573	1190	1090	1235	1160
4000	3	6.1	573	1235	1135	1280	1205
	6	6.7	630	1140	1040	1185	1115
4500	3	6.7	630	1185	1085	1230	1160
	6	7.1	667	1110	1010	1155	1090
5000	3	7.5	705	1125	1025	1170	1105
Approximate total water, gal. per cu. yd.				33*–35	31*–33	36*–38	34*–36

No. 4 to 1½-In. Graded Coarse Aggregate

Gravel			Stone		
Dry Rodded Weight, lb. per cu. ft.	Bulk Specific Gravity	Pounds of Gravel	Dry Rodded Weight, lb. per cu. ft.	Bulk Specific Gravity	Pounds of Stone
9	10	11	12	13	14
103.8	2.62	1990	99.4	2.65	1905
104.2	2.63	2000	101.3	2.70	1940
104.6	2.64	2005	103.1	2.75	1980
105.0	2.65	2015	105.0	2.80	2015
105.4	2.66	2020	106.9	2.85	2050
105.8	2.67	2030	108.8	2.90	2085
106.2	2.68	2035	110.6	2.95	2120

* For 3-in. and 6-in. slumps respectively.

TABLE 12C. PAVEMENT CONCRETE DESIGN MIXES

QUANTITIES OF INGREDIENTS TO PRODUCE GIVEN STRENGTHS AIR-DRY WEIGHTS FOR 27 CU. FT. OF CONCRETE

					Graded Natural Sand Fineness Modulus 2.70, Bulk Specific Gravity 2.65 Pounds of Sand			
					With Gravel		With Stone	
28-Day Strength, p.s.i.		Slump, in.	Cement Minimum		Regular Concrete	Air-Entrained Concrete	Regular Concrete	Air-Entrained Concrete
Flexural	Compressive		Bags	Quantity, lb.				
1	2	3	4	5	6	7	8	9
500	2900	2	4.3	404	1250	1140	1305	1185
550	3700	2	5.2	489	1180	1065	1230	1115
600	4300	2	5.8	545	1135	1015	1185	1070
650	4800	2	6.4	602	1085	970	1135	1020
700	5100	2	6.8	639	1055	940	1105	990
Approximate total water, gal. per cu. yd.					30	27	33	30

No. 4 to 1½-In. Graded Coarse Aggregate

Gravel			Stone		
Dry Rodded Weight, lb. per cu. ft.	Bulk Specific Gravity	Pounds of Gravel	Dry Rodded Weight, lb. per cu. ft.	Bulk Specific Gravity	Pounds of Stone
10	11	12	13	14	15
103.8	2.62	2185	99.4	2.65	2095
104.2	2.63	2195	101.3	2.70	2135
104.6	2.64	2205	103.1	2.75	2170
105.0	2.65	2210	105.0	2.80	2210
105.4	2.66	2220	106.9	2.85	2250
105.8	2.67	2230	108.8	2.90	2290
106.2	2.68	2235	110.6	2.95	2330

Problem

Given:

No. 4 to ¾-in. graded stone

3500 p.s.i. concrete strength (non-air-entraining), 3-in. slump

105.0 lb. dry rodded weight of stone

2.80 * specific gravity of stone

2.65 * specific gravity of sand

2.70 * fineness modulus of sand

* Commonly assumed values where accurate measurements are not available.

To determine mix

1. Enter Table 12A.

> Strength: 3500 p.s.i. (Col. 1); 3-in. slump (Col. 2)
> Cement: Read 6.0 bags (Col. 3) or 564 lb. (Col. 4)
> Sand: 1285 lb. (Col. 7)

Enter Col. 13, specific gravity 2.80. Read 1930 lb. stone (for dry rodded weight of 105.0 lb.). Water: 40 gal. (bottom of Col. 7), *including surface water in aggregates.*

If dry rodded weight is less, increase the sand and decrease the stone weights; if it is more, decrease the sand and increase the stone weights as follows:

2. Given same problem as above but dry rodded weight of stone is 107.0 lb. instead of 105.0 lb. for specific gravity of 2.80.

Increase stone weight:

$$= (107.0 - 105.0) \times 2.80 \times 0.115 \times 62.4 = 40 \text{ lb.}$$

Decrease sand weight:

$$= (107.0 - 105.0) \times 2.65 \times 0.115 \times 62.4 = 38 \text{ lb.}$$

Therefore,

> *Dry sand* weight will be $1285 - 38 = 1247$ lb.
>
> *Dry stone* weight will be $1930 + 40 = 1970$ lb.

For a given specific gravity of coarse aggregate lying between any two given in Cols. 10 and 13, make a straight-line interpolation for corresponding dry rodded weights in Cols. 9 and 12 and weights in Cols. 11 and 14.

If fineness modulus of sand is less than 2.70, decrease the sand and increase the stone weights; if it is more than 2.70, increase the sand and decrease the stone weights as follows:

3. Given problem 2 above, except fineness modulus of sand is 2.50 instead of 2.70, correct as follows:

Decrease sand weight:

$$= (2.70 - 2.50) \times 2.65 \times 62.4 \times 1.65 = 55 \text{ lb.}$$

Increase stone weight:

$$= (2.70 - 2.50) \times 2.80 \times 62.4 \times 1.65 = 58 \text{ lb.}$$

Then:

> Cement 6.0 bags (564 lb.) ⎤
> Sand $1247 - 55 = 1192$ lb. ⎟ Correction to be made for surface water
> Stone $1970 + 58 = 2028$ lb. ⎟ in aggregates. (See example below.)
> Water 40 gal. ⎦

4. Given problem 3 above, except:

4½-in. slump

Interpolate in Cols. 3, 4, and 7 as follows:

Cement $\left(\dfrac{4.5'' - 3''}{6'' - 3''}\right) \times (611\# - 564\#) + 564 = 588$ lb. or 6.25 bags

Sand $1285\# \left(\dfrac{4.5'' - 3''}{6'' - 3''}\right) \times (1285\# - 1245\#) = 1265 - 38 - 55$

$= 1172$ lb.

Stone $1930 + 40 + 58 = 2028$ lb.

Water 40 gal., *including surface water in aggregates*

5. *Field Corrections of Batch Weights for Surface Water in Aggregates*

	Uncorrected Weight, lb.	Water Correction, lb.	Corrected Field Weights, lb.
Cement	588	0	588
Sand	1172	+50	1222
Stone	2028	+20	2048
Water	333 (40 gal.)	−70	263 (32 gal.)

Check actual batch for slump of 4½ in. and make further correction if necessary.

Refer to *Bulletin* 11 of the National Crushed Stone Association by Goldbeck and Gray for the method of designing regular mixes used in these tables. They are suggested for field check or trial and are not intended to replace laboratory design mixes.

CONCRETE REINFORCEMENT

TABLE 13. DIAMETERS AND SECTIONAL AREAS OF WELDED WIRE FABRIC

(Area in square inches per foot of width for various spacings of wire)

AS&W Co. Steel Wire Gage Numbers	Wire Diam-eter, in.	Center-to-Center Spacing, in.							
		2	3	4	6	8	10	12	16
0000	.3938	.731	.487	.365	.244	.183	.146	.122	.091
000	.3625	.619	.413	.310	.206	.155	.124	.103	.077
00	.3310	.516	.344	.258	.172	.129	.103	.086	.065
0	.3065	.443	.295	.221	.148	.111	.089	.074	.055
1	.2830	.377	.252	.189	.126	.094	.075	.063	.047
2	.2625	.325	.216	.162	.108	.081	.065	.054	.041
¼″	.2500	.295	.196	.147	.098	.074	.059	.049	.037
3	.2437	.280	.187	.140	.093	.070	.056	.047	.035
4	.2253	.239	.159	.120	.080	.060	.048	.040	.030
5	.2070	.202	.135	.101	.067	.050	.040	.034	.025
6	.1920	.174	.116	.087	.058	.043	.035	.029	.022
7	.1770	.148	.098	.074	.049	.037	.030	.025	.018
8	.1620	.124	.082	.062	.041	.031	.025	.021	.015
9	.1483	.104	.069	.052	.035	.026	.021	.017	.013
10	.1350	.086	.057	.043	.029	.021	.017	.014	.011
11	.1205	.068	.046	.034	.023	.017	.014	.011	.009
12	.1055	.052	.035	.026	.017	.013	.010	.009	.007
13	.0915	.039	.026	.020	.013	.010	.008	.007	.005
14	.0800	.030	.020	.015	.010	.008	.006	.005	.004
15	.0720	.024	.016	.012	.008	.006	.005	.004	.003
16	.0625	.018	.012	.009	.006	.005	.004	.003	.002

From American Steel & Wire Co. Handbook.

TABLE 14. MINIMUM BEAM WIDTHS IN INCHES *

Bar No.	Size of Bar	No. of Bars in Single Layer of Reinforcement							Add for Each Additional Bar
		2	3	4	5	6	7	8	
4	$\frac{1}{2}''\,\phi$	6″	$7\frac{1}{2}''$	9″					$1\frac{1}{2}''$
5	$\frac{5}{8}''\,\phi$	6″	8″	$9\frac{1}{2}''$	11″	$12\frac{1}{2}''$			$1\frac{5}{8}''$
6	$\frac{3}{4}''\,\phi$	$6\frac{1}{2}''$	$8\frac{1}{2}''$	$10\frac{1}{2}''$	12″	14″			$1\frac{7}{8}''$
7	$\frac{7}{8}''\,\phi$	7″	9″	$11\frac{1}{2}''$	$13\frac{1}{2}''$	16″	18″	20″	$2\frac{3}{16}''$
8	$1''\quad\phi$	$7\frac{1}{2}''$	10″	$12\frac{1}{2}''$	15″	$17\frac{1}{2}''$	20″	$22\frac{1}{2}''$	$2\frac{1}{2}''$
9	$1''\quad\square\dagger$	8″	11″	14″	17″	20″	23″	26″	3″
10	$1\frac{1}{8}''\,\square\dagger$	$8\frac{1}{2}''$	12″	15″	$18\frac{1}{2}''$	22″	$25\frac{1}{2}''$	$28\frac{1}{2}''$	$3\frac{3}{8}''$
11	$1\frac{1}{4}''\,\square\dagger$	9″	$12\frac{1}{2}''$	$16\frac{1}{2}''$	20″	24″	27″	$31\frac{1}{2}''$	$3\frac{3}{4}''$

* Where specially anchored bars are used, haunch width may be narrowed.
† Numbers 9, 10, and 11 are round bars equivalent in area to obsolete 1″, $1\frac{1}{8}''$, and $1\frac{1}{4}''$ square bars.

TABLE 15. PROPERTIES OF REINFORCING BARS AND HOOK DIMENSIONS

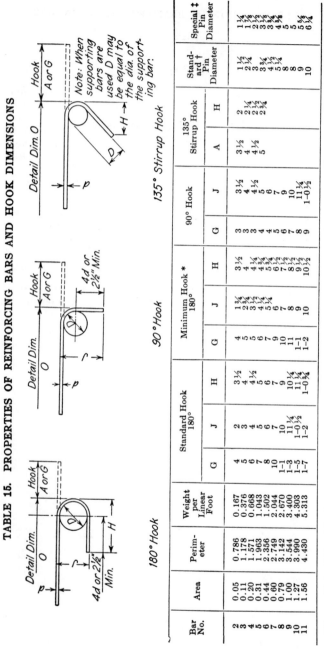

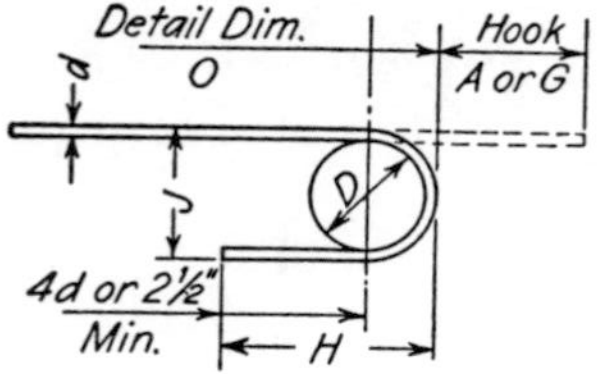
Detail Dim.
O
Hook
A or G
d
J
D
4d or 2½"
Min.
H

180° Hook

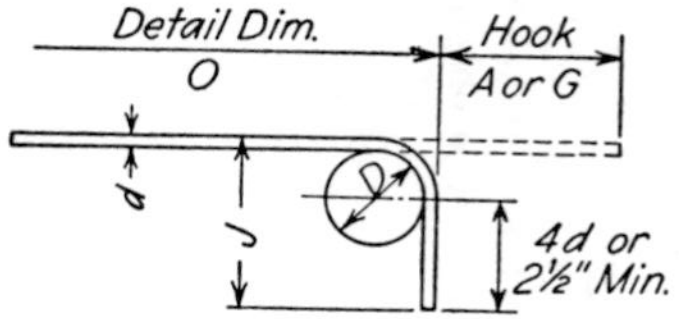
Detail Dim.
O
Hook
A or G
d
J
D
4d or
2½" Min.

90° Hook

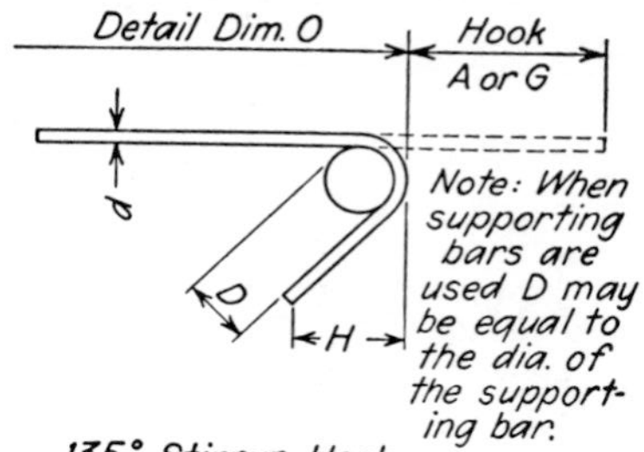
Detail Dim. O
Hook
A or G
d
D
H
Note: When supporting bars are used D may be equal to the dia. of the supporting bar.

135° Stirrup Hook

Bar No.	Area	Perimeter	Weight per Linear Foot	Standard Hook 180°			Minimum Hook * 180°			90° Hook		135° Stirrup Hook		Standard † Pin Diameter	Special ‡ Pin Diameter
				G	J	H	G	J	H	G	J	A	H		
2	0.05	0.786	0.167	4	2	3½	4	1¾	3½	3	3½	3½	2	1½	1¼
3	0.11	1.178	0.376	5	3	4	5	2¾	4	3	4	4	2¼	2¼	1⅞
4	0.20	1.571	0.668	6	4	4½	5	3½	4¼	3	4½	4½	2½	3	2½
5	0.31	1.963	1.043	7	5	5	6	4¼	4¾	4	5	5	2¾	3¾	3⅛
6	0.44	2.356	1.502	8	6	6	7	5¼	5¾	4	6			4½	3¾
7	0.60	2.749	2.044	10	7	7	9	6	6½	5	7			5¼	4⅜
8	0.79	3.142	2.670	1-1	10	9	10	7	7½	6	9			8	5
9	1.00	3.544	3.400	1-3	11¼	10¼	11	8	8½	7	10			8	5
10	1.27	3.990	4.303	1-5	1-0½	11¼	1-1	9	9½	8	11¼			9	5⅝
11	1.56	4.430	5.313	1-7	1-2	1-0¾	1-2	10	10½	9	1-0½			10	6¼

* For special conditions only; do not use for hard grade steel.
† For Standard Hook 180° and 90°.
‡ For Special Hook 180° and Stirrup Hook 135°.
Hooks in accordance with ACI Specifications.

TABLE 16. AREA OF STEEL PER FOOT OF WIDTH

SPACING OF BARS

BAR No.	4"	4½"	5"	5½"	6"	6½"	7"	7½"	8"	8½"	9"	9½"	10"	10½"	11"	11½"	12"
2 *	0.15	0.13	0.12	0.11	0.10	0.09	0.08	0.08	0.07	0.07	0.07	0.06	0.06	0.06	0.05	0.05	0.05
3	0.33	0.29	0.26	0.24	0.22	0.20	0.19	0.18	0.17	0.16	0.15	0.14	0.13	0.13	0.12	0.11	0.11
4	0.59	0.52	0.47	0.43	0.39	0.36	0.34	0.31	0.29	0.28	0.26	0.25	0.23	0.22	0.21	0.20	0.20
5	0.92	0.82	0.74	0.67	0.61	0.57	0.53	0.49	0.46	0.43	0.41	0.39	0.37	0.35	0.33	0.32	.031
6	1.33	1.18	1.06	0.96	0.88	0.82	0.76	0.71	0.66	0.62	0.59	0.56	0.53	0.51	0.48	0.46	0.44
7	1.80	1.60	1.44	1.31	1.20	1.11	1.03	0.96	0.90	0.85	0.80	0.76	0.72	0.69	0.66	0.62	0.60
8	2.36	2.09	1.88	1.71	1.57	1.45	1.35	1.26	1.18	1.11	1.05	0.99	0.94	0.90	0.86	0.82	0.78
9 †	3.00	2.67	2.40	2.18	2.00	1.85	1.71	1.60	1.50	1.41	1.33	1.26	1.20	1.14	1.09	1.04	1.00
10 †	3.80	3.37	3.04	2.76	2.53	2.34	2.17	2.02	1.90	1.79	1.69	1.60	1.52	1.45	1.38	1.32	1.27
11 †	4.69	4.17	3.75	3.41	3.13	2.89	2.68	2.50	2.34	2.21	2.08	1.97	1.87	1.79	1.70	1.63	1.56

* Bar No. 2, plain round; all others, deformed bars.

† Bars Nos. 9, 10, and 11 are round bars equivalent in weight and area to obsolete 1″, 1⅛″, and 1¼″ square bars.

LOAD TESTS

Permanent measurable deflections are a sign of weakness.

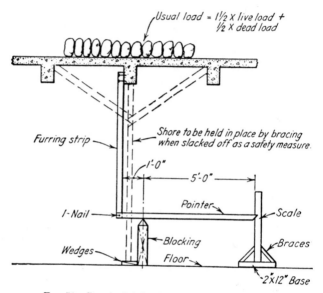

Fig. 51. Standard deflection magnifier for load tests.

Note. When the expense of safety shoring is too great, men conducting a load test may be protected by using a roller as a test load, the roller being towed from some safe distance. Level shots to measure deflection can be taken.

Load to be left in position 24 hours before removal. If the maximum deflection exceeds the deflection computed by accepted formulas, and if after the removal of load the recovery is less than 75% of the maximum deflection, the structure will have failed to pass the test.

See also A.C.I. code par. 202, or local code that may be applicable.

CONCRETE INGREDIENTS—DISCUSSION

Portland Cement, Non-Air-Entraining. ASTM Designation: C-150

Type I: For use in general concrete construction when special-property cements are not required.

Type II: For use in general concrete construction exposed to moderate sulfate action or where moderate sulfate action or moderate heat of hydration is required. Some type II cements produce concrete of slightly less strength at 28 days than type I.

Type III: For use when high early strength is required.

Type IV: * For use when a low heat of hydration is required.

Type V: * For use when high sulfate resistance is required.

Portland Cement, Air-Entraining. ASTM Designation: C-175

Types IA, IIA, and IIIA are similar to portland cement ASTM Designation C-150 except that air-entraining agents have been interground with the cement clinker. See *Data Book*, Vol. 2, *Specifications and Costs.*

Warning: Do not use an air-entraining cement in concrete whose proportions have been designed for regular cement. The difference in sand and water content will cause a critical reduction in design strength and durability. It is important that the inspector check to determine whether portland cement is regular or is air-entraining. To make a simple field check, obtain a wide-mouthed bottle, preferably one having straight sides, and pour into it some of the cement in question to one-third of its depth. Then add water to more than two-thirds the bottle depth and shake vigorously for about 15 to 20 seconds. Let it stand for 10 minutes. If the resulting foam disappears during this time, the cement is regular; otherwise it is probably air-entraining. This check is an indication only and should be

Fig. 52. Air-pressure meter complete with chamber for sample, pressure gage, and hand pump. Used for testing air content of concrete by measuring the decrease in volume of concrete under pressure.

* Types IV and V are usually manufactured on special order only.

verified by a laboratory test or by determining at the point of discharge the air content of the concrete in which the cement is used. One method of doing this is by means of an air-pressure meter (see Fig. 52).

Natural Cement.   ASTM Designation: C-10

Natural cement differs from portland cement in that a natural instead of an artificial mixture is used and in that it calcines at a lower temperature.

It is used largely as a blend with portland cement in concrete and produces about 2% air content in concrete compared to about 1% in non-air-entraining portland cement. It is not generally used for structural concrete.

Fine aggregate (see specification requirements) may be natural sand or sand manufactured from blast-furnace slag or approved native crushed stone such as trap rock, dolomitic limestone, or gravel; it should be clean and reasonably free from organic matter, uniformly graded, and otherwise in conformity with ASTM Designation C-33. Manufactured sand usually produces concrete that is somewhat harsher than that made with natural sand, but workability is improved in air-entrained concrete.

Check for proof of grading and quality. See specifications and colorimetric tests and bottle tests described in another part of this book.

Sands should be sharp or hard, rounded particles free from flat, shaly or partly disintegrated grains.

Beach sands or other unwashed salt-impregnated sands should not be used in reinforced concrete.

Coarse aggregate (see specification requirements) should be derived from hard, strong, and durable native stone such as trap rock, dolomites, limestones, granites, calcareous and silicious gravels as well as blast-furnace slag having a compact weight of not less than 70 lb. per cu. ft. Where blast-furnace slag is used, the stock piles should be kept wet.

Reject aggregate which is dirty or which contains soft, shaly, slaty, and thin, elongated stones.

Check with specifications for limiting sizes and gradation. Large stone, sometimes called plums, in concrete usually is prohibitive because of high labor cost in handling.

Water used in concrete must be clean and free from injurious amounts of acids, alkalies, or organic substances. The use of salt water for mixing concrete is not recommended.

Air-Entrained Concrete

Air-entrained concrete can be produced by using air-entraining portland cement or one of the various types of portland cements to which has been added at the mixing plant a suitable air-entraining agent. This should be done only where facilities and engineering supervision are available for accurate control.

The advantages of air-entrained concrete are considered to be:

1. Prevention of disintegrating effects of chlorides and alternate freezing and thawing.

2. High durability; unaffected by the action of salt spread on pavements to melt ice; elimination of surface scaling.

3. Increased resistance to the aggressive action of sulfates and sea waters.

4. Increased workability and cohesiveness.

5. Reduced segregation and bleeding tendency.

6. Produces a more homogeneous and durable concrete and better appearing structures.

Air in mass concrete should be not less than 3% nor more than 6%, preferably 4% to 5%. Corrections should be made for increased air by reducing the volume of sand. (Ordinary non-air-entraining cement contains about 1% air which is not usually corrected for.) About 4 gal. less water per cubic yard is required. Proper gradation of fine aggregate is important.

Air-entrained concrete ordinarily will have a slight decrease in strength from that of common portland cement concrete. However, close control should be maintained and air content should be determined frequently by means of an air-pressure meter (see Fig. 52), in which the entrained air and concrete in a closed container of known capacity can be measured by the reduction in volume caused by a known applied pressure. This must be done to prevent deviations in strength. Concrete cylinders should be made and tested to check strengths.

There are several types of pressure meters for determining the air content in plastic concrete, all of which operate on the same theory. The concrete is placed in a container of known volume, and water is forced by air pressure into the concrete. The volume of air displaced is indicated on a dial gage in terms of percentage of total air, which is read directly on the gage after calibration of the gage.

SEELYE STEVENSON VALUE & KNECHT
Consulting Engineers
101 PARK AVENUE   NEW YORK 17, N. Y.

CONCRETE PLANT INSPECTORS DAILY REPORT

Client _____ Date _____

Project _____ Report No. _____

Producer _____

Contractor _____

Section Poured _____ Strength Requirements: at 7 days _____

at 28 days _____

		Water-Cement	Gals.
Nominal Mix 1:____·____		Ratio _____	per Sack _____
Dry Wts.	Lb.	Lb.	Lb.
per Cu. Yd. _____ Cement _____		Fine Agg. _____	Coarse Agg. _____
Type &		Type &	Type &
Cement Brand _____	F. A. Source _____		C. A. Source _____

Admixture, Kind & Amount _____ Air Content _____ Per Cent _____

FIELD TESTS AND COMPUTATIONS FOR ACTUAL BATCH PROPORTIONS

PER CENT PASSING U. S. STANDARD SIEVES

Sieve Coarse Agg.	2″	1½″	1″	¾″	⅜″	No. 4	F.M.
Sieve Fine Agg.	No. 4	8	16	30	50	100	F.M.

Time	Moisture %	Free Water Gals. per Cu. Yd.	Actual Batch Wts.	Added Water Gals.	Actual W/C Gals. per Sack	Method of Batching / Weather
F. A.	_____	_____	_____			Average A.M. Temp. P.M.
C. A.	_____	_____	_____	____	____	Slump Ordered
F. A.	_____	_____	_____			Time Arrived Plant
C. A.	_____	_____	_____	____	____	
F. A.	_____	_____	_____			Time Left Plant
C. A.	_____	_____	_____	____	____	Total Hours
						No. Cu. Yds.
						Inspector

Respectfully Submitted

SEELYE STEVENSON VALUE & KNECHT

SEELYE STEVENSON VALUE & KNECHT

Consulting Engineers

101 Park Avenue   New York 17, N. Y.

CONCRETE INSPECTION

Project _____ Date _____

Supplier _____ Truck No. _____

Contractor _____

Nominal Mix: 1 _____

Required Strength,

28 days _____ p.s.i.

Class or

Type _____

	Dry Wt. per Cu. Yd.	Wet Batch Wt. per Cu. Yd.
Cement		
Sand		
Gravel		
Stone		
Water		
Admixture		
Moisture Fine Agg.		
Per Cent Coarse Agg.		
Load No. _____ Cu. Yds.		
Rated Truck Capacity: _____ Cu. Yds.		

The above batch weights are certified as correct.

_____

Inspector

Firm Name _____

FIG. 53. Truck ticket for transit mix. Plant inspector fills out one for each truck. Driver turns in copy to field inspector before load is accepted.

SEELYE STEVENSON VALUE & KNECHT

Consulting Engineers

101 PARK AVENUE   NEW YORK 17, N. Y.

REPORT ON CONCRETE STRUCTURES
FIELD INSPECTION
(Short Form)

Report No. _____   Date _____

Job _____   Temp. _____

Reported to _____

WORK INSPECTED

	Location or Station	Reinforcement and Forms	Concrete (cu. yds. placed)	
			This day	TOTAL To date
Footings				
Columns				
Beams				
Slabs				
Walls				

Admixture _____

Air content _____

Aggregate inspected _____

Slump tests made _____

Test cylinders made _____

Frost protection checked _____

Curing _____

Remarks _____

(orders to Contractor, notes on concrete)

_____
Inspector

Engineer

REPORT ON CONCRETE TEST SPECIMENS *

FIELD DATA
To be filled in by maker of specimens

Project, name and symbol:	District:

No. specimens in shipment:	Type of specimens: ☐ Cylinders: ☐ Cores: ☐ Beams: Diameter, in.	Symbols:

Date placed:	Date extracted (*cores*):	Date shipped:	Monolith extracted from:

Cubic yards represented:	Mixture (*by weight*):	Slump, in.:	Sand. Total aggregate ratio % by vol.

Theoretical unit wt.: lb./cu. ft.	Actual Unit Weight: lb./cu. ft.	Calculated air content: %	Net actual *W/C*: gal./bag

Admixture type and amount: %	*Cement Factor*	
	Theoretical: bags/cu. yd.	Actual: bags/cu. yd.

Cement

Cement Spec. SS-C- Type:	Brand:
Mill name and location:	Car number:

Type and Source of Aggregate

Fine aggregate:	Coarse aggregate:

By Whom Prepared (*Inspector*)

LABORATORY DATA
To be filled in by laboratory

Specimen No.	Date Tested:	☐ Flexural Strength, p.s.i. ☐ Compressive Strength, p.s.i.				
		7-Day	14-Day	28-Day	90-Day	180-Day

Date specimens received:	Temperature of specimens when received: °F.

Remarks:

* *From War Department Corps of Engineers, North Atlantic Division.*

Engineer

REPORT ON CONCRETE STRUCTURES
FIELD INSPECTION
(*Long Form*)

Report No. _____   Date _____

Job _____   Temp. _____

Reported to _____

WORK INSPECTED

	Location or Station	Reinforcement and Forms	Concrete	Yardage	
				Today	Total to date incl.
Footings					
Columns					
Beams					
Slabs					
Walls					

Cement: tested and sealed _____

Coarse aggregate: size, appearance, cleanliness, soundness _____

Fine aggregate: grading, silt content by bottle sediment test _____

Forms: dimensions, oil, cleanliness, tightness _____

Reinforcement: inserts, recesses, concrete coverage for protection _____

Slump tests: cylinders and/or test beams _____

Construction joints _____

Mixing: proportioning, water content, time of mixing _____

Concrete placing, vibration or rodding _____

Finishing _____

Protection vs. frost _____

Curing _____

Form stripping and reshoring _____

Air content_____

_____
Inspector

_____
Engineer

CEMENT SHIPPING REPORT *

_____

Gentlemen:

Shipments of portland cement indicated have been mill inspected and sealed for your account.

Car Number	Seal	Contents Bbl.	Bin	Brand	Destination

Reported to:

_____
Inspector

* *Adapted from Haller Testing Laboratories, Inc.*

MASONRY

CHECK LIST FOR INSPECTORS

Inspectors' Equipment

Complete set of plans, specifications, approved samples and shop drawings.

Set of sieves of specified sand sizes.

Plumb bob and line.

6-foot rule.

Procedure in Inspection

Prepare and ship samples of brick, concrete block, clay tile, sand lime bricks, cement, and sand lime to laboratory for test.

Perform sieve tests on sand for mortar at site.

Inspect brick. Discard underburned brick (sometimes called salmon brick), which is pale in color if a red brick. Compare brick with specifications. Face brick can best be checked from approved sample, or sample panel.

See that joints are according to specification.

If the engineer has built up a sample of wall, see that this is followed.

Check thickness of joints, type of pointing, and mortar against specifications.

Check lime against lime memorandum on p. 77, particularly as to length of time after slaking.

Do not permit laying of brick in weather cold enough to freeze mortar. See specifications.

Check bonding of brickwork.

No voids permitted in interior of wall.

Check wall for plumbness and level courses.

All flashings, weep holes, and sills built in as required by plans and specifications.

Lift brick up that are laid. There should be sufficient suction to lift mortar with them.

All brick, except face brick, to be thoroughly wet before laying, and brick laid at a time when freezing may be expected shall be protected from the formation of ice.

Follow specifications for back plastering; if not covered in specifications, face brick to be back-plastered before setting back-up brick.

All joints to be completely filled with mortar. Mortar to be troweled on end of brick and brick shoved into place.

Set brick accurately to line when first laid; the bond is broken when a brick is moved after partial set.

See that the type of joint specified is carried out. See that a jointing tool is used that produces the joint specified and does not tend to draw the mortar away from the brick.

Around windows and doorways, brickwork to be kept back from frames a sufficient distance to permit a calked joint.

Steel members imbedded in masonry and not indicated to be fireproofed with concrete shall be "buttered" with not less than ½ in. of setting mortar.

Note that brickwork before setting of the mortar has no transverse or arching strength; therefore, see that no more weight of raw brickwork is laid at one time than the lintels, piers, or other supporting members can carry without excessive deflection.

Where putlog holes are left through the wall for support of outside scaffolding, see that the holes are filled with mortar for the full thickness of the wall after they have served their purpose.

LIME FOR MORTAR AND MASONRY

Lime is produced in two forms as follows:

1. High-calcium quicklime, which is sent to the job in powdered form of two different kinds: pulverized or granular, labeled as quicklime. One has no particular advantage over the other.

This lime is slaked by adding water similar to the method of preparing lump lime, and must be allowed to age 3 to 7 days.

One ton of quicklime will produce approximately 80 cu. ft. of stiff lime putty.

2. Hydrated lime, which is lime containing water in chemical combination. It is a calcium hydroxide and comes on the job labeled hydrated masons' lime. This lime also comes in two different kinds: (a) Ordinary hydrated lime. This product should be soaked in water for not less than 24 hours before using. (b) Pressure hydrated lime. This lime can safely be put in the mixer without any treatment whatever. It is used exactly the same as cement.

STRUCTURAL – BRICK MASONRY

COMMON (Header bond)
¾ brick   Bond course every 6th row

Stretcher or running bond, similar but without headers, except every other course at corner.

COMMON (Flemish bond)
¾ brick   Bond course every 6th row

ENGLISH
(Closer)

FLEMISH (Double stretcher)

ENGLISH (Cross)

FLEMISH

FLEMISH (Cross)

FLEMISH (Diagonal)

SPECIAL BRICK SIZES

Norman: 12" length × 2¼" × 3¾"
Roman: 12" length × 1⅝" × 3¾"
Baby Roman: " × 1⅝" × 3¾"
Two Brick Type: 5" high × 8" × 3¾"
Oversize: 2¾" × 3¾" × 8"

BRICK JOINTS
3" = 1'–0"

Struck   Weathered
Raked   Strip-   "V"   Concave   Flush   Beaded
          ped   or   or
          nosed   plain cut

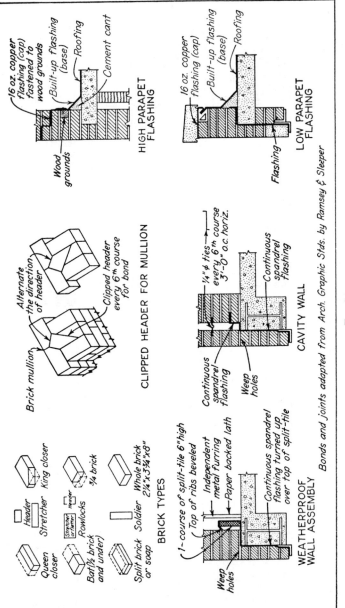

HIGH PARAPET FLASHING

16 oz. copper flashing (cap) fastened to wood grounds
(Built-up flashing (base)
Roofing
Cement cant
Wood grounds

LOW PARAPET FLASHING

16 oz. copper flashing (cap)
Built-up flashing (base)
Roofing
Flashing

CLIPPED HEADER FOR MULLION

Brick mullion
Alternate the direction of header
Clipped header every 6th course for bond

CAVITY WALL

¼"∅ ties—every 6th course 3'-0" o.c. horiz.
Continuous spandrel flashing
Continuous spandrel flashing
Weep holes

BRICK TYPES

Queen closer
Header
Stretcher
King closer
Rowlocks
¾ brick
Whole brick 2¼"x3¾"x8"
Bat(½ brick and under)
Split brick or soap
Soldier

WEATHERPROOF WALL ASSEMBLY

1 - course of split-tile 6" high (Top of ribs beveled)
Independent metal furring
Paper backed lath
Continuous spandrel flashing turned up over top of split-tile
Weep holes

Bonds and joints adapted from Arch. Graphic Stds. by Ramsey & Sleeper

FIG. 54.

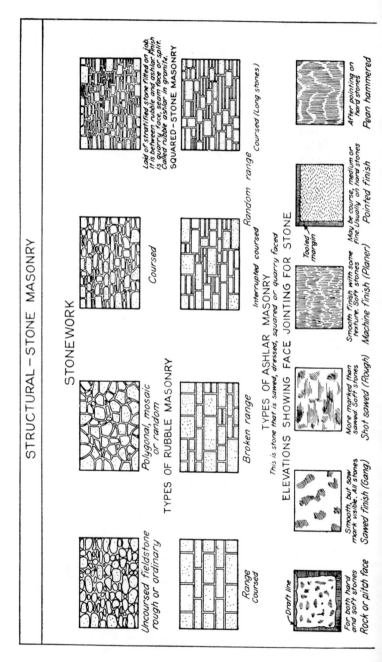

STRUCTURAL-STONE MASONRY

STONEWORK

TYPES OF RUBBLE MASONRY

Uncoursed fieldstone rough or ordinary

Polygonal, mosaic or random

Coursed

Interrupted coursed

Random range

Coursed (Long stones)

Range Coursed

Broken range

Laid of stratified stone fitted on job. It is between rubble and ashlar. Finish is quarry face, seam face or split. Called rubble ashlar in granite. SQUARED-STONE MASONRY

TYPES OF ASHLAR MASONRY
This is stone that is sawed, dressed, squared or quarry faced

ELEVATIONS SHOWING FACE JOINTING FOR STONE

For both hard and soft stones
Rock or pitch face

Smooth, but saw mark visible. All stones visible.
Sawed finish (Gang)

More marked than sawed. Soft stones
Shot sawed (Rough)

Smooth finish with some texture. Soft stones
Machine finish (Planer)

May be coarse, medium or fine. Usually on hard stones
Pointed finish

After pointing on hard stones
Pean hammered

Tooled margin

Draft line

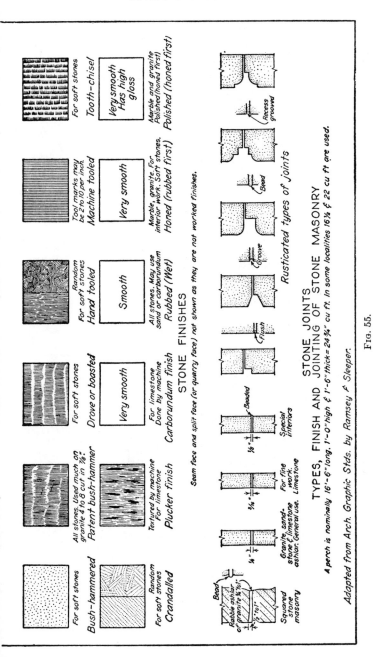

STONE FINISHES

Bush-hammered	Crandalled	Patent bush-hammer	Plucker finish	Drove or boasted	Hand tooled	Machine tooled	Tooth-chisel
For soft stones	For soft stones Random	All stones. Used much on granite 4 to 8 cut in ⅜".	Textured by machine For limestone	For soft stones	Random For soft stones	Tool marks may be 2 to 10 per inch.	For soft stones

Squared stone masonry	Carborundum finish	Rubbed (Wet)	Smooth	Very smooth	Honed (rubbed first)	Very smooth	Polished (honed first)	Very smooth Has high gloss
	For limestone Done by machine	All stones. May use sand or carborundum			Marble, granite. For interior work. Soft stones.		Marble and granite Polished (honed first)	

Seam face and split face (or quarry face) not shown as they are not worked finishes.

STONE JOINTS

½" ±⅛" Granite, sand-stone & limestone ashlar. General use.

¾" For fine work. Limestone

⅛" ± Special interiors — Beaded

Flush

Rusticated types of joints — Groove, Bead, Recess grooved

Rabble ashlar or granite ¾/½" — Bead — ½" to 1"

A perch is nominally 16'-6" long, 1'-0" high & 1'-6" thick = 24¾" cu ft. In some localities 16½ & 22 cu ft are used.

TYPES, FINISH AND JOINTING OF STONE MASONRY

Adapted from Arch. Graphic Stds. by Ramsey & Sleeper.

FIG. 55.

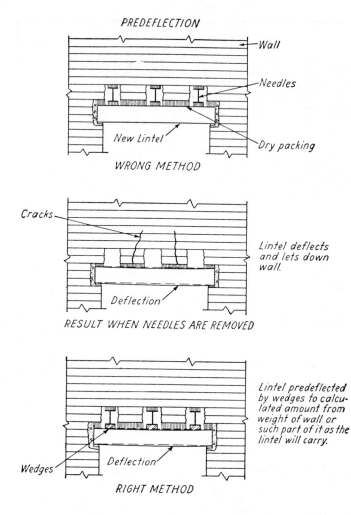

PREDEFLECTION

Wall

Needles

New Lintel

Dry packing

WRONG METHOD

Cracks

Lintel deflects
and lets down
wall.

Deflection

RESULT WHEN NEEDLES ARE REMOVED

Lintel predeflected
by wedges to calcu-
lated amount from
weight of wall or
such part of it as the
lintel will carry.

Wedges   Deflection

RIGHT METHOD

All structural materials are elastic and because of this
property will deflect under load. If the weight of masonry
is allowed to deflect a beam supporting it, cracks will occur
in the masonry. Therefore, lintels or beams inserted to support
masonry already in place should be predeflected to avoid
letting down the masonry and causing cracks.

Fig. 56.

BRICK

DISCUSSION

Face brick is specified for use where distinctive brickwork finish is intended. It is usually packed in straw or cardboard separators for protection against damage. Finish is smooth or textured as specified. There are three grades (ASTM, C-216).

Type FBX: narrow variation in size, and generally in straight shades of color.

Type FBS: wider color ranges and greater variation in size.

Type FBA: generally special sizes; includes swelled or clinker brick and greater dimensional tolerances.

Ceramic glazed face brick (enameled brick, ASTM, C-126) is also used for distinctive brickwork but has a glazed finish. Usually it is packed in cartons to prevent damage. Its name is derived from the fact that ceramic clay is applied to the surface of fire-clay brick for architectural effect. It is made in glossy or matte finish.

Common brick (building brick) is used where the superior architectural effect of face brick is not called for. It may be either hand stacked or dumped as specified. Grade SW is used for resistance to extreme frost action or for work below grade; Grade MW, for exterior or interior use above grade in normal climates; Grade NW, as back-up for interior work only. (ASTM, C-62.)

Sand-Lime Building Brick. (Not a burned product.) Grade and use are similar to those for common brick and sometimes back-up; usually white (ASTM, C-73).

Paving brick is tough, durable, and evenly burned. When broken it shows uniformly dense structure free from lime, air pockets, and marked laminations. Check with specifications. (See ASTM, C-7.)

Sewer brick is made specially for sewers and their structures carrying sanitary and industrial wastes. Grades SA, MA, and NA are to be used as specified (or see ASTM, C-32). Visual inspection shall show freedom from cracks, warpage, stones, pebbles, or particles of lime affecting serviceability.

General. Brick, except sand-lime brick, are burned from clay, shale, or fire clay, or mixtures thereof. Fire-clay brick burns to buffs and grays; shale brick burns to red with variations in color to browns and blacks.

Underburned (salmon) brick is too soft for ordinary use and should be rejected. It is salmon colored but this light color alone does not necessarily indicate underburned brick as some clays may produce a similar color. Salmon brick gives a dull sound when struck and is likely to disintegrate when wet and subjected to freezing. See Fig. 81.

Efflorescence on laid-up brickwork may be caused by the leaching-out of salts in brick or mortar; no cure is known. The use of calcium chloride or other salts as an antifreeze and a retarder in mortar may increase efflorescence.

Waterproofing. No waterproofing or application should be put on a brick wall until the wall is dry.

Wetting of Brick. Make an oil crayon circle around a 25-cent piece and apply 20 drops of water in the circle. If the brick absorbs the water within 1½ minutes, the bricks should be wetted before laying; otherwise they will absorb water from mortar needed for its proper hydration.

Silicone exterior masonry water repellent can be applied to finished brickwork and stucco walls for repelling of water and protection of masonry joints, preventing efflorescence, and retarding staining. Application by spray using a low-pressure tree spray with rubber washers and hose replaced by Neoprene is recommended in preference to application by brush. It is recommended that the waterproofing mixture contain at least 3% of silicone resin solids.

CONCRETE BUILDING UNITS

Plant Inspection

Concrete mix should be checked against specifications (cement shall be not less than 4 sacks to 54 cu. ft. of loose aggregate). Check gradation between sand and coarse aggregate by trial to produce dense concrete required. For instance, exterior blocks should be dense and smooth; interior blocks which are to receive plaster should be rough.

Check compacting unit (tamping, vibration or both) by visual observation of uniformity of units and proper stripping.

Check specification to see whether high-pressure or low-pressure steam curing is required.

Low-pressure steam curing: Air-cure for 2 hours; then in closed curing units, steam chamber at 170° F. ± 5° for 8 hours; then cure in temperature above 40° F. for 7 days before delivery.

High-pressure steam curing: Total curing cycle 12 hours: 2 hours to raise from atmospheric pressure to 145 lb.; hold at 145 lb. for 8 hours, then reduce atmospheric pressure in 2 hours. Steam to be at 370° F. No further treatment is necessary in all temperatures.

Low-grade concrete and cinder blocks can be made without steam curing, but the process is not recommended.

IDENTIFICATION OF BUILDING STONE

Granite is a coarse-grained, hard, igneous rock in which the different minerals give a speckled appearance.

True granite contains the following elements:

Quartz—a clear, hard crystal.

Feldspar, which looks like a yellowish tooth.

Hornblende—hard, black, shiny.

Mica—thin, flaky, transparent.

Pyrite, which looks like a yellowish metal.

Bastard granite contains some but not all of these crystals.

Both granites are excellent building materials, although too much pyrite might cause stain and a possible breaking down of the stone by weathering.

Gneiss may be either sedimentary or igneous rock which has been metamorphosed, that is, compressed and worked under sufficient pressure and heat so that the structural changes were by plastic flow rather than by cracking.

In gneiss, the interlocking minerals are for the most part visible to the naked eye. The gneisses are banded. Gneiss is a satisfactory building material. Its constituents are the same as those of granite.

Gneisses merge into schists as the texture becomes finer.

Schists with a large percentage of mica are known as mica schists. As a building material they are subject to cleavage.

Basalt is a heavy, dark, igneous rock ranging from dark gray to black and very fine in texture. As commonly used, basalt includes all dark, basic, volcanic rocks. It makes an excellent road material and concrete aggregate and, in some cases, building stone.

Bluestone is the commercial name for a dark, bluish-gray sandstone, quarried in New York State. It is an arkose, i.e., composed of the same constituents as granite (feldspar, mica, and quartz), from which it probably was formed by disintegration and later consolidation. It is noted for its fine grain and high crushing strength, up to 1900 p.s.i. The term has also been applied locally to other rocks, among which are dark blue slate and blue limestone.

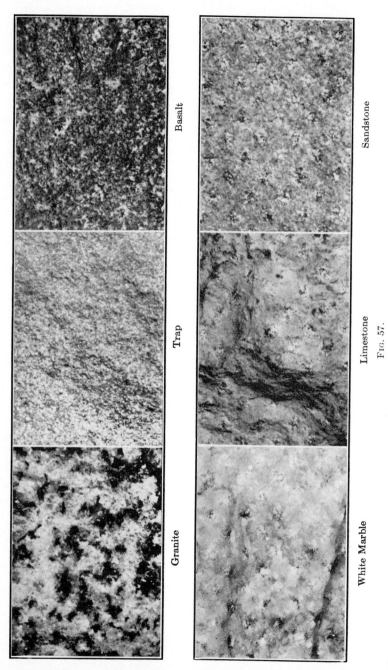

Basalt

Sandstone

Trap

Limestone

Fig. 57.

Granite

White Marble

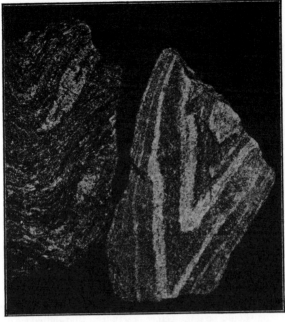

Schist   Gneiss

FIG. 58.

Limestone is calcium carbonate rock of sedimentary origin. It is somewhat vulnerable and may be distinguished from magnesium carbonate limestone by the fact that it effervesces under a dilute solution of acid, which is not the case with the dolomite. Individual grains cannot be distinguished. Limestone is soft, easily worked, and a reasonably good building stone but vulnerable.

Sandstone, as its name implies, is made up of sand cemented with silica or lime. In general, the grains are distinguishable. Its reliability as a building material can be ascertained only after investigation; for instance, brownstone is a sandstone which has not always proved reliable.

Slates are metamorphosed shale and have cleavage planes along which the stone is split for commercial purposes. These cleavage planes occur at an angle to the bed planes. Slates are a satisfactory building material, particularly for roofs.

Shale comes from silt and clay and occurs in beds which tend to "shale" off. It is softer than limestone and unreliable as a building material.

Caution: Sedimentary stone should be laid on natural beds.

Definition: Porphyritic texture means a texture in which the larger minerals appear to be embedded in a matrix.

STRUCTURAL STEEL

CHECK LIST FOR INSPECTORS

The following is based on the assumption that steel has been inspected in shop. If this has not been done, steel should be completely checked against shop details and for correct sections.

Inspectors' Equipment

Complete set of erection plans and specifications.

Details should not be necessary unless shop inspection was not made or unless necessary to show special field connections.

Steel tape.

6-ft. rule.

Plumb bob.

Rivet-testing hammer.

Steel handbook.

Necessary coveralls, helmet, gloves, etc.

Calipers, gages, etc.

Procedure in Inspection

Members should be checked for damage in shipment, such as bent plates, connection angles or members themselves, and condition of paint. This checking should be done before erection so that damaged pieces may be rejected or rectified by straightening or reinforcing.

Anchor bolts should be checked as to size, location, elevation, and plumbing.

Base plates and grillages should be checked for correct work, level, and proper grouting. In general, they should be leveled up so as to carry load direct to foundations or walls.

Columns resting on base plates, grillages, or girders and column splices should be checked for proper bearing of milled surfaces. Where column sections change in nominal section and milled fillers are used, they should be carefully inspected.

Minor corrections may be made with steel shims.

Plumbing of columns should be checked to specified tolerance before any riveting or permanent bolting of floors is done.

As erection proceeds, inspector should match pieces against erection plans to see that proper piece is in correct position. Usually material is properly marked, but where there is any doubt, section of member should be checked.

The inspector should make sure that rivets or turned bolts are used where called for on erection plans or specifications. If there is any question

TABLE 17. STRUCTURAL STEEL SECTIONS

Amer. Std. Channel Sect.

D	Wt	S	t	d	b
18 [	58	74.5	.70	18	4¼
	51.9	69.1	.60	18	4⅛
	45.8	63.7	.50	18	4
	42.7	61.0	.45	18	4
15 [	50	53.6	.72	15	3¾
	40	46.2	.52	15	3½
	33.9	41.7	.40	15	3⅜
13 * [	50	48.1	.79	13	4⅜
	40	41.7	.56	13	4⅛
	35	38.6	.45	13	4⅛
	31.8	36.5	.38	13	4

D	Wt	S	t	d	b
12 [	30	26.9	.51	12	3⅛
	25	23.9	.39	12	3
	20.7	21.4	.28	12	3
10 [	30	20.6	.67	10	3
	25	18.1	.53	10	2⅞
	20	15.7	.38	10	2¾
	15.3	13.4	.24	10	2⅝

D	Wt	S	t	d	b
9 [	20	13.5	.45	9	2⅝
	15	11.3	.28	9	2½
	13.4	10.5	.23	9	2⅜
8 [	18.75	10.9	.49	8	2½
	13.75	9.0	.30	8	2⅜
	11.5	8.1	.22	8	2¼
7 [	14.75	7.7	.42	7	2¼
	12.25	6.9	.31	7	2¼
	9.8	6.0	.21	7	2⅛

D	Wt	S	t	d	b
6 [	13.0	5.8	.44	6	2⅛
	10.5	5.0	.31	6	2
	8.2	4.3	.20	6	1⅞
5 [	9.0	3.5	.32	5	1⅞
	6.7	3.0	.19	5	1¾
4 [	7.25	2.3	.32	4	1¾
	5.4	1.9	.18	4	1⅝
3 [	6.0	1.4	.36	3	1⅝
	5.0	1.2	.26	3	1½
	4.1	1.1	.17	3	1⅜

Amer. Std. Beam Sect.

D	Wt	S	t	d	b
24 I	120	250.9	.80	24	8
	105.9	234.3	.62	24	7⅞
	100	197.6	.75	24	7¼
	90	185.8	.62	24	7⅛
	79.9	173.9	.50	24	7
20 I	95	160.0	.80	20	7¼
	85	150.2	.65	20	7
	75	126.3	.64	20	6⅜
	65.4	116.9	.50	20	6¼

D	Wt	S	t	d	b
18 I	70	101.9	.71	18	6¼
	54.7	88.4	.46	18	6
15 I	50	64.2	.55	15	5⅝
	42.9	58.9	.41	15	5½
12 I	50	50.3	.69	12	5½
	40.8	44.8	.46	12	5¼
	35	37.8	.43	12	5⅛
	31.8	36.0	.35	12	5

D	Wt	S	t	d	b
10 I	35	29.2	.59	10	5
	25.4	24.4	.31	10	4⅝
8 I	23.0	16.0	.44	8	4⅛
	18.4	14.2	.27	8	4
7 I	20.0	12.0	.45	7	3⅞
	15.3	10.4	.25	7	3⅝
6 I	17.25	8.7	.47	6	3⅝
	12.5	7.3	.23	6	3⅜

D	Wt	S	t	d	b
5 I	14.75	6.0	.49	5	3¼
	10.0	4.8	.21	5	3
4 I	9.5	3.3	.33	4	2¾
	7.7	3.0	.19	4	2⅝
3 I	7.5	1.9	.35	3	2½
	5.7	1.7	.17	3	2⅜

Nomenclature

D = nominal depth in inches.
Wt = weight per foot in pounds.
S = Section modulus.

t = web thickness in inches.
d = actual depth in inches.
b = flange width in inches.

I = American Standard Section.
* Car building and shipbuilding.

TABLE 17. STRUCTURAL STEEL SECTIONS (Continued)

Bethlehem and Carnegie Structural Sections

D	Wt	S	t	d	b
36 WF	300	1105.1	.94	36¾	16⅝
	280	1031.2	.88	36½	16⅝
	260	951.1	.84	36¼	16½
	245	892.5	.80	36	16½
	230	835.5	.76	35⅞	16½
	194	663.6	.77	36½	12⅛
	182	621.2	.72	36⅜	12⅛
	170	579.1	.68	36⅛	12
	160	541.0	.65	36	12
	150	502.9	.62	35⅞	12
33 WF	240	811.1	.83	33½	15⅞
	220	740.6	.77	33¼	15¾
	200	669.6	.71	33	15¾
	152	486.4	.63	33½	11⅝
	141	446.8	.60	33¼	11½
	130	404.8	.58	33⅛	11½

D	Wt	S	t	d	b
24 WF	160	413.5	.66	24¾	14⅛
	145	372.5	.61	24½	14
	130	330.7	.56	24¼	14
	120	299.1	.56	24¼	12⅛
	110	274.4	.51	24⅛	12
	100	248.9	.47	24	12
	94	220.9	.52	24¼	9
	84	196.3	.47	24⅛	9
	76	175.4	.44	23⅞	9
21 WF	142	317.2	.66	21½	13⅛
	127	284.1	.59	21¼	13
	112	249.6	.53	21	13
	96	197.6	.57	21⅛	9
	82	168.0	.50	20⅞	9
	73	150.7	.45	21¼	8¼
	68	139.9	.43	21⅛	8¼
	62	126.4	.40	21	8¼

D	Wt	S	t	d	b
14 WF	426	707.4	1.87	18¾	16¾
	398	656.9	1.77	18¼	16⅝
	370	608.1	1.65	18	16½
	342	559.4	1.54	17½	16⅜
	314	511.9	1.41	17¼	16¼
	287	465.5	1.31	16¾	16⅛
	264	427.4	1.20	16½	16
14 WF	320	492.8	1.89	16¾	16¾
	246	397.4	1.12	16¼	16
	237	382.2	1.09	16⅛	15⅞
	228	367.8	1.04	16	15⅞
	219	352.6	1.00	15⅞	15⅞
	211	339.2	.98	15¾	15¾
	202	324.9	.93	15⅝	15¾
	193	310.0	.89	15½	15¾
	184	295.8	.84	15⅜	15⅝
	176	281.9	.82	15¼	15⅝
	167	267.3	.78	15⅛	15⅝
	158	253.4	.73	15	15½
	150	240.2	.69	14⅞	15½
	142	226.7	.68	14¾	15½

D	Wt	S	t	d	b
12 WF	190	263.2	1.06	14⅜	12⅝
	161	222.2	.90	13⅞	12½
	133	182.5	.75	13⅜	12⅜
	120	163.4	.71	13⅛	12⅜
	106	144.5	.62	12⅞	12¼
	99	134.7	.58	12¾	12¼
	92	125.0	.54	12⅝	12⅛
	85	115.7	.49	12½	12⅛
	79	107.1	.47	12⅜	12⅛
	72	97.5	.43	12¼	12
	65	88.0	.39	12⅛	12
12 WF	58	78.1	.36	12¼	10
	53	70.7	.34	12	10
	50	64.7	.37	12¼	8⅛
	45	58.2	.34	12	8
	40	51.9	.29	12	8
	36	45.9	.30	12¼	6⅝
	31	39.4	.27	12⅛	6½
	27	34.1	.24	12	6½
	22	25.3	.26	12¼	4
	19	21.4	.24	12⅛	4
	16.5	17.5	.23	12	4
	14	14.8	.20	11⅞	4

D	Wt	S	t	d	b
8 WF	67	60.4	.57	9	8¼
	58	52.0	.51	8¾	8¼
	48	43.2	.40	8½	8⅛
	40	35.5	.36	8¼	8⅛
	35	31.1	.31	8⅛	8
	31	27.4	.29	8	8
	28	24.3	.29	8	6½
	24	20.8	.24	7⅞	6½
	20	17.0	.25	8⅛	5¼
	17	14.1	.23	8	5¼
	15	11.8	.24	8⅛	4
	13	9.9	.23	8	4
	10	7.8	.17	7⅞	4
8 C	34.3	28.9	.37	8	8
6 WF	25	16.8	.32	6⅜	6
	20	13.4	.26	6¼	6
6 WF	15.5	10.1	.24	6	6
	16	10.1	.26	6¼	4
	12	7.24	.23	6	4
	8.5	5.07	.17	5⅞	4

D	Wt	S	t	d	b
30 WF	210	649.9	.77	30⅜	15⅛
	190	586.1	.71	30⅛	15
	172	528.2	.65	29⅞	15
	132	379.7	.61	30¼	10½
	124	354.6	.58	30⅛	10½
	116	327.9	.56	30	10½
	108	299.2	.55	29⅞	10½
27 WF	177	492.8	.72	27¼	14⅛
	160	444.5	.66	27⅛	14
	145	402.9	.60	26⅞	14
	114	299.2	.57	27¼	10⅛
	102	266.3	.52	27⅛	10
	94	242.8	.49	26⅞	10

D	Wt	S	t	d	b
18 WF	114	220.1	.59	18½	11⅞
	105	202.2	.55	18⅜	11¾
	96	184.4	.51	18⅛	11¾
	85	156.1	.53	18⅜	8⅞
	77	141.7	.47	18⅛	8¾
	70	128.2	.44	18	8¾
	64	117.0	.40	17⅞	8¾
	60	107.8	.42	18¼	7½
	55	98.2	.39	18⅛	7½
	50	89.0	.36	18	7½
16 WF	96	166.1	.53	16⅜	11½
	88	151.3	.50	16⅛	11½
	78	127.8	.53	16⅜	8⅝
	71	115.9	.49	16⅛	8½
	64	104.2	.44	16	8½
	58	94.1	.41	15⅞	8½
	50	80.7	.38	16¼	7⅛
	45	72.4	.35	16⅛	7
	40	64.4	.31	16	7
	36	56.3	.30	15⅞	7

D	Wt	S	t	d	b
14 WF	136	216.0	.66	14¾	14¾
	127	202.0	.61	14⅝	14¾
	119	189.4	.57	14½	14⅝
	111	176.3	.54	14⅜	14⅝
	103	163.6	.49	14¼	14⅝
	95	150.6	.46	14⅛	14½
	87	138.1	.42	14	14½
14 WF	84	130.9	.45	14⅛	12
	78	121.1	.43	14	12
	74	112.3	.45	14¼	10⅛
	68	103.0	.42	14	10
	61	92.2	.38	13⅞	10
	53	77.8	.37	14	8
	48	70.2	.34	13¾	8
	43	62.7	.31	13⅝	8
	38	54.6	.31	14⅛	6¾
	34	48.5	.29	14	6¾
	30	41.8	.27	13⅞	6¾

D	Wt	S	t	d	b
10 WF	112	126.3	.75	11⅜	10⅜
	100	112.4	.68	11⅛	10⅜
	89	99.7	.61	10⅞	10¼
	77	86.1	.53	10⅝	10¼
	72	80.1	.51	10½	10⅛
	66	73.7	.46	10⅜	10⅛
	60	67.1	.41	10¼	10⅛
	54	60.4	.37	10⅛	10
	49	54.6	.34	10	10
10 WF	45	49.1	.35	10⅛	8
	39	42.2	.32	10	8
	33	35.0	.29	9¾	8
	29	30.8	.29	10¼	5¾
	25	26.4	.25	10⅛	5¾
	21	21.5	.24	9⅞	5¾
	19	18.8	.25	10¼	4
	17	16.2	.24	10⅛	4
	15	13.8	.23	10	4
	11.5	10.5	.18	9⅞	4

D	Wt	S	t	d	b
6 C	25	15.7	.31	6	6
	20	12.9	.25	6	6
5 WF	18.9	9.5	.31	5	5
	18.5	9.9	.26	5⅛	5
	16	8.5	.24	5	5
4 C	13	5.2	.25	4	4
4 B	13	5.45	.28	4⅛	4
	10	4.16	.22	4	4

Nomenclature

D = nominal depth in inches.
Wt = weight per foot in pounds.
S = section modulus.
t = web thickness in inches.
d = actual depth in inches.
b = flange width in inches.
B = Bethlehem Steel Co. Section.
C = U. S. Steel Corp. Section.

Properties of Angles

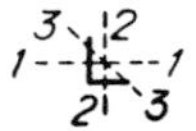

Size	S 1–1	Area	r 1–1	r 3–3	Size	S 1–1	S 2–2	Area	r 1–1	r 2–2	r 3–3
2½ x 2½ x ¼	.39	1.19	.77	.49	2½ x 2 x ¼	.38	.25	1.06	.78	.59	.42
3 x 3 x ¼	.58	1.44	.93	.59	3 x 2½ x ¼	.56	.40	1.31	.95	.75	.53
3 x 3 x 5/16	.71	1.78	.92	.59	3½ x 3 x ¼	.78	.59	1.56	1.11	.91	.63
3½ x 3½ x ¼	.79	1.69	1.09	.69	4 x 3 x ¼	1.0	.60	1.69	1.28	.90	.65
3½ x 3½ x 5/16	.98	2.09	1.08	.69	4 x 3 x 5/16	1.2	.73	2.09	1.27	.89	.65
4 x 4 x ¼	1.1	1.94	1.25	.80	5 x 3½ x 5/16	1.9	1.0	2.56	1.61	1.03	.77
4 x 4 x 5/16	1.3	2.40	1.24	.79	5 x 3½ x ⅜	2.3	1.2	3.05	1.60	1.02	.76
5 x 5 x ⅜	2.4	3.61	1.56	.99	5 x 3½ x 7/16	2.6	1.4	3.53	1.59	1.01	.76
6 x 6 x ⅜	3.5	4.36	1.88	1.19	6 x 4 x ⅜	3.3	1.6	3.61	1.93	1.17	.88
6 x 6 x 7/16	4.1	5.06	1.87	1.19	6 x 4 x 7/16	3.8	1.9	4.18	1.92	1.16	.87
6 x 6 x ½	4.6	5.75	1.86	1.18	6 x 4 x ½	4.3	2.1	4.75	1.91	1.15	.87

TABLE 18. WIRE AND SHEET METAL GAGES IN DECIMALS OF AN INCH

Gage No.	United States Standard Gage * — Weight, oz. per sq ft.	United States Standard Gage * — Approximate Thickness, in.	The United States Steel Wire Gage — Steel Wire except Music Wire	American or Brown & Sharpe Wire Gage — Non-Ferrous Sheets and Wire	New Birmingham Standard Sheet & Hoop Gage — Iron and Steel Sheets and Hoops	British Imperial or English Legal Standard Wire Gage — Wire	Birmingham or Stubs Iron Wire Gage — Strips, Bands, Hoops and Wire
	(Uncoated Steel Sheets and Light Plates)			Thickness, in.			
7/0's			.4900		.6666	.500	
6/0's			.4615	.5800	.625	.464	
5/0's			.4305	.5165	.5883	.432	.500
4/0's			.3938	.4600	.5416	.400	.454
3/0's			.3625	.4096	.500	.372	.425
2/0's			.3310	.3648	.4452	.348	.380
0			.3065	.3249	.3964	.324	.340
1			.2830	.2893	.3532	.300	.300
2			.2625	.2576	.3147	.276	.284
3	160	.2391	.2437	.2294	.2804	.252	.259
4	150	.2242	.2253	.2043	.250	.232	.238
5	140	.2092	.2070	.1819	.2225	.212	.220
6	130	.1943	.1920	.1620	.1981	.192	.203
7	120	.1793	.1770	.1443	.1764	.176	.180
8	110	.1644	.1620	.1285	.1570	.160	.165
9	100	.1495	.1483	.1144	.1398	.144	.148
10	90	.1345	.1350	.1019	.1250	.128	.134
11	80	.1196	.1205	.0907	.1113	.116	.120
12	70	.1046	.1055	.0808	.0991	.104	.109
13	60	.0897	.0915	.0720	.0882	.092	.095
14	50	.0747	.0800	.0641	.0785	.080	.083
15	45	.0673	.0720	.0571	.0699	.072	.072
16	40	.0598	.0625	.0508	.0625	.064	.065
17	36	.0538	.0540	.0453	.0556	.056	.058
18	32	.0478	.0475	.0403	.0495	.048	.049
19	28	.0418	.0410	.0359	.0440	.040	.042
20	24	.0359	.0348	.0320	.0392	.036	.035

Gage No.	Gage No.						
21		.0329	.0318	.0285	.0349	.032	.032
22		.0299	.0286	.0253	.0313	.028	.028
23		.0269	.0258	.0226	.0278	.024	.025
24		.0239	.0230	.0201	.0248	.022	.022
25		.0209	.0204	.0179	.0220	.020	.020
26		.0179	.0181	.0159	.0196	.018	.018
27		.0164	.0173	.0142	.0175	.0164	.016
28		.0149	.0162	.0126	.0156	.0148	.014
29		.0135	.0150	.0113	.0139	.0136	.013
30		.0120	.0140	.0100	.0123	.0124	.012
31	7	.0105	.0132	.0089	.0110	.0116	.010
32	6.5	.0097	.0128	.0080	.0098	.0108	.009
33	6	.0090	.0118	.0071	.0087	.0100	.008
34	5.5	.0082	.0104	.0063	.0077	.0092	.007
35	5	.0075	.0095	.0056	.0069	.0084	.005
36	4.5	.0067	.0090	.0050	.0061	.0076	.004
37	4.25	.0064	.0085	.0045	.0054	.0068	
38	4	.0060	.0080	.0040	.0048	.0060	
39			.0075	.0035	.0043	.0052	
40			.0070	.0031	.0039	.0048	

* U. S. Standard Gage is officially a weight gage, in oz. per sq. ft. as tabulated. The approximate thickness shown is the "Manufacturers' Standard" of the American Iron and Steel Institute, based on steel as weighing 501.81 lb. per cu. ft. (489.6 true weight plus 2.5% for average over-run in area and thickness). The A.I.S.I. standard nomenclature for flat rolled carbon steel is as follows:

Widths, in.	Thicknesses, in.							
	0.2500 and Thicker	0.2499 to 0.2031	0.2030 to 0.1875	0.1874 to 0.0568	0.0567 to 0.0344	0.0343 to 0.0255	0.0254 to 0.0142	0.0141 and Thinner
To 3½ incl.	Bar	Bar	Strip	Strip	Strip	Strip	Sheet	Sheet
Over 3½ to 6 incl.	Bar	Bar	Strip	Strip	Strip	Sheet	Sheet	Sheet
Over 6 to 12 incl.	Plate	Strip	Strip	Strip	Sheet	Sheet	Sheet	Black Plate / Sheet
Over 12 to 32 incl.	Plate	Sheet	Sheet	Sheet	Sheet	Sheet	Sheet	
Over 32 to 48 incl.	Plate	Sheet	Sheet	Sheet	Sheet	Sheet	Sheet	Sheet
Over 48	Plate	Plate	Plate					—

as to what connection is to be used, inspector should check with engineer's office.

Rivets should be checked for size and tightness. The alignment of holes should be checked before driving. Where they are not true, holes should be reamed and larger rivets driven. If rivet is tight and has full concentric head, it should be passed.

In no case should the following be allowed:

> Burning of holes with torch.
> Enlarging unfair holes with drift pins.
> Tightening of rivet by calking of head.

Rivets should be tested with small hammer. Strike rivet head with several good blows of hammer to see if it can be "floated" or moved up and down. Defective rivets should be marked with chalk. When a loose rivet is removed, it may loosen adjoining rivets. In small groups, it may be necessary to remove all the rivets in group. However, as a rivet shrinks in cooling, a slight vibration is not cause for condemning a rivet. Sufficient temporary bolts should be used to hold pieces tight together while riveting.

Bolted connections should be reasonably tight but should not be turned up so as to strip thread. Where washer, lock washer, lock nuts, etc., are called for, they should be checked.

Beams on walls should be checked for proper wall bearing and anchorage.

Inspector should cooperate with the erector in safeguarding structure from accidents during erection. He should see that derrick base is secured from horizontal thrust of boom in any direction. Steel carrying derricks should be strong enough and have sufficient connections for erection stresses involved. The erectors should be warned against such dangerous practices as lifting too heavy a load for the strength of counter ties of derrick, booming out too far and splicing of boom. Guying and bracing of steel in process of erection against wind pressure are important. Shrinkage of a wet rope should be allowed for.

Painting should be done according to specifications. Where shop paint has been removed during shipment, repainting should be done before erection. Field paint should be of different color from shop paint. All steel should be free from rust and scale and should be dry. Painting should not be permitted in freezing weather.

Inspector should be familiar with design of building if possible. In any event, he should confer with the engineer to see whether there are any special connections which should be watched. If, in the opinion of inspector, any part of the structure does not appear structurally sound, he should notify engineer.

_____

Engineer

REPORT ON STRUCTURAL STEEL—RIVETED OR BOLTED FIELD INSPECTION

_____

Report No. _____

Job _____ Date _____

Reported to _____ Temp. _____

_____

Marks of Members Erected	No. Erected	DATE			Marks of Members Erected	No. Erected	DATE		
		Plumbed	Riveted or Bolted	Accepted			Plumbed	Riveted or Bolted	Accepted

Worked from approved erection plans and specifications _____

The fact that shop inspection has been made has been verified _____

All steel accepted has been inspected and approved as follows with special attention to the following:

All members have been checked against plans for piece mark and location

_____

Column bases, leveled and grouted _____

Columns plumbed _____

Riveting where called for on plans _____ Quality _____

Bolting quality _____

Painting _____

Every column splice has been inspected for true bearing _____

Ends of beams on seat connections are within $1\frac{1}{16}$ in. maximum of face of supporting members _____

Remarks (rejections, corrections, etc.) _ _____

_____

Inspector

Engineer

REPORT ON STRUCTURAL STEEL—RIVETED OR BOLTED

Shop Inspection, Part I

Report No. _____

Job _____ Date _____

Reported to _____ Where Inspected _____

Approved drawings used for inspection, shop drawings, erection plan, joint details _____

Steel inspected for:

Surface defects, folds, twists, straightness _____

All sections called for on plans _____

Connections agree with details and for correct location _____

All members requiring bearing ends have full square-milled bearing _____

Stiffeners are full in contact at both ends for plate girders and at the ends shown in contact for seats and rolled sections _____

All skewed connecting angles and plates have been bent hot _____

Rivets are tight and of correct diameter _____

The ends of beams bearing on seat connections will be not more than $1\frac{1}{16}$ in. maximum from the face of column or supporting member _____

Not more than 2 of the rivets are punched more than $\frac{1}{16}$ in. off for any connections and not more than $\frac{1}{4}$ in. in any case _____.

Material has been properly cleaned before painting _____.

Painting is according to specifications or drawings _____

Sample of shop coat paint has been taken for analysis _____

Inspector has marked every member after accepting same _____

No member has been shipped without inspector's mark except _____

Inspector has marked on plans and column schedule all members accepted

Members have been assembled to insure proper alignment and fit, and freedom from twists, bends, and open joints between the component parts _____

Inspector will be able to state in final report that every member has been covered _____

Special requests have been attended to _____

Remarks (rejections, corrections, attention to warning notes, etc.) _____

Inspector

Engineer

REPORT ON STRUCTURAL STEEL

Shop Inspection, Part II
(For both riveted and welded steel)

Report No. _____

Job _____ Date _____

Reported to _____ Where Inspected _____

| Material | Required | Being Fabricated | Finished | Shipments | | | Weights |
				Date	R.R.	Car No.	

Remarks _____

_____

Shipped this report _____

Previous _____

Total to date _____

Inspector

Mill _____
Client _____
Project _____

Engineer _____

Date _____
Report _____
Sheet _____ of _____
Specification _____

STEEL MILL REPORT *

Requirements

Quantity	Description of Material	Weight	Heat Number	Yield Point P.s.i.	Ultimate P.s.i.	% Elongation In.	% Reduction Area	Bend	Chemical Tests			
									Car.	Man.	Phos.	Sul.

Total this report _____
Previous _____
Total to date _____

The above tests do / do not fulfil A.S.T.M. Spec. _____, _____ Grade

_____
Inspector

* Adapted from *Haller Testing Laboratories, Inc.*

WELDING

CHECK LIST FOR INSPECTORS

See also "Check List for Inspectors, Structural Steel," p. 88.

Extra Equipment for Structural Welded Job

Welding gage.
Chipping hammer.
Wire brush.
Protective shield.

Procedure in Inspection

Qualifications of welder. If there is any question as to his qualifications, he should be required to make test pieces for inspector.

Conformity of electrodes to specifications or correct usage. See p. 107. For current actually used, see Fig. 68.

Condition and capacity of welding equipment.

Quality of welds for overlap, color, porosity, slag inclusions, undercutting, uniformity, and workmanlike appearance.

Fitting up of members for tightness. In fillet welds when the gap exceeds $\frac{1}{16}$ in., size of weld should be increased.

Sequence of welding in order to minimize residual stresses.

Condition of any tack welds which are to be fused with final welds. If any of these are not satisfactory, they should be removed.

Cleaniness of work, as good welds cannot be made on dirt, rust, or slag. In a multiple pass weld, slag must be chipped and wire-brushed to shiny surface before next pass is made.

Weather conditions, as welding should not be done in temperature less than 0° F., or when surfaces are wet from condensation, rain, snow, or ice. Welder should be properly protected from wind. At temperatures between 0° F. and 32° F., surfaces must be heated. Material $1\frac{1}{2}$ in. thick or over should be 70° F. minimum.

Conformity to approved plans for the following details:

Cross-sectional size, length, location, and omission. They should not be increased arbitrarily as longer welds sometimes introduce more restraint than calculated.

Operator at work at frequent intervals. If welding is not being properly done, he should be corrected. An experienced welder knows when he is making a good weld. He also knows whether equipment is working properly.

AMERICAN WELDING SOCIETY ARC AND GAS WELDING SYMBOLS

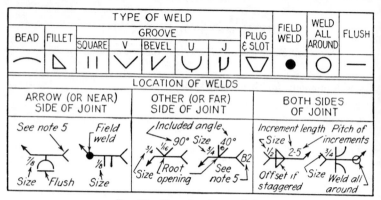

Fig. 59. Shop drawing symbols.

1. The side of the joint to which the arrow points is the arrow (or near) side, and the opposite side of the joint is the other (or far) side.

2. Arrow side and other side welds are same size unless noted.

3. Symbols apply between abrupt changes in direction of joint or as dimensioned (except where all around symbol is used).

4. All welds are continuous and of user's standard proportions unless noted.

5. Tail of arrow used for specification reference (tail may be omitted when reference not used), e.g., C.A. = automatic shielded carbon arc, S.A. = automatic submerged arc.

6. In joints in which one member only is to be grooved arrow points to that member.

7. Dimensions of weld sizes, increment lengths, and spacings in inches.

COMMON WELDING PROCESSES

Figures 60 and 61 indicate common welding processes and the action of the shielded arc electrode. In the electric arc welding process a metal electrode is melted and fuses with contiguous metal surfaces to be joined. The welding heat is obtained from the electric arc formed between the electrode and the parts to be welded. The temperature of the arc is approximately 10,000° F.

In the electric arc process if the direction of flow of current is through ground lead, into work, into electrode, into work lead, and back to machine, the circuit is known as electrode negative (straight polarity). With the electrode positive (reverse polarity) the direction of the flow of current is reversed. In alternating-current welding the direct-current generator is replaced by a transformer. Direct current with electrode positive (reverse polarity) is used for structural work except where deep penetration is required. The type of electrodes affects the polarity, as electrodes can be used only as shown in Table 20, p. 107, on account of the material of the covering.

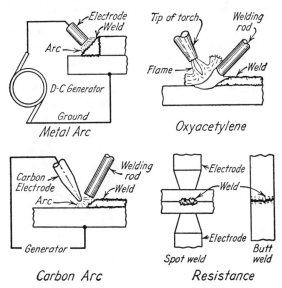

Fig. 60. Welding processes. *From H. Malcolm Priest, Practical Design of Welded Steel Structures, American Welding Society.*

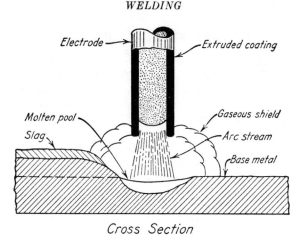

Cross Section

FIG. 61. Shielded arc electrode. *From H. Malcolm Priest, Practical Design of Welded Steel Structures, American Welding Society.*

WELDERS' QUALIFICATION TEST USING FILLET WELDS *

Take two bars 5 in. by ½ in. by 4 in., and weld as indicated in Fig. 62 in the desired position, that is, flat, horizontal, vertical, or overhead. Turn plates over and break with a blow by a sledge hammer. The weld should break cleanly along the center line, showing a clean cross section of weld material. Visual inspection of the weld and its fracture readily reveals any improper fusion between the weld and base metal or any lack of soundness.

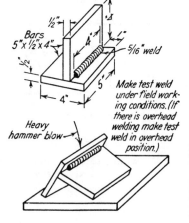

FIG. 62.   Test for weld soundness.

* A quick check-up for use by inspectors. For formal qualification tests, see *American Welding Society Handbook.*

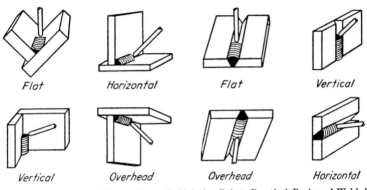

Fig. 63. Welding positions. *From H. Malcolm Priest, Practical Design of Welded Steel Structures, American Welding Society.*

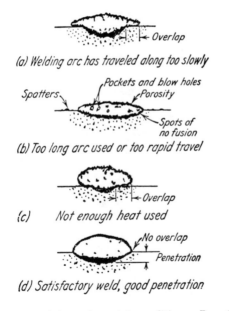

(a) Welding arc has traveled along too slowly

(b) Too long arc used or too rapid travel

(c) Not enough heat used

(d) Satisfactory weld, good penetration

Fig. 64. Weld characteristics under certain conditions. *From Gilbert D. Fish, Arc-Welded Steel Frame Structures, McGraw-Hill Book Company.*

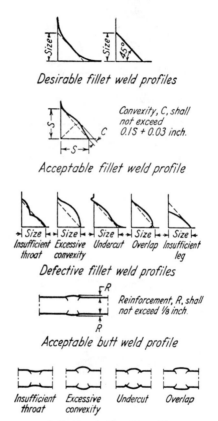

Desirable fillet weld profiles

Convexity, C, shall not exceed 0.1S + 0.03 inch.

Acceptable fillet weld profile

| Insufficient throat | Excessive convexity | Undercut | Overlap | Insufficient leg |

Defective fillet weld profiles

Reinforcement, R, shall not exceed 1/8 inch.

Acceptable butt weld profile

| Insufficient throat | Excessive convexity | Undercut | Overlap |

Defective butt weld profiles

Fig. 65. Illustrations of acceptable and defective welds as contained in A.W.S. Code. *From Specifications for Design, Fabrication and Erection of Structural Steel for Buildings by Arc and Gas Welding, 1942, American Institute of Steel Construction.*

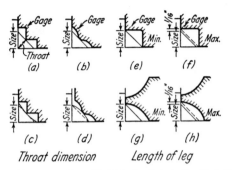

FIG. 66. Fillet weld gages. *From H. Malcolm Priest, Practical Design of Welded Steel Structures, American Welding Society.*

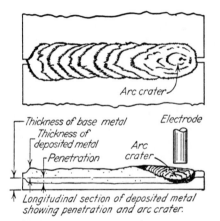

FIG. 67. Weld penetration and arc crater. *From Gilbert D. Fish, Arc-Welded Steel Frame Structures, McGraw-Hill Book Company.*

TABLE 19. MAXIMUM SIZE OF ELECTRODES

Type	POSITION				
	Flat	Horizontal	Vertical	Overhead	
Fillet	¼ in.	¼ in.	³⁄₁₆ in.	³⁄₁₆ in.	*Note:* Maximum size of fillet weld in one pass is ⁵⁄₁₆ in., except that vertical welds can be ½ in.
Butt	¼ in.	³⁄₁₆ in.	³⁄₁₆ in.	³⁄₁₆ in.	

Electrodes for a single pass fillet weld and for root pass of a multilayer weld shall be of proper size to insure thorough fusion and penetration with freedom from slag incursions, but shall not exceed $\frac{5}{32}$ in. diameter for butt welds, vertical and overhead fillet welds.

Read off electrode container recommended current. Check vs. current being used.

To find current being used, time rate of electrode burn-off, find current from chart on following page.

EXAMPLE. Given $\frac{5}{32}$ electrode and burn-off rate of 12 in. in 70 seconds.

Enter chart (Fig. 68) at 70 seconds, proceed across to intersection with $\frac{5}{32}$ in. curve, drop vertical to ampere scale, read 150 amperes.

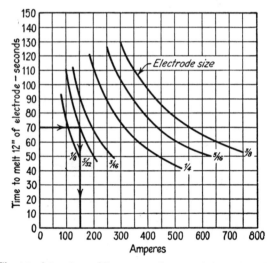

FIG. 68. Chart to determine welding current by rate of electrode melt-off. *From Procedure Handbook of Arc Welding—Design and Practice, The Lincoln Electric Company.*

Welding of Steel Wire Fabric. To join the ends of sheets of steel wire fabric, such as circular reinforcement for precast concrete pipe, first remove all rust by an electric wire brushing tool. Check that ends of circular wires are in contact and lapped 8 to 10 diameters, say not less than $1\frac{1}{2}$ to 2 in. Electric-arc-weld both fillets full. Multiple-pass-weld, if necessary. Low-current electrodes requiring not more than 80 amperes are recommended to avoid undercutting damage to base metals. Be sure not to exceed ampere range directed by manufacturer. Check specifications. Deposit enough metal (A.W.S. 6012) to develop strength of wire, usually 70,000 p.s.i.

MILD STEEL ARC-WELDING ELECTRODES (A233-1948)

TABLE 20. ELECTRODE CLASSIFICATION

Electrode Classification Number	Type of Coating or Covering	Capable of Producing Satisfactory Welds in Position Shown *	Type of Current
E45 Series. Minimum Tensile Strength of Deposited Metal in Non-Stress-Relieved Condition 45,000 P.S.I.			
E4510 E4520	Sulcoated or light coated	F, V, OH, H H-fillets, F	Not specified, but generally d.c., straight polarity (electrode negative)
E60 Series. Minimum Tensile Strength of Deposited Metal in Non-Stress-Relieved Condition 62,000 P.S.I. (or higher)			
E6010	High cellulose sodium	F, V, OH, H	For use with d.c., reversed polarity (electrode positive) only
E6011	High cellulose potassium	F, V, OH, H	For use with a.c. or d.c., reversed polarity (electrode positive)
E6012	High titania sodium	F, V, OH, H	For use with d.c., straight polarity (electrode negative), or a.c.
E6013	High titania potassium	F, V, OH, H	For use with a.c. or d.c., straight polarity (electrode negative)
E6015	Low hydrogen sodium	F, V, OH, H	For use with d.c., reversed polarity (electrode positive) only
E6016	Low hydrogen potassium	F, V, OH, H	For use with a.c. or d.c., reversed polarity (electrode positive)
E6020	High iron oxide	H-fillets, F	For use with d.c., straight polarity (electrode negative), or a.c., for horizontal fillet welds; and d.c., either polarity, or a.c., for flat-position welding
E6030	High iron oxide	F	For use with d.c., either polarity, or a.c.

* The letters indicate welding positions as follows:

F = flat.
H = horizontal.   V = vertical   } For electrodes $\frac{3}{16}$ in. and under, except in
H-fillets = horizontal fillets.   OH = overhead } classification E6015 and E6016, $\frac{5}{32}$ in. and under.

Electrodes E6010 and E6011 produce welds of x-ray quality.

Electrodes E6012 and E6013 produce welds of lesser penetration and therefore are used where poor fit is encountered and they bridge a gap better.

Electrodes E6015 and E6016 are so-called low-hydrogen electrodes. They are used where hardenability or crack sensitivity is a problem.

Electrodes E6020 and E6030 are downhand electrodes. They give greatest penetration and speed in the flat position.

The operating characteristics and uses of each classification is described in detail in **A.W.S.** Spec. A233.

―――――――――――――――――――

Engineer

REPORT ON STRUCTURAL STEEL—WELDED
Shop Inspection, Part I
(See p. 97 for Part II.)

Report No. _____

Job _____ Date _____

Reported to _____ Where inspected _____

Approved drawings used for inspection, shop drawings, erection plan, joint details _____

Steel inspected for surface defects, fold, twists, straightness _____

All sections are as called for on plans _____

Connections agree with details and for correct locations _____

All members requiring bearing ends have full square-milled bearing _____

Stiffeners are full in contact at both ends for plate girders and at the ends shown in contact for seats and rolled sections _____

All skewed connecting angles and plates have been bent hot _____

The ends of beams bearing on seat connections will be not more than $1\frac{1}{16}$ in. maximum from the face of column or supporting member _____

Material has been properly cleaned before painting _____

Painting is according to specifications or drawings _____

Sample of shop coat paint taken for analysis _____

Inspector has marked every member after accepting same _____

No member has been shipped without inspector's mark except _____

Inspector has marked on plans and column schedule all members accepted

Inspector will be able to state in final report that every member has been covered _____

Every weld inspected for size _____ length _____

location _____ and quality_____

Every welder has marked every weld group for identification _____

All welders' qualifications checked _____ Authority _____

Number of welders _____ Names _____

Make and capacity of machines _____

Kind of current _____

Make, grade, style No., and size of electrodes _____

Special requests have been attended to _____

Remarks (rejections, corrections, etc.) _____

―――――――――――――――――――

Inspector

_____
Engineer

REPORT ON STRUCTURAL STEEL—WELDED

FIELD INSPECTION

_____

Project _____ Date _____

Welding permit No. _____ Report No. _____

Welding contractor _____

Description of work _____

	Erected during this Period	Erected to Date	Plumbed	Welded	Accepted
Columns					
Beams					

Shop welded or riveted _____

Weather and temperature _____

Checked against approved typical details and erection plans _____

_____

Machines _____

Electrodes * _____ Sizes _____ Polarity _____

Weld sizes _____ No. of layers or beads _____ Symbol _____

Positions employed: horizontal _____ flat _____ vertical _____

overhead _____

All welders' qualifications checked _____ Authority _____

Has every welder marked joint with index number? _____

Has inspector kept complete record of welding? _____

Has every weld been checked for size? _____ Length? _____

Location? _____ Quality? _____ Workmanship? _____

Number of individual welds made: _____ Accepted: _____

Rewelded: _____

Reasons for rejections and rewelding, method of correction of defective welds, and name and index numbers of welders making such defective welds:

_____

_____

_____

* See p. 107.

Inspector has marked on plans all joints accepted including column splices using separate prints where plans cover two or more tiers _____

Before welding was the steel properly cleaned, and free from corrosion, water, oil, scale, dirt, paint, etc.? _____

Were proper methods employed when setting up the work to insure tight fit without displacement of component parts after welding, together with full penetration of the weld metal to the root of the joints? _____

Was inspector in full attendance at all times while welds or fusion was being made in the passing of metal from the electrode to the base metal? ____

Was each completed weld carefully examined for defects and irregularities such as: undercutting, overlaps, lack of fusion at edges, lack of penetration, place cracks adjacent to or behind weld, water cracks and cracks in weld metal, slag inclusions? _____

Remarks _____

_____

Joints welded and accepted _____

Inspector has marked every weld group after accepting them. _____

_____

_____
Inspector

CORROSION OF METALS

1. Corrosion is caused by the direct exposure of a metal to an acid, as when pipes are imbedded in cinder concrete.

2. Corrosion of metals is caused by a process similar to that in a galvanic bath, in which two metals separated in the galvanic series are connected by an electrolyte, usually water. This causes the solution (corrosion) of one of the metals and the protection, or building up, of the other. The metal that corrodes is known as the anode (the positive pole of a battery), and the protected metal as the cathode (the negative pole of a battery). Every metal has a place in the "galvanic series." The following is a condensed table. The farther apart in the table any two metals are, the greater the galvanic action between them.

Galvanic Series (Condensed)

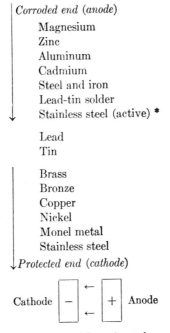

Corroded end (*anode*)

Magnesium
Zinc
Aluminum
Cadmium
Steel and iron
Lead-tin solder
Stainless steel (active) *

Lead
Tin

Brass
Bronze
Copper
Nickel
Monel metal
Stainless steel

Protected end (*cathode*)

Cathode | − | ← + | Anode
←

Direction of flow of metal

* Active stainless steel has its natural protective film removed.

Refer to job specifications for insulation of galvanic metals in contact direct or through a liquid.

If the anodic metal is large in area relative to the cathodic, the corrosive action will be relatively small, and vice versa. Thus, whereas steel fasteners joining Monel sheets are not practicable, Monel rivets are satisfactory for joining steel sheets.

3. Corrosion may also be caused by so-called "concentration cells." When two metals, or a metal and a non-metal, are in nominal contact, there are always minute spaces between the surfaces. In these confined spaces, galvanic action may be set up which will cause the corrosion of one of the metal surfaces. The hoops of a wood-stave water tank are very susceptible to corrosion from this cause owing to the spaces under them and small leaks at the joints which form concentration cells.

4. To prevent corrosion:

Avoid contact of metals with corrosive, especially acid, substances. Seal the edges of weld back-up plates by welding or otherwise.

Thoroughly paint exposed steel at all points.

Report all leaks, especially in wood-stave tanks with steel hoops.

Do not paint over rusted steel; clean it thoroughly.

Clean corroded reinforcing bars.

Reject corroded material of any sort.

IDENTIFICATION OF METALS

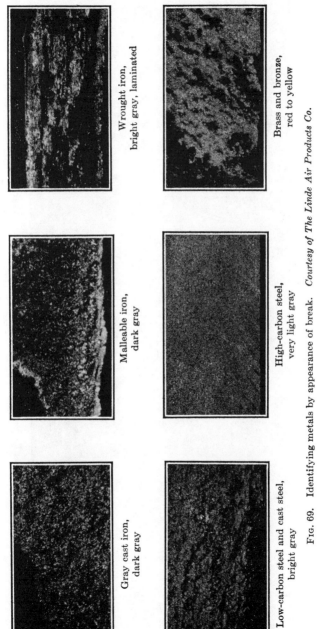

Wrought iron, bright gray, laminated

Brass and bronze, red to yellow

Malleable iron, dark gray

High-carbon steel, very light gray

Gray cast iron, dark gray

Low-carbon steel and cast steel, bright gray

Fig. 69.   Identifying metals by appearance of break.   *Courtesy of The Linde Air Products Co.*

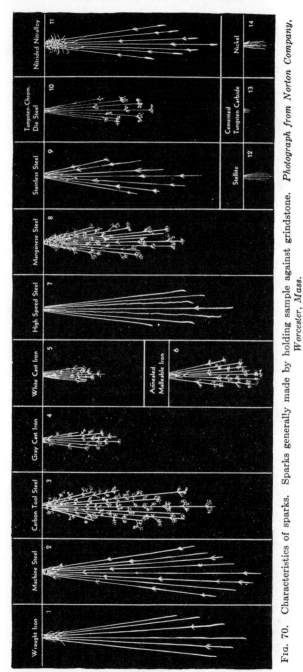

FIG. 70. Characteristics of sparks. Sparks generally made by holding sample against grindstone. *Photograph from Norton Company, Worcester, Mass.*

CHARACTERISTICS OF SPARKS

Metal	Volume of Stream	Length of Stream, Inches	Color of Stream Close to Wheel	Color of Streaks Near End of Stream	Quantity of Spurts	Nature of Spurts
1. Wrought iron	Large	65	Straw	White	Very few	Forked
2. Machine steel	Large	70	White	White	Few	Forked
3. Carbon tool steel	Moderately large	55	White	White	Very many	Fine, repeating
4. Gray cast iron	Small	25	Red	Straw	Many	Fine, repeating
5. White cast iron	Very small	20	Red	Straw	Few	Fine, repeating
6. Annealed mall. iron	Moderate	30	Red	Straw	Many	Fine, repeating
7. High speed steel	Small	60	Red	Straw	Extremely few	Forked
8. Manganese steel	Moderately large	45	White	White	Many	Fine, repeating
9. Stainless steel	Moderate	50	Straw	White	Moderate	Forked
10. Tungsten-chromium die steel	Small	35	Red	Straw *	Many	Fine, repeating *
11. Nitrided Nitralloy	Large (curved)	55	White	White	Moderate	Forked
12. Stellite	Very small	10	Orange	Orange	None	
13. Cemented tungsten carbide	Extremely small	2	Light orange	Light orange	None	
14. Nickel	Very small **	10	Orange	Orange	None	
15. Copper, brass, aluminum	None				None	

* Blue-white spurts.   ** Some wavy streaks.

BRIDGES

Reports When under Construction

Structural steel	see pp. 95 to 98.
Concrete	see pp. 70 to 75.
Piles	see p. 139.
Timber	see p. 151.
Other items	

Field Data Required for Rating Existing Bridges if Plans Not Available

Sizes of all members.

All span and panel point dimensions.

Sketches of all joints including dimensions and sizes of bolts, rivets, pins, connection angles, washers, etc.

Data for dead-load computations such as material and thickness of floor construction.

Live loads from using railroad or proper highway department.

INSPECTION OF EXISTING BRIDGES *

Waterway. First show the area of the structure in square feet in the space provided.

Conditions in the streambed should be noted as to (1) adequacy of channel afforded by the existing structure; (2) probability of scour that may endanger the footings; and (3) presence of obstructions, such as drift logs, stumps, or old piers, that may be diverting the current so as to cause undermining of the footings. Also note any undergrowth or obstructions that can be removed to increase the adequacy of the waterway or to lessen the fire hazard of timber structures. Lastly, note whether stream-bank protection is necessary to keep the channel properly confined and thus to avoid endangering the bridge foundations or the end fills. Also note if there are any indications of unusual corrosiveness at the site.

Piers and Abutments. The type and material used should be listed.

Timber Piles. Piles supporting timber bridges should be inspected carefully at the ground line, where decay first sets in. A ¾-in. hexagonal steel bar about 4 ft. long, with one end sharpened to a long tapering point and the other end provided with a chisel face, is a very useful tool in such examinations. It can be jabbed into a pile to disclose deterioration not apparent on the surface and to determine the extent of sap rot. Piles in which the diameter of sound material has been reduced to 6 in. or less should be marked with yellow keel for replacement.

---

* From Toncan Culvert Manuf. Assoc.

Steel Tubular Piers. Steel tubular piers should be carefully examined for corrosion in rivets or bolt heads connecting the cylindrical sections. (The filling material in such steel cylinders is usually inferior and without strength in itself.) Also note whether there has been appreciable movement of the tubes due to impact of heavy loads on the structure; if so, additional footings or bracing may be needed. Note whether the steel tubes are out of plumb and if so whether this is due to undermining, to lack of proper bracing, or to inadequate support below. Examine base of tubes for exposed piling caused by scour.

Concrete Substructures. The pier shafts should be examined for damage from drift or ice. Examine exposed footings for rock pockets due to improper placement of concrete. Note extent of any undermining. Look for cracks, and note whether they are caused by unequal settlement, contraction, or fill pressure. Check abutments and adequacy of wing walls. Recommend placement of riprap and rock slope protection where necessary.

Concrete Structures

Culverts. Examine barrel and wing walls of culverts to find any harmful cracks due to settlement that should be grouted to prevent deterioration of the reinforcing steel. Also examine floor of barrel to note any upheaval which may cause failure of side walls due to excessive fill pressure; especially note this in culverts under high fills.

Beam-and-Slab Spans. Note condition of railing for damage by collision; sight alignment of railing for indication of settlement of the structure. On heavily traveled roads, the handrail should be kept clean in order to provide proper visibility for night driving and, if conditions warrant, should be painted with a cement wash coat. Examine beams for cracks that may be due to clogged expansion joints, settlement, or fill pressure at either end of bridge. Note any surface checking in deck, railing, curbs, or sidewalks that may allow water to seep in and cause disintegration by freezing action.

Steel Structures

On steel trusses note first the general alignment of the span to see whether the end posts and top chords are straight and in line. Any buckling indicates that the structure has been overloaded. Especially note this for light construction. Kinks in any one member may have been caused by damage in shipment, in erection, or by collision; the inspector should make sure that any such kinks are not due to overstress.

Examine members for reduction of area due to rust and corrosion.

For all pin-connected trusses, note whether eyebars in the same member are taking equal tension. Overloading or lack of proper camber adjustment may cause one eyebar to take all the stress, leaving the others loose on the pin. Especially note this for the diagonal and hip vertical members. Observe the structure under heavy loading, and note whether there is any excessive deflection or bowing in or out of the diagonal eyebar members which would indicate lack of proper counterbracing. Note condition of end shoes and rollers to see whether proper expansion is being provided for and whether the rollers are free to move.

Timber Structures

Timber Trusses. In inspecting a timber truss, first see if it has any noticeable sag. If sag is present note whether it is due to failure of splices, improper adjustment of vertical rods, or crushing of diagonal members. Examine all splices for splitting or cracking of the shear tables. Sound the rods with a hammer to note whether each is carrying the same amount of tension, and examine condition of caps and ends of diagonal members for signs of crushing. If the structure is very old, it will be advisable to use a $\frac{3}{8}$-in. auger bit to test out the center of the top and bottom chord members for heart rot at all panel points and splices; the floor beams at contact with bottom chords should also be bored. Decay will be found first at contact points and where rods go through timber members.

Other points to check on covered trusses will be the condition of the roof and housing. Be sure to examine truss bearings over the pier caps and the condition of caps over pier piling for crushing, and bore with auger bit where there is any doubt as to their soundness. Note whether all bolts through splices, packing blocks, and cross bracing are tight and in good order. The substructure of timber trestles should be examined as directed under "Piers and Abutments." Caps should be examined for any crushing over the posts or piling. Decay will always be found first at bearing contacts, and a testing bar or auger bit should be used on all doubtful timbers. A thorough boring test must be made on all timbers that have been in place more than 6 years.

Note condition of bulkheads at each end of the bridge for decay, height, and proper retention of approach fill. Check sway and longitudinal bracing to note whether any members are broken or decayed and whether additional bracing is required. In examining the superstructure, first go under the bridge, examine each span, and note (1) whether stringers are crushing, cracking, or splitting, (2) whether they have full bearing over the caps, and (3) whether bridging between stringers is in place. Note condition of under side of decking, and see whether all bolts are properly tightened or have become loose due to shrinkage of timbers.

Second, examine deck and handrail from roadway. Especially on high bridges, sound handrail posts with testing bar at contact with felloe guard, stringers, and railing to see that members are not badly decayed. Handrail should be kept painted for protection against decay and to provide visibility for night traffic; all decayed members must be replaced. Timber handrails require repainting about every 3 years.

FIELD DATA FOR NEW SMALL BRIDGES

The following bridge inspection report on p. 121 is devoted to data that should be gathered in the field for the replacement of an existing small bridge with a new structure. All data requested in the heading is self-explanatory; however, it should be emphasized that, if the existing structure is noticeably too small or too large, then the area to be drained, expressed as drainage area in acres, should be as accurate as possible. Likewise, the correct value for c should be shown for use in the Talbot formula.

Fill in the data requested for the respective type; however, if a decision as to proper selection has not been made, it is advisable to list the data for both pipe and arches since very little extra time will be required to develop the additional information.

It is important that the profile of the stream bed and road and location sketch be as accurate as possible. Be sure to indicate on the location sketch any suggested desired change in location for the new structure.

EXISTING BRIDGE INSPECTION REPORT *

Bridge No. _____ Route No. _____ _____ Miles from _____ Toward_____
Bridge over _____ Type _____ Date built _____
Overall length _____ Width between rails _____ Vertical clearance _____
Load capacity _____ Approaches_____ Date inspected _____
     H-10, H-15, or H-20
     H 20-S16 or H 15-S12

OBSERVATIONS

Waterway Area ____ sq. ft.	Piers and Abutments Type _____	Concrete Structures and Floors	Steel Construction Type _____	Timber Spans and Floors
Adequacy ___	Undermining ___	Cracking ___	Condition of paint ___	Condition of paint ___
Scour ___	Settlement ___	Scaling ___	Corrosion ___	Decay ___
Obstructions ___	Cracking ___	Scour ___	Joints ___	Wear (floors) ___
Undergrowth ___	Disintegration ___	Settlement ___	Loose rivets ___	Structural defects† ___
Channel shifting ___	Decay (lumber) ___	Disintegration ___	Camber ___	Crushing at joints ___
Revetments ___	Corrosion (steel) ___	Waterproofing ___	End shoes ___	Splices ___
Other features ___	Other defects ___	Other defects ___	Other defects ___	Camber ___
	Piling foundations? ___		Wear ___	Other defects ___

Make above observations for each part of structure, and note with (√) mark to indicate "OK" or "None." For items needing explanation mark with circle with a number inserted to refer to corresponding remark listed below. Amplify remarks with sketches on second sheet when necessary.

REMARKS

(Use second sheet when space below is not sufficient; also, list causes of all defects such as cracking and scaling of concrete whenever possible.)

_____
_____
_____
_____

RECOMMENDATIONS

(Furnish data on p. 121 when total replacement is recommended)

Item	Estimated Cost	
	Maintenance	Improvements

Note. List under maintenance and "Recommendations" all necessary channel clearing, revetments, bank protection, channel changes, stream-bed pavements, riprap work, underpinning or other foundation protection, shoulder and slope protection, repairs to concrete work, painting, waterproofing, preservative treatments, repairs to roadway surfaces, repairs and renewals to timber and piling, and all other maintenance and repair work of whatever nature.

                 Inspector

* *From Toncan Culvert Manuf. Assoc.*
† Under "Structural defects" note any tendency to warp, split, crack, etc.

FIELD DATA FOR NEW STRUCTURE *

Location _____ Width of roadway between outside of **rails**
　　　　　Same—New
　or shoulders _____
Drainage area _____ Acres　Talbot formula factor c = _____
Distance between stream bed and roadway _____
Slope of embankment _____ Type recommended _____
　　　　　　　　　　　　　　　　　　　　　　　　Pipe—Arch

ADDITIONAL DATA FOR PIPE STRUCTURE

Waterway area required _____ sq. ft.　Live load _____
Cover over pipe _____ No. of pipes _____ Diameter _____
Slope or skew _____ Center line length _____
Headwalls or riprap _____ Type of material in stream bed _____
_____ Slope of stream _____ %

ADDITIONAL DATA FOR ARCH STRUCTURE

Waterway area required _____ sq. ft.　Live load _____
Cover over arch _____ No. of arches _____ Span _____
Rise _____ Slope or skew _____
Center-line length _____ Head walls or riprap _____
Bearing power of soil _____ Type of material in stream bed _____
Recommended material for abutments, piers, and walls _____
Depth of abutments and piers below stream bed _____
Slope of stream _____ %

Profile of stream and road.

Note. Indicate normal and flood level of stream.

Location sketch

Show angle of skew of structure with centerline of roadway.

_____
　　　　　　　　　　　Inspector

* *From Toncan Culvert Manuf. Assoc.*

PAINTING

CHECK LIST FOR INSPECTORS

TREATMENT OF SURFACES FOR PAINTING

General Conditions

All surfaces to be painted shall be thoroughly dry.

No exterior painting to be done in rainy, damp, or frosty weather.

Permit no interior painting until surfaces have become thoroughly dry. (By artificial heating if necessary.)

Allow no painting on metal surfaces to be welded. If such surfaces have been painted, paint is to be removed.

All surfaces must be of material in compliance with specifications. Surfaces must be checked for shop coat where called for in specifications.

Surface Preparation

Metal Surfaces. Remove dirt and mud by brushing and/or washing.

Remove grease and oil with benzine, naphtha, or turpentine.

Rust and scale to be removed with wire brush, steel scraper, or sand blasting.

Mill scale to be removed by burning.

Old paint to be removed by burning, scraping or paint remover.

Before painting over prime coat, check and reprime where necessary.

Before priming new galvanized metal wash with copper sulfate solution to remove grease and chemicals.

Before hot asphaltic applications, heat metal.

Where phosphoric acid treatment is specified, immerse material in caustic soda solution at 200° F. to remove grease and oils; rinse in hot water; immerse in 5% sulfuric acid pickle, then rinse in hot water.

Wood Surfaces. Remove dirt and dust with brush and rag.

Stop out all knots and sap streaks with shellac.

Putty nail holes, cracks, and other depressions after primer coat has thoroughly dried. Tint putty to match finish.

Old paint to be removed by sanding, wire brushing, scraping, or burning.

Floors to be sanded or scraped.

Open-grained woods to be varnished to be given first an application of wood paste filler thinned with turpentine.

Masonry Surfaces. Dust, dirt, and excess material to be removed with stiff bristle or wire brush.

Remove salts from brickwork with zinc sulfate water solution, and brush off surface when dry.

All masonry surfaces to be allowed thorough period for curing.

Porous block to be primed with casein paste or resin sealer.

Cement floors to be prepared by acid etching with muriatic acid to improve adhesion; acid to be washed off and floor dried before painting.

Stucco and concrete to be cleaned with stiff fiber brush; traces of oil to be removed with abrasive stone or, if general, by light sand blasting. Sealer to be added to the paint.

Smooth dense concrete surfaces to be roughened by light sand blasting, muriatic acid etching, or rubbing with abrasive stone to improve adhesion.

Where cement paints are used on exterior concrete the surface to be thoroughly wetted before application.

Plaster Surfaces. Allow 30 days for drying before painting.

Apply prime coat of sealer to clean dry surface.

Check prime coat for fading caused by hot spots (incomplete mixing of hydrated lime) and suction spots (thin spots and inadequate troweling).

SPREAD FOUNDATIONS

CHECK LIST FOR INSPECTORS

The inspector should determine from plans the type of soil on which the foundation design is based and check against actual conditions.

Shallow pipe borings under each footing should be made if there is a question about the underlying soils.

If there is any question in regard to soil bearing capacity, the inspector should notify the engineer, who may according to his judgment revise the size of footings or require the footings to be carried deeper. A soil test may be required.

Keep footings clear of water when concrete is poured.

In clay soil, footings to be free of water at all times. The excavation should be back-filled to the top of the footing with thoroughly tamped dry clay as soon as forms are removed to prevent water from reaching the bottom of the footing.

In clay soil, if there is danger of water entering the excavation, a tile drain above the footing is advisable.

Soil to be original strata and below loam or vegetation. Remove disturbed soil before concreting.

Bottom elevation of footing to be at least the elevation called for on plan. If necessary, owing to soil condition, elevation may be lowered for suitable bearing.

Keep record of actual elevation of footings installed.

Check slope between footings when elevations differ from plans or when determined in the field. This slope should not be more than 2 horizontal to 1 vertical for compact soils but should be fixed by the engineer.

Conditions which may require sheeting where it is impossible to keep minimum slope should be watched.

Possible undermining of existing foundations should be checked.

Footings should be of size shown on plans.

Concrete for footing. See "Instructions to Inspectors, Concrete."

Report any unusual conditions, such as change of soil, springs, poor rock, or danger of slides.

Footings not to be built on frozen ground, or where there is danger of frost heave during construction.

In back-filling trenches or pits, see that the soil is thoroughly tamped, preferably with a mechanical tamper to minimize settlement.

When cranes for hoisting heavy loads are placed adjacent to foundation walls, see that the walls are adequately shored. Check the shoring of walls before back-filling.

Provide relief holes for water in walls and slabs where hydrostatic head may develop before concrete has set or where water conditions may require.

Be on the lookout for cracks or bulges in walls of adjoining property.

FOUNDATIONS ON SOIL

Method of conducting a load test, N. Y. City code.   See also Fig. 71.

Procedure. Apply sufficient load uniformly on platform to produce a center load of four times the proposed "design load per square foot."

Center load equals load of platform times $\dfrac{b}{a + b}$.

Read settlement every 24 hours until no settlement occurs in 24 hours.

Add 50% more load and read settlement every 24 hours until no settlement occurs in 24 hours.

Settlement under proposed load should not show more than ¾ in., or increment of settlement under 50% overload should not exceed 60% of settlement under proposed load.

If the above limitations are not met, repeat test with reduced load.

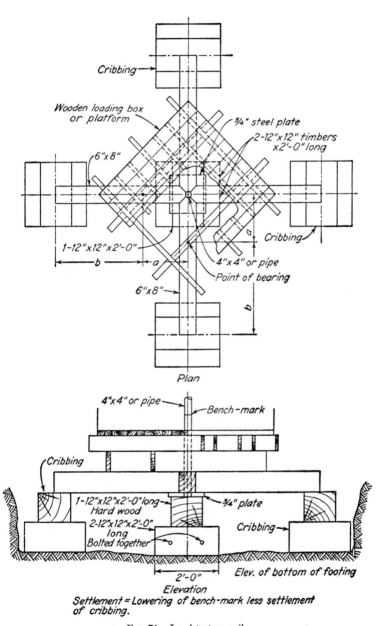

FIG. 71. Load test on soil.

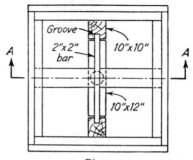

Plan

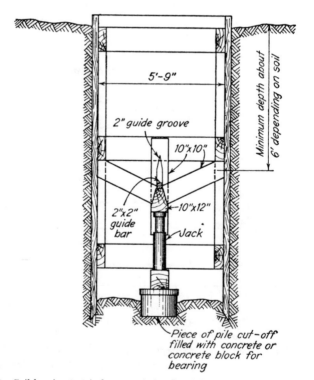

FIG. 72. Soil bearing test in bottom of pit. Lateral pressure from diagonal struts prevents form from being lifted out of the pit. *From "Underpinning" by Prentice and White.*

PRESUMPTIVE BEARING CAPACITIES OF SOILS *

a. Satisfactory bearing materials shall be ledge rock in its natural bed, natural deposits of gravel, sand, compact inorganic silt, or clay, or any combination of these materials. These bearing materials shall not contain an appreciable amount of organic matter or other unsatisfactory material, nor shall they be underlaid by layers of such unsatisfactory materials of appreciable thickness.

b. Fill material, mud, muck, peat, organic silt, loose inorganic silt, and soft clay shall be considered as unsatisfactory bearing materials and shall be treated as having no presumptive bearing value.

c. The maximum allowable presumptive bearing values for satisfactory bearing materials shall, except for pile foundations (see section C26-405.0, c), in the absence of satisfactory load tests or other evidence, be those established in the classification in Table 21.

TABLE 21. CLASSIFICATION OF SUPPORTING SOILS

Class	Material	Maximum Allowable Presumptive Bearing Values in Tons per Square Foot
1	Hard sound rock	60
2	Medium hard rock	40
3	Hardpan overlaying rock	12
4	Compact gravel and boulder-gravel formations: Very compact sandy gravel	10
5	Soft rock	8
6	Loose gravel and sandy gravel; compact sand and gravelly sand; very compact sand-inorganic silt soils	6
7	Hard, dry consolidated clay	5
8	Loose coarse to medium sand; medium compact fine sand	4
9	Compact sand-clay soils	3
10	Loose fine sand; medium compact sand-inorganic silt soils	2
11	Firm or stiff clay	1.5
12	Loose saturated sand-clay soils; medium soft clay	1

See Tables 29 and 30 for relation of compaction to spoon blows.

* Paragraph C26-377.0, New York City Building Code.

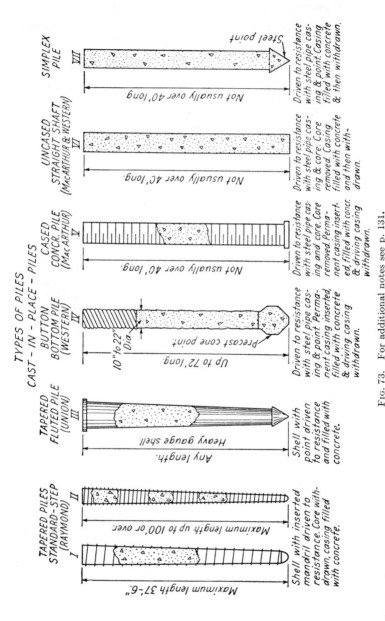

TYPES OF PILES
CAST - IN - PLACE - PILES

I — II. TAPERED PILES STANDARD-STEP (RAYMOND)
Maximum length 37'-6". — Maximum length up to 100' or over.
Shell with inserted mandril driven to resistance. Core withdrawn, casing filled with concrete.

III. TAPERED FLUTED PILE (UNION)
Any length. — Heavy gauge shell.
Shell with point driven to resistance and filled with concrete.

IV. BUTTON BOTTOM PILE (WESTERN)
Up to 72' long. — 10" to 22" Dia. — Precast cone point.
Driven to resistance with steel pipe casing & point. Permanent casing inserted, filled with concrete & driving casing withdrawn.

V. CASED CONCR. PILE (MacARTHUR)
Not usually over 40' long.
Driven to resistance with steel pipe casing and core. Core removed. Permanent casing inserted, filled with concr. & driving casing withdrawn.

VI. UNCASED STRAIGHT SHAFT (MacARTHUR & WESTERN)
Not usually over 40' long.
Driven to resistance with steel pipe casing & core. Core removed. Casing filled with concrete and then withdrawn.

VII. SIMPLEX PILE
Not usually over 40' long. — Steel point.
Driven to resistance with steel pipe casing & point. Casing filled with concrete & then withdrawn.

Fig. 73. For additional notes see p. 131.

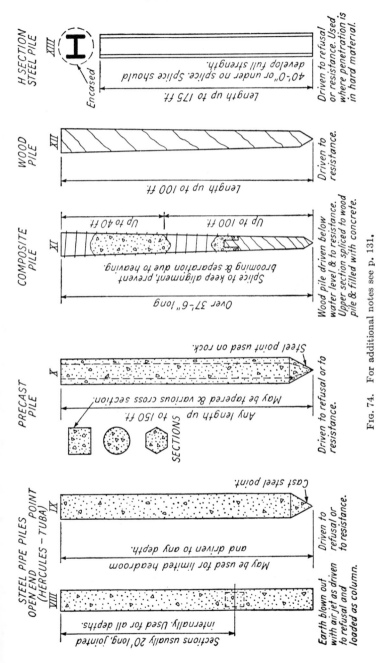

STEEL PIPE PILES OPEN END POINT (HERCULES – TUBA)

VIII
Sections usually 20' long, jointed
Earth blown out with air jet as driven internally. Used for all depths to refusal and loaded as column.

IX
May be used for limited headroom and driven to any depth.
Cast steel point.
Driven to refusal or to resistance.

PRECAST PILE

X
Steel point used on rock.
May be tapered & various cross section.
Any length up to 150 ft.
SECTIONS
Driven to refusal or to resistance.

COMPOSITE PILE

XI
Splice to keep alignment, prevent brooming & separation due to heaving.
Up to 40 ft.
Up to 100 ft.
Over 37'-6" long
Wood pile driven below water level & to resistance. Upper section spliced to wood pile & filled with concrete.

WOOD PILE

XII
Length up to 100 ft
Driven to resistance.

H SECTION STEEL PILE

XIII
Encased
40'-0" or under no splice. Splice should develop full strength.
Length up to 175 ft
Driven to refusal or resistance. Used where penetration is in hard material.

Fig. 74. For additional notes see p. 131.

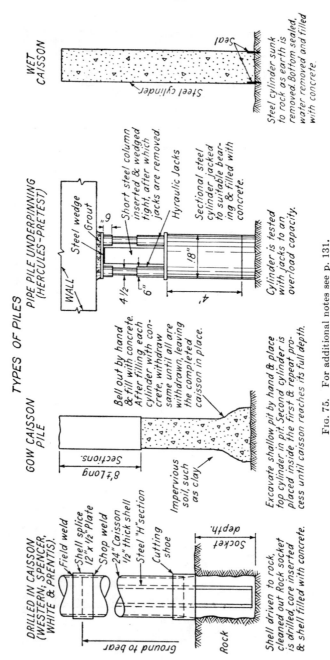

TYPES OF PILES

WET CAISSON

Steel cylinder sunk to rock as earth is removed. Bottom sealed, water removed and filled with concrete.

Seal

Steel cylinder

PIPE PILE UNDERPINNING (HERCULES-PRETEST)

WALL

Steel wedge
Grout

Short steel column inserted & wedged tight, after which jacks are removed.

Hyraulic Jacks

Sectional steel cylinder jacked to suitable bearing & filled with concrete.

Cylinder is tested with jacks to an overload capacity.

9"
4½"
6"
18"
4'

GOW CAISSON PILE

8' Long
Sections.

Bell out by hand & fill with concrete. After filling each cylinder with concrete, withdraw same until all are withdrawn, leaving the completed caisson in place.

Impervious soil, such as clay

Excavate shallow pit by hand & place top cylinder in pit. Second cylinder is placed inside the first & repeat process until caisson reaches its full depth.

DRILLED IN CAISSON (WESTERN; SPENCER, WHITE & PRENTIS).

Field weld
Shell splice 12" x ½" Plate
Shop weld
24" Caisson ½" thick shell
Steel "H" section
Cutting shoe

Ground to bear

Socket depth.

Rock

Shell driven to rock, cleaned out Rock socket is drilled, core inserted & shell filled with concrete.

Fig. 75.   For additional notes see p. 131.

PILE NO.	TYPES OF PILES — NOTES
I, II, III, V, XI	Precautions are required to prevent collapsing of shell when driving adjoining piles.
VI, VII	With uncased piles, precautions should be taken to prevent damage due to driving adjoining piles because pile has no sheet casing around it.
IV, V	When shell is inserted inside driving casing and casing withdrawn, soil must be relied upon to grip pile as firmly as if it had been driven without casing.
VIII	Concrete steel pipe piles sometimes driven open ended to predetermined depth and filled with concrete. After concrete has set pile is driven to required resistance. This is done so as not to disturb adjoining wall and foundation and also when driving cast-in-place piles to prevent heaving.
X	Precast piles are used for marine structure, require heavy handling equipment.
XI, XII	Cut off untreated wood pile below permanent water level. Creosoted wood piles used with cut off above permanent water level for 50 year life. Precautions against overdriving should be taken. Vulnerable to marine borers if not treated with creosote.
I TO IX	The piles have less give under hammer than a pile of more flexible material such as wood or concrete and consequently if driven to the same resistance have a greater safety factor.
XIII	Steel H sections should not be used through cinders, ash fill or normally active rust producing material without adequate protection.
CAISSONS	These are usually used when sinking foundation to considerable depths with heavy loads. This is done by wood sheeting, steel sheeting and steel cylinders. In case of water condition, operations carried on under compressed air.

PILE DRIVING

CHECK LIST FOR INSPECTORS AND DATA

Procedure in Inspection

Inspector should first determine from specifications the type of pile to be used, should familiarize himself with specifications, and should have approved drawings for his use in field.

Check steam or compressor capacity against pile hammer manufacturer's rating.

Condition and lengths of pile or pile shells before driving.

Type of pile hammer and size. Weight of striking part or ram and stroke (see Tables 22A, 22B, 23, and 24) or energy per blow and rated strokes per minute.

Plumbing of pile or mandrel before driving.

Lateral tolerance of pile; limit 3 in. from horizontal location.

Plumbness of pile; limit 2% of pile length.

Inspect pile shell just before concrete is poured, by lowering a droplight or large flashlight to observe collapse or tearing of shell or the presence of sand, mud, or water.

Inspector to plot lengths of piles below cutoff.

Inspector to check boring record to assure that the ends of piles are in solid material and that there is no compressible soil stratum below them.

Check concrete mix from specifications or drawings.

Protection of concrete against freezing.

Pile caps not laid on frozen ground.

Piles cut off at the proper elevation.

After tops of bearing piles have been cut off see that the heads are brush painted twice with hot creosote oil and then heavily coated with asphalt roofing cement.

Check diameters of points and cutoffs of timber piles against specifications.

If untreated piles are partly above water lines call your superior's attention to it.

Check nearby structures for damage from vibration.

Driving Control

The *Engineering News* formulas for allowable bearing loads where piles are driven to practical refusal are:

(*a*)   Drop hammer: $P = \dfrac{2WH}{S + 1}$

(*b*)   Single-acting power hammer: $P = \dfrac{2WH}{S + 0.1}$ *

(*c*)   Double-acting hammer: $P = \dfrac{2E}{S + 0.1}$ *

where P = allowable load, in pounds; W = weight of ram, in pounds; H = height of fall or stroke, in feet; and S = average penetration per blow under last 5 blows, in inches.†

* The trend of engineering thought is that the application of the *Engineering News* formula for steam hammers is not on the side of safety for small penetrations per blow (less than $\frac{1}{4}$ in.). Hence, recommended values are suggested in Tables 22A, 22B, and 24, based on a modification of the formula, in which the dividing factor is changed from $S + 0.1$ to $S + 0.3$.

The weight of the ram should be substantially greater than the weight of the pile.

† Five blows is commonly specified, but because of the high speed of a power hammer it is suggested that the average penetration per blow during the last 20 seconds of driving is more practicable.

The reason for the difference in the formulas (a) and (b) or (c) is the extra speed of the power hammer, which affects the time of consolidation of earth between blows. Both (a) and (b) are gravity-type hammers, while (c) is steam or air driven.

Examples. Given W = 2000 lb., H = 15 ft. 0 in., S = 0.5 in. Required P, using drop hammer.

$$P = \frac{2 \times 2000 \times 15}{0.5 + 1} = 40,000 \text{ lb.}$$

Given W = 5000 lb., H = 3 ft. 0 in., S = 0.4 in. Required P, using single-acting steam hammer.

$$P = \frac{2 \times 5000 \times 3}{0.4 + 0.1} * = 60,000 \text{ lb.}$$

Comments. The field engineer's checking criterion is the number of strokes per minute, rather than the steam pressure, and also penetration. If steam pressure falls off, the number of blows per minute cannot be delivered and the penetration falls off. Therefore, time the driving speed of steam hammers to be sure that the manufacturer's rated speed in strokes per minute is maintained at all times to attain allowable bearing loads indicated by formulas. See that sufficient steam or air capacity is provided.

Load Tests

Conduct as follows. A suitable balanced platform shall be built on top of pile which has been in place for at least 2 days. If it is a concrete pile, the concrete should be thoroughly hardened. Place initial load equal to the proposed pile load using heavy material such as pig iron. Increase this load 25% after 12 hours, and 25% after 24 hours, thus the total load is 150% of proposed load.

Allow final load to remain at least 48 hours. Take readings before and after placing of each load and 12 and 24 hours after placing final load.

The total net settlement deducting rebound after removing load should not be more than 0.01 in. per ton of total test load.

* See footnote p. 132.

TABLE 22A. SINGLE-ACTING POWER HAMMERS—VULCAN IRON WORKS
(ALLOWABLE LOADS BY FORMULA (b) AND RECOMMENDED VALUES)

Hammer Size	Weight of Ram	Stroke (feet)	Blows per Minute	Penetration per Blow in Inches									
				Loads in 1000 Pounds									
				0.1	0.2	0.3	0.4	0.5	0.6	0.7	0.8	0.9	1.0
2	3000	2.42	70	73	48	36	29	24	20	18	16	14	13
1	5000	3.00	60	150	100	75	60	50	43	37	33	30	27
0	8000	3.05	50	244	162	122	97	81	69	60	54	48	44
OK	9300	3.25	50	302	202	152	121	101	86	75	67	60	55
Recommended values *				50%	60%	66.5%	71.5%	75%	78%	80%	82%	83.5%	84.5%

TABLE 22B. SINGLE-ACTING POWER HAMMERS—McKIERNAN-TERRY
(ALLOWABLE LOADS BY FORMULA (b) AND RECOMMENDED VALUES)

Hammer Size	Weight of Ram	Stroke (feet)	Blows per Minute	Penetration per Blow in Inches									
				Loads in 1000 Pounds									
				0.1	0.2	0.3	0.4	0.5	0.6	0.7	0.8	0.9	1.0
S-3	3000	3.00	65	90	60	45	36	30	26	22	20	18	16
S-5	5000	3.25	60	162	108	81	65	54	46	41	36	32	30
S-8	8000	3.25	55	260	173	130	104	87	74	65	58	52	47
S-10	10000	3.25	55	325	217	162	130	108	93	81	72	65	59
S-14	14000	2.67	60	375	250	187	150	125	107	94	83	75	68
Recommended values *				50%	60%	66.5%	71.5%	75%	78%	80%	82%	83.5%	84.5%

* See footnote p. 132.

TABLE 23. VALUES OF *E* FOR McKIERNAN-TERRY DOUBLE-ACTING PILE HAMMERS *

SIZE OF HAMMER	FT-LB. BLOW AT GIVEN STROKES PER MINUTE		SIZE OF HAMMER	FT-LB. BLOW AT GIVEN STROKES PER MINUTE	
	Strokes per Min.	Ft-Lb. per Blow = *E*		Strokes per Min.	Ft-Lb. per Blow = *E*
7	225	4,150	9B2	100	3,700
	195	3,720		105	4,200
	170	3,280		110	4,750
				115	5,350
9B3	145	8,750		120	5,940
	140	8,100		130	7,000
	135	7,500		140	8,200
	130	6,800			
			10B2	100	10,700
10B3	105	13,100		105	12,000
	100	12,000		110	13,500
	95	10,900		115	15,000
	90	9,550			
			11B2	100	15,600
11B3	95	19,150		105	17,250
	90	18,300		110	18,900
	85	17,500		115	20,500
	80	16,700		120	22,000

* *E* = energy of ram for various driving speeds.

TABLE 24. ALLOWABLE LOADS ON PILES IN THOUSANDS OF POUNDS USING MAXIMUM *E* (DOUBLE-ACTING HAMMER)

PENETRATION PER BLOW IN INCHES	RECOMMENDED VALUES, %	SIZE OF HAMMER						
		7	9B3	10B3	11B3	9B2	10B2	11B2
0.1	50	41.5	87.5	131.0	191.5	82.0	150.0	220.8
0.2	60	27.6	58.3	87.3	127.6	54.6	100.0	147.2
0.3	66.5	20.7	43.7	65.5	95.7	41.0	75.0	110.4
0.4	71.5	16.6	35.0	52.4	76.6	32.8	60.0	88.3
0.5	75	13.8	29.1	43.6	63.8	27.3	50.0	73.6
0.6	78	11.8	25.0	37.4	54.7	23.4	42.9	63.2
0.7	80	10.3	21.8	32.7	47.8	20.5	37.5	55.3
0.8	82	9.2	19.4	29.1	42.5	18.2	33.3	49.1
0.9	83.5	8.3	17.5	26.2	38.3	16.4	30.0	44.1
1.0	84.5	7.5	15.9	23.8	34.8	14.9	27.3	40.1

SHEET PILING

Purpose: To support the sides of an excavation and prevent loss of ground.

Types: 1. Wood sheeting.
           2. Reinforced-concrete sheeting.
           3. Interlocking steel sheeting.

Wood sheeting may be tongue and groove, splined, or Wakefield sheeting, made with three planks bolted together with the center one offset to form a tongue and groove. Wood sheeting is limited to about 20 ft., but the excavation can be made deeper than this by driving a second set of sheeting planks within the first set.

Reinforced concrete is used where it is to be left in place as part of the permanent structure.

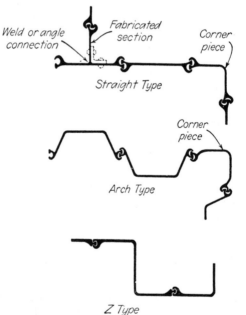

FIG. 76. Steel sheeting.

Steel sheeting can be driven in lengths up to about 50 ft. Sheeting of all types should be driven progressively in stages around the excavation to minimize the danger of breaking the lock between sheets. Driving should be stopped at once if the sheet meets an obstruction, especially in the case of wood sheeting.

See that sheet piling is driven straight and to the required lines and grade, with penetrations to firm subsoil.

Steel sheet piling along excavations to be carefully interlocked and timber sheeting tightly closed to prevent loss of ground.

Soldier-beam and timber-lagged sheeting should be watched to avoid ground slippage.

Vertical sheeting permits little water to pass through it; consequently the difference in hydrostatic pressure may cause boils in the bottom of the excavation. If steel sheeting can be driven into an impervious stratum to cut off the water, it will prevent boils. Sheeting should be held in place with wales and braces or shores. In excavations of more than a few feet the braces or shores should be prestressed with wedges or otherwise to prevent movement.

Small pits may be carried to great depths with horizontal sheeting because the earth arches around them. Great care must be taken, however, not to loosen ground, as that may break the arch and cause a collapse. On long excavations, horizontal sheeting has been used successfully between vertical H beams called "soldier beams."

SPUD PILES

Where it is found difficult to drive a pile of the type that is being used on the job on account of boulders or other obstructions, a so-called "spud pile" will sometimes solve the difficulty. This is a rugged steel mandrel, strong enough to break up or push aside the obstructions. The mandrel, in driving, may be forced sideways or "weaved," to push obstructions out of the way, but such forcing, if carried to extremes, can break or bend the mandrel.

RED LIGHTS—PILES

Buckling and heaving of cast-in-place piles is caused by bulking of the soil due to driving closely spaced piles in a group. The shells for an entire group are commonly driven before any are filled, and buckling can be detected by inspection. The practice of driving an inner shell to correct a buckling should not be allowed. A buckled shell should be withdrawn. A successful method for avoiding buckling is called "pre-excavation." An open-end pipe, somewhat smaller than the pile, is first driven and the soil in it is removed by compressed air or other means. The pipe is then withdrawn and the pile is driven down in the hole. If the shell has been filled before the adjoining shell is driven, a rise of the wet concrete in the shell indicates buckling, and the pile should be replaced if the rise is significant.

Heaving, or lifting of a filled pile out of the ground, may result from the same cause. Unless the heaving is very slight, the pile should be rejected.

Injury to wood piles. Crushing or brooming of pile head or, in precast concrete piles, the cracking or disintegrating of concrete makes it impossible to drive piles properly as this dissipates the energy of the blow of hammer.

Possible telescoping or crushing of the middle of wooden piles is indicated by sudden loss of resistance.

Possible deflection of the foot of pile. This happens when a pile hits a slanting surface of rock and then drives easier as the result of the splitting or sliding of the bottom.

A sudden change from hard to easy driving in wood piles should be regarded with great suspicion. If withdrawn, the pile is usually found to be shattered. Excavations have disclosed many old wood piles shattered or sheared by overdriving.

Hard driving, and especially hammering down the concrete in a pipe pile to form a "pedestal," causes vibration that may be dangerous to surrounding structures.

In driving open-end pipes through sand, the sand tends to arch in the pipe and form a solid plug which makes the pipe drive like a closed-end pipe or solid pile and greatly adds to the difficulty of driving. The sand should therefore be blown or dippered out frequently.

Water jetting of piles should be prohibited (1) where there is a risk of undermining a nearby structure; (2) where clayey and silty soils would be disturbed; (3) where piles depend on friction in such soils. Sometimes it is impossible to drive piles in sand except by jetting.

Borings should be inspected for soft strata below the level where piles fetch up, as reliance in pile-driving formulas for allowable loads may be unsafe.

_____
Engineer

REPORT ON PILE DRIVING

FIELD INSPECTION

Report No. _____

Job _____ Date _____

Reported to _____

Hammer data _____

Foot-ing No.	Pile No.	Pene-tration	No. Blows Last In.	No. Strokes per Min.	Bearing Capacity	Ap-proved	Re-jected	Re-marks

See field drawing No. _____ for field location of piles in this report _____

_____
Inspector

ROPE AND CABLE—STRENGTHS

TABLE 25. WEIGHT AND STRENGTH OF MANILA AND SISAL ROPE *

Diameter, in.	Circumference, in.	Approx. Feet per Lb.	Ultimate Breaking Strength of Manila Rope (Min. Government Allowance), lb.	Safe Working Strains, lb.	Ultimate Breaking Strength of Sisal Rope (Min. Government Allowance), lb.	Safe Working Strains, lb.
1/4	3/4	50.0	600	120	480	96
3/8	1 1/8	24.4	1,350	270	1,080	216
1/2	1 1/2	13.3	2,650	530	2,120	424
3/4	2 1/4	6.00	5,400	1,080	4,320	864
1	3	3.71	9,000	1,800	7,200	1,440
1 1/2	4 1/2	1.67	18,500	3,700	14,800	2,960
2	6	.930	31,000	6,200	24,800	4,960

* Adapted from American Manufacturing Company.

WIRE ROPE 6 x 19 STANDARD HOISTING—PLOW STEEL *

Diameter, in.	3/8	7/16	1/2	9/16	5/8	3/4	7/8	1	1 1/8	1 1/4	1 3/8	1 1/2	1 5/8	1 3/4	1 7/8	2	2 1/4	2 3/4
Breaking strength, tons of 2000 lb.	5.31	7.19	9.35	11.8	14.5	20.7	28.0	36.4	45.7	56.2	67.5	80.0	93.4	108.0	123.0	139.0	174.0	254.0

* From John A. Roebling's Sons Company.

VARIETIES OF KNOTS

A great number of knots have been devised, of which only a few are illustrated, but those selected are the most frequently used. See Fig. 77.

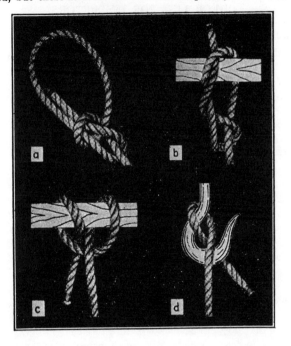

Fig. 77. *From American Manufacturing Company.*

a. Bowline. Makes a slip-proof loop. Popular because it is easy to untie.

b. Timber hitch. For securing a line to logs or planks. For lifting or dragging.

c. Clove hitch. For attaching rope to a fixed object, or small rope to a larger one.

d. Blackwall hitch. A temporary hook tie. More secure with two turns around hook.

WOOD AND TIMBER CONSTRUCTION

CHECK LIST FOR INSPECTORS

Inspectors' Equipment

Complete set of final structural plans, specifications, and approved shop details.

Copy of rules for stress grade of lumber.

6-foot rule.

Plumb bob.

Moisture meter.

Procedure in Inspection

Grade of lumber checked. Material should be stamped with grade shown on plans or called for in specifications. The inspector should familiarize himself with rules for grading of lumber to be used so that he may check grading if from appearance it looks incorrect.

Selection of already graded lumber checked. Select beams so as to avoid slope of grain in lower third of beam steeper than 1:20. Slope of grain in tension member of truss not to be steeper than 1:20. Avoid knots in lower edge of beams. By utilizing elsewhere or inverting pieces which do not conform, these results should be attained without waste.

Imperfections that may have occurred after grading, such as broken fibers due to transportation, decay, and moisture content, which should not be more than 20%, to be checked. Moisture content may be checked with moisture meter if available; otherwise inspector will have to accept manufacturer's certificate of moisture at time of grading plus visual inspection.

Increased checks, loose knots, and warping due to unsatisfactory seasoning watched.

Sizes, lengths and spacing of all members checked.

Bearing and anchorage of beam, girder, or joists on masonry checked.

Plumbing, base, cap, and splice details of columns, especially checking bearing at ends, checked.

All special details shown on plans carefully followed.

Correct fabrication of built-up member such as laminated members and trusses. All members with bolts and ring connectors should be fabricated with standard tools and strictly according to instructions furnished by manufacturer of same.

Drilling and grooving of ring connector members. Any material that is incorrectly drilled or grooved must be rejected, as it is impossible to correct it.

Tightness of bolts in bolted or connected work. These should be tightened up so hard that washer makes a slight impression in wood surface but not so as to tear fibers. After construction until seasoning, bolts should be given a periodic inspection for tightness and at the same time timber should be inspected for further checking. This particularly applies to ring connectors or keyed work, as ring or key tends to rotate as bolts loosen.

Alignment, bearing, or connection of trusses after erection. Trusses should be straight and in a vertical position, and bearing or connection in accordance with plans.

Gluing of glued laminated members. This is usually done in a shop with proper facilities. Inspector should check to see that specifications are followed exactly with special attention to the following: type and quality of glue, mixing of glue, amount of glue used, method of applying, moisture content of lumber, curing of members, and temperatures of manufacturing space. In field, watch for tendency of laminations to separate.

Retouching of cut, preserved members, see specifications.

Lumber seasons to a moisture content in balance with the surrounding air, and shrinks in seasoning somewhat more or less than $\frac{1}{32}$ in. per in. of width, depending on the temperature and humidity of the air.

Access to fabricated wood trusses should be available after erection, and they should be inspected periodically. When the air in which the trusses are in use is warm and dry, the first inspection should be within a month after erection; under less rapid conditions of seasoning, a longer time can intervene, but it should not be more than 3 or 4 months. Nuts should be tightened on bolts when necessary, until the wood members have become thoroughly seasoned, in order to keep the adjacent faces in contact.

Splitting in seasoning occurs more extensively at ends of members than at intermediate points. As connections are near ends of web members, and web members are usually smaller than chord members, it may be possible to obtain seasoned material for them when it cannot be obtained for chord members.

Wood dries faster through ends of pieces than through sides. Moisture-retaining end coatings tend to balance end and side drying and reduce end splitting. They should be applied at a mill as soon as possible after cutting, and again whenever a piece is recut at a fabricating plant or elsewhere and a fresh end cut is exposed. "End Seals for West Coast Lumber," published by the West Coast Lumbermen's Association, gives detailed information on this subject.

Separate splice plate for each row of connectors in tension and compression splices, and saw kerfs between longitudinal rows of connectors at the ends of axially loaded members, obviate the splitting from shrinkage which might occur in a wide member. Holes should be bored at ends of saw kerfs to prevent the start of splits or checks.

Long end distances, longer if possible than design data requires, and straight end cuts without re-entrant angles (at a right angle to the member if possible), reduce end splitting.

RED LIGHTS

All timber construction is subject to shrinkage.

1. To avoid checking, paint ends of planks exposed to air.

2. Ends of wide timbers which are prevented from shrinking laterally by bolts, as truss members, tend to check and split; therefore, use two

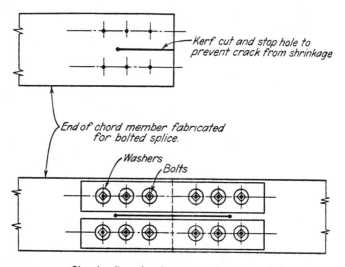

Kerf cut and stop hole to prevent crack from shrinkage

End of chord member fabricated for bolted splice.

Washers
Bolts

Chord splice, showing separate splice plates for each row of bolts and kerf cuts and stop holes in main members.

Fig. 78. Timber details to prevent shrinkage cracks. *Adapted from West Coast Lumberman's Assoc. Timber Fabrication Division.*

narrow fish plates in preference to one wide one, and in the ends of wide truss members make a saw cut terminating in a drilled hole to permit some shrinkage without splitting. (See Fig. 78.)

3. Timber connectors act as keys rather than bolts and will rotate as shrinkage slacks off tension in the bolts. Trusses with ring connectors should be checked and tightened periodically for several years; they are very dangerous otherwise.

4. Wood floor joists or roof beams framing into steel girders should have tops ¼ in. above steel to allow for shrinkage and prevent humps in the floor or roof.

IDENTIFICATION OF TIMBER

DISCUSSION

Spruce. The following varieties are in common use as stress grade lumber.

Eastern or Red Spruce. Grows in New England and south to North Carolina. See illustration.

Engelmann Spruce. Rocky Mountain region.

Sitka Spruce. Pacific Northwest. The wood is light, soft, even-grained, easily worked, and strong in proportion to its weight.

Redwood (*Sequoia*) is soft, light, straight-grained, and very durable. It grows to great size and age. It is used in general building and especially for siding and shingles on account of its durability. It is also used for millwork because of its comparative freedom from swelling and shrinking due to atmospheric changes after it is seasoned.

Cedar. There are so many woods popularly known as cedar that the word means little, botanically; but all the so-called cedars are light in weight, soft, even-grained, and resistant to decay, though in varying degrees.

The principal species that go by the name of cedar are:

1. Eastern or Red Cedar. Kentucky and Tennessee.
2. Northern White Cedar or Arbor Vitae. New England and Lake states.
3. Southern White Cedar. Atlantic Coast.
4. Western Red Cedar or Giant Arbor Vitae. Pacific Northwest and Rocky Mountain states.
5. Port Orford Cedar. Oregon.
6. Incense Cedar. Southern Oregon.
7. Alaska or Yellow Cedar. Oregon to Alaska.

Cedar is used principally for posts, for shingles, and for special purposes such as canoe building.

HARDWOODS—RED OAK

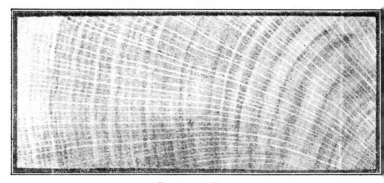

Transverse Section

Radial Section

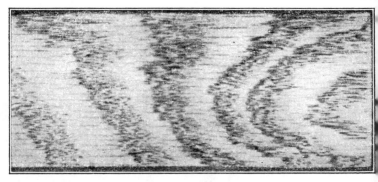

Tangential Section

This illustration is representative of the oaks, which are all very strong and suitable for the manufacture of anything from piles to furniture. The wood is very heavy; the white oak is the most resistant to decay. Largely used for church trusses.

Fig. 79.

Tidewater Red Cypress { light with dark stain, decay resistant.

Eastern Hemlock { light, brash, not easily dressed, slightly yellowish brown color.

Eastern Spruce { whitish gray, soft texture, light and strong.

Larch { fine-grained, light wood coming into wider use for joists and planks.

Douglas Fir { the western counterpart of Long Leaf Yellow Pine, heavy for soft wood, distinctly reddish in color. Summer wood dark red, very hard. Splits easily but has good resistance against decay. One of the strongest soft woods. Known locally as Douglas spruce and Oregon pine.

Short Leaf Yellow Pine { distinctly yellowish in color. Summer wood same color as spring wood. Coarse graining gives ornamental appearance when cut on tangential plane.

Long Leaf Yellow Pine { counterpart of western fir except that it is yellow with a reddish cast. Summer rings dark colored, very dense. Wood gets its great strength from this feature. Used for wood trusses and high-class timber construction.

Eastern White Pine { very soft, whittling wood with pungent odor, excellent for timber, color white to a slight pinkish.

Greenheart { heavy, tough with distinct greenish tint; resistant to *Limnoria*, used for piling and dock work.

Fig. 80. Identification of common soft woods by comparison.

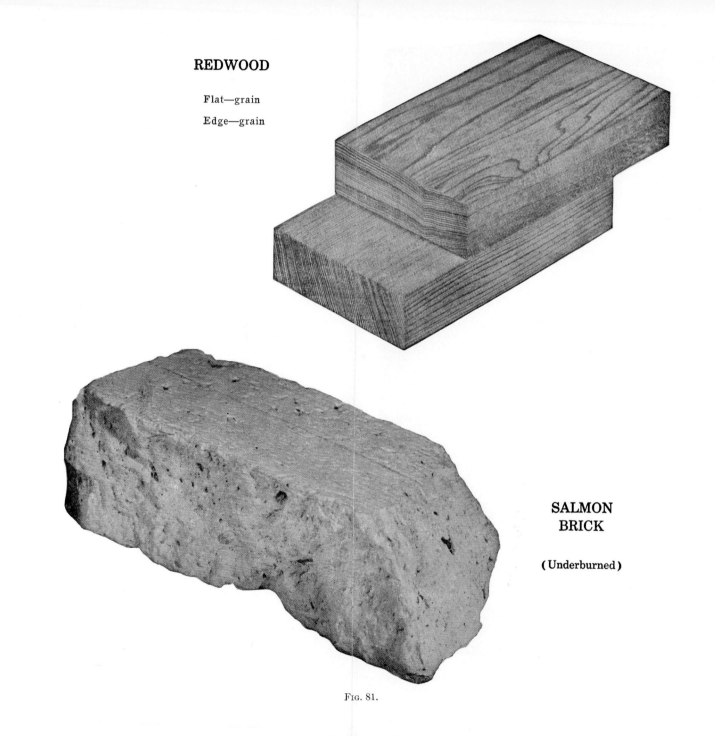

REDWOOD

Flat—grain

Edge—grain

**SALMON
BRICK**

(Underburned)

Fig. 81.

HARDWOODS—MAPLE

Transverse Section

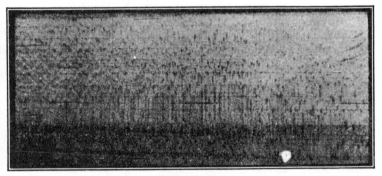

Radial Section

Tangential Section

This illustration is representative of the maples, an excellent flooring and furniture material but not used very much as structural timber.

Fig. 82.

TIMBER

WOOD JOISTS—NET SECTION

TABLE 26. SECTION MODULI = $bd^2/6$

Nom. Size	Actual Size	S	Nom. Size	Actual Size	S	Nom. Size	Actual Size	S	Nom. Size	Actual Size	S	Nom. Size	Actual Size	S	Nom. Size	Actual Size	S
2 × 4	1⅝ × 3⅝	3.56	3 × 4	2⅝ × 3⅝	5.75	4 × 4	3⅝ × 3⅝	7.94	6 × 6	5½ × 5½	27.7	8 × 8	7½ × 7½	70.3	10 × 10	9½ × 9½	143
2 × 6	1⅝ × 5⅝	8.57	3 × 6	2⅝ × 5⅝	13.8	4 × 6	3⅝ × 5⅝	19.1	6 × 8	5½ × 7½	51.6	8 × 10	7½ × 9½	113	10 × 12	9½ × 11½	209
2 × 8	1⅝ × 7½	15.3	3 × 8	2⅝ × 7½	24.6	4 × 8	3⅝ × 7½	34.0	6 × 10	5½ × 9½	82.7	8 × 12	7½ × 11½	165	10 × 14	9½ × 13½	289
2 × 10	1⅝ × 9½	24.4	3 × 10	2⅝ × 9½	39.5	4 × 10	3⅝ × 9½	54.5	6 × 12	5½ × 11½	121	8 × 14	7½ × 13½	228	10 × 16	9½ × 15½	380
2 × 12	1⅝ × 11½	35.8	3 × 12	2⅝ × 11½	57.9	4 × 12	3⅝ × 11½	79.9	6 × 14	5½ × 13½	167	8 × 16	7½ × 15½	300	10 × 18	9½ × 17½	485
2 × 14	1⅝ × 13½	49.4	3 × 14	2⅝ × 13½	79.7	4 × 14	3⅝ × 13½	110.0	6 × 16	5½ × 15½	220	8 × 18	7½ × 17½	383	10 × 20	9½ × 19½	602

_____
Engineer

REPORT ON WOOD PRESERVATION *

PLANT INSPECTION

Report for _____

Material _____

Project _____

Producer _____

Contractor _____ Specs. _____ _____

Treatment No. _____ Report No. _____ Date _____

Charge No. _____ Preservative _____

Board feet _____ Treatment specified _____ Process _____ _____

Lineal feet _____ Net retention _____

Cubic feet_____ Condition of _____

Steam _____ hours at __ pounds maximum pressure ____ °F. maximum temperature

Vacuum _____ hours at __ inches maximum pressure ____ °F. minimum temperature

Air _____ hours at __ pounds maximum pressure _____

Preservative _____ hours at __ pounds maximum pressure ____ °F. average temperature

Vacuum _____ hours at __ inches maximum mercury ____ °F. minimum temperature

Special operation _____

Penetration _____ Specific gravity or preservative _____

_____

No. Pieces	Size	Length	Total Treated	Total to Date

Remarks:

The above preservative and treatment fulfills the specification.

_____
Inspector

* *Adapted from Haller Testing Laboratories, Inc.*

SOILS

TABLE 27. EXPLORATION AND SAMPLING METHODS *

Method	Material in Which Used	Penetration Method	Sampling Method	Type of Sample	Purpose or Value
Rod sounding or jet probing	All soils except hardpan or boulders	Driving 1 in. steel rod or ¾ in. jet pipe with hand pump	No sample		To obtain depth of muck or soft strata. Location ledge or boulders. Otherwise valueless.
Wash borings		Washing inside 2½ in. driven casing with chopping bit on end of 1 in. extra heavy pipe	Sample recovered from sediment in wash water	Disturbed—sedimentary, coarse grains only	Depth to ledge or boulders; otherwise valueless. Results deceptive and dangerous.
Dry sample boring		Open-end pipe or split spoon sampler driven into soil	Disturbed but not separated	Density data from penetration of spoon. Fairly reliable and inexpensive.
Special sampling devices	Cohesive soils	Driven casing or auger boring	By special sampling spoon or device	Undisturbed	To obtain samples for laboratory study
Auger boring	Cohesive soils. Cohesionless soils above ground water	Soil, wood or post hole; auger rotated by hand or machine and withdrawn	Sample recovered from soil brought up by auger	Disturbed but better than wash samples	To locate soil strata and ground water. Roads, airfields, canals, and railroads. Samples for visual inspection and soil profile.
Well or churn drilling	All soils including boulders, rock, and gravel	Churn drilling by power	Bailed sample of churned material or use of "clay socket"	"Clay socket" or "dry"	Occasionally used for foundations. "Bailed" samples worthless.
Rotary drilling		Rotating bit	From circulating liquid	Fluid	Samples worthless
Core drill borings	Large boulders and solid rock	Diamond, shot, or saw-tooth cutters	Cores cut and recovered	Rock cores ⅞ in. and over in diameter	Best method to obtain type and condition of rock
Test pits and caissons; may be dug by clamshell or back-hoe	All soils; below ground water use pneumatic caisson or lower water table	Excavate by hand or power; pit over 6 ft. sheeted or lagged	Bulk sample by hand; undisturbed sample with spoon, tube, or special device	Disturbed or undisturbed	Most satisfactory method; should supplement others. To obtain undisturbed sample cohesionless soil. Soil can be inspected in natural condition.
Geophysical, seismic, electric resistance, electric potential	No samples, vibrations. Mostly patented methods.	Continuous vibration or impulse from dynamite explosion.		Device to register	Primary exploration will indicate earth, loose rock, or solid rock. Interpretation uncertain.

* Adapted from "Low Dams" by Natural Resources Comm., based on Harvard Grad. Eng. School Pub. 208 by H. A. Mohr.

TABLE 28. SIZE OF SAMPLES

Visual inspection and record, 1 qt. mason jar.
California bearing ratio, 125 lb.
Soil stabilization, 125 lb.
Physical constants and mech. analysis, 5–15 lb.
Aggregates for construction (concrete), 35 lb.
Moisture-density (Proctor tests), 10–35 lb.
Undisturbed sample, 12″ to 2′ long x 2″ to 5″ diam.
Rock core, usually $\frac{7}{8}$″ to $1\frac{1}{2}$″ diam.

Note. Seal undisturbed samples in tube with paraffin so structure and moisture content are not disturbed. Place bulk (disturbed) samples in bag or container tight enough so fines will not be lost.

Boring Spoon Penetrations. The number of blows per foot of penetration of spoon under a definite weight falling a certain height gives important information in classifying soils. Easy driving of sampling spoons usually indicates a high percentage of water and consequently danger of settlement.

TABLE 29. SOIL COMPACTION RELATED TO SPOON BLOWS WITH 2½-IN. SPOON

FOR SAND

Descriptive Term	Blows per Foot	Remarks
Loose	15 or less	These figures approximate for medium sand,
Compact	16 to 50	300-lb. hammer, 18-in. fall. Coarser soil re-
Very compact	50 or more	quires more blows; finer material, fewer blows.

FOR CLAY

Very soft	Push to 2	Molded with relatively slight finger pressure.
Soft	3 to 10	
Stiff	11 to 30	Molded with substantial finger pressure; might be removed by spading.
Hard	30 or more	Not molded by fingers, or with extreme difficulty; might require picking for removal.

From New York City Building Code Section C26-377.0. See also Table 21.

TABLE 30. COMPACTION RELATED TO SPOON BLOWS WITH 2-IN. SPOONS *

FOR MEDIUM SAND

Approximate Blows per Foot

Descriptive Term	148-Lb. Hammer† 30-In. Fall	300-Lb. Hammer 18-In. Fall	150-Lb. Hammer 18-In. Fall
Loose	15 or less	10 or less	25 or less
Medium compact	16 to 30	11 to 25	25 to 45
Compact	30 to 50	26 to 45	45 to 65
Very compact	50 or more	45 or more	65 or more

FOR CLAY

Very soft	Push to 3	Push to 2	Push to 5
Soft	4 to 12	3 to 10	5 to 15
Stiff	12 to 35	10 to 25	15 to 40
Hard	35 or more	25 or more	40 or more

* Prepared by Department of Housing and Buildings of the City of New York supplementing Section C-26-3770, Presumptive Bearing Capacities of Soils. See also Table 21.

† Used by most boring contractors.

Coarser soils require more blows; finer material, fewer blows. A variation of 10% in the weight of the hammer will not materially affect the values in the tables.

The use of any specific size of spoon, weight and fall of hammer is not mandatory in the Code. However, any other size spoon or weight of hammer exceeding the 10% variation from weight of hammer specified in the tables shall not be accepted until sufficient data have been submitted for investigation and approval.

SPACING AND DEPTH OF BORINGS AND TEST PITS OR TEST HOLES

Highways.* At 100 ft. stations plus additional necessary at culverts, bridges, weak zones, wide cuts and fills, muck deposits, borrow pits, and sources of base material. Depth not less than 3 ft. below subgrade. Locate ground water table, seepage sources, and direction of flow

Airfields.† At 100-ft. to 1000-ft. spacing on center line, edge of pavement, and edge of shoulders. Depth not less than 4 to 6 ft. below subgrade in cut or ground surface in fill. Not less than twice diameter of tire contact area nor less than frost penetration. Locate ground water table and seepage data.

* A.S.T.M. D-420, C.A.A. Specs.
† P.R.A., U.S.E.D., A.A.F., C.A.A.

Make field load-bearing tests at time of survey (from 5 to 10 usual for each airfield).

Bridges, Dams, and Piers.* Borings spaced as needed to bedrock or well below foundation level. Make borings at least 20 ft. into solid rock. Make 1 or more borings at each pier 50 ft. minimum into solid rock. Use open-pit exploration on land and in shallow water. Make soil bearing tests and pile loading tests.

Building Foundations, Towers, Chimneys, etc.* Borings spaced not over 50 ft. center to center. Depth 15 ft. to 20 ft. minimum below foundation level. Initial borings to depth = 2 × width loaded area.

Core borings into rock greater than minimum design depth of rock required. Supplement borings with test pits, load tests, and test piles.

Sewers, Drains, and Tunnels. On center line recommend 100- to 200-ft. intervals, except in rock, 10 to 50 ft., to determine the irregular rock surface for pay lines and to make certain of adequate rock cover for tunnels, also to avoid risk of preglacial gorges. Depth of boring below anticipated or preliminary subgrade in good soil or rock should be 5 ft. or more to allow for latitude in final design affecting subgrade elevations; in poor or unstable soils, explore to necessary depth requisite for determining kind of foundation such as bearing piles or replacing poor soil with other consolidating materials.

* Man. Eng. Practice 8, A.S.C.E.

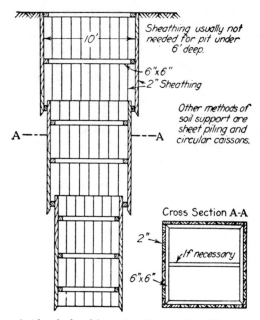

FIG. 83. Test pit (sheathed and braced). *Krynine, Soil Mechanics, McGraw-Hill Book Company.*

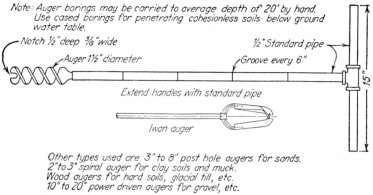

FIG. 84. Soil auger.

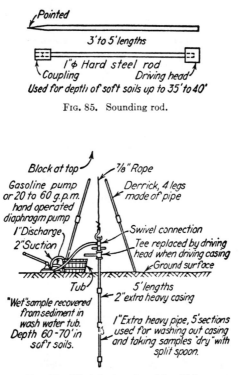

FIG. 85. Sounding rod.

FIG. 86. Wash boring rig. *After Mohr.*

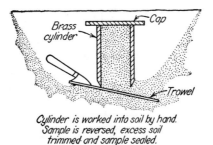

Cylinder is worked into soil by hand.
Sample is reversed, excess soil
trimmed and sample sealed.

FIG. 87. Shallow sampling, cohesionless soil (sand). *Krynine, Soil Mechanics,*
McGraw-Hill Book Company.

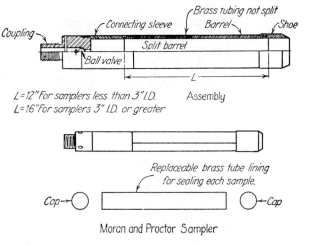

L=12" For samplers less than 3" I.D.   Assembly
L=16" For samplers 3" I.D. or greater

Moran and Proctor Sampler

FIG. 88.   Deep sampler, cohesive soils.

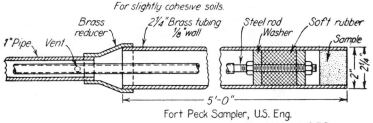

For slightly cohesive soils.

Fort Peck Sampler, U.S. Eng.

Note: A suggested sampler is the one used by Providence U.S.E.D., "Clay Sampler Type C"- consists of 4¾" diameter brass tube ¹⁄₁₆" thick plus a piston.

FIG. 89.   Piston-type sampler, cohesive soils.

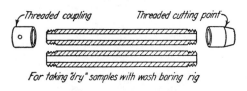

For taking "dry" samples with wash boring rig

FIG. 90.   Split spoon sampler.

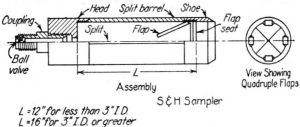

L = 12" for less than 3" I.D.
L = 16" for 3" I.D. or greater
This sampler disturbs soil. Freezing and core drilling have been
used with success for undisturbed samples.

Fɪɢ. 91. Deep sampler, cohesionless soil. *Krynine, Soil Mechanics, McGraw-Hill
Book Company.*

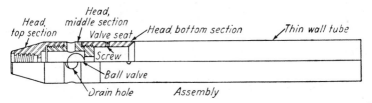

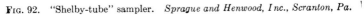

Fɪɢ. 92. "Shelby-tube" sampler. *Sprague and Henwood, Inc., Scranton, Pa.*

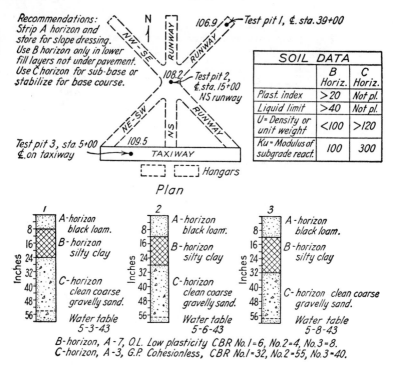

FIG. 93. Plan and log of test pits for airfield.

CHECK LIST—BORING INSPECTION

1. Note depth of hole and elevation at which each sample was taken.

2. See that boring plan and specification are up to date.

3. Record boring in log as per Fig. 94.

4. Take all samples with a sample spoon. No sample out of casing or washwater.

5. Record ground-water level daily. Pull casing up 1 ft. at night.

6. Undisturbed samples, if called for, to be taken by pushing sampler into ground. Do not drive.

7. Clean-out casing before each sample is taken.

8. Samples to be taken below the bottom of the casing, and in undisturbed soil.

9. Casing not to be driven until the soil has been washed out below for the depth the casing is to be driven.

10. Record number of blows per foot and size of hammer in driving sample spoon.

11. Check washwater for change in type of soil that the chopping bit is going through.

BORING LOG † (TYPICAL STRUCTURES)

Location _Antigua_ _ _ _ _ _ _   Structure _Hangar (See key plan)_ Sheet No. _1_ of _2_.

Boring No. _1_.Datum _ _ _ _ _ _   Boring Inspector _Smith_ _ _   Date _1-3-41_ _ _

Stratification				Casing		Sample or Spoon		*	Miscellaneous Data
Elevation	Depth	Legend	Description of Materials (Type, Color, & Consistency)	Blows	Penetration	Blows	Penetration	Sample No.	Length of hole 26'-0" / Rock 5'-0" / Weight of hammer 300 lbs / Aver. fall of hammer 30" / El. of ground water +68.4
73.7	0		Surface						Remarks**
			Brown sandy loam						Few roots
			Trace of gravel			6	12"	1D	Dry and friable
71.2	2'6"								
				8	12"	32	18"	2D	Fairly firm
			Fine brown sand	10	12"				Cohesionless
66'	G.W.		Trace of gravel	16	12"				Resistance
				16	12"	28	12"	3D	increases with
64.7	9'0"								depth.
			Firm, hard, yellow,						Becomes plastic
62.2	11'6"		silty clay.	18	12"	20	18"	4D	when worked.
			Compact gravel, silt,	380	12"	60	3"	5D	Chips of black slate
52.7	21'0"		and sand "Hardpan"						embedded in silt.
			Buff-colored						Casing and rods
			limestone.					6C	refused at 21'-0"
			Hard 80% core						Bottom of hole
47.7	26'0"		recovery.						at 26'-0."

Note: Additional data may include: Key plan with contours, stations coordinates, and building outline; Benchmarks, date, drilling rig, casing dia.; length and diameter of sampler; Atterberg Limits, Mech. Analysis, density, water content.

* Write sample number at corresponding depth, designate dry samples by D, wash samples by W, undisturbed samples by U, and rock cores by C.
** When drilling cores in rock record the percentage of recovery in each foot of penetration.

Fig. 94.

† *Caribbean Architect-Engineer.*

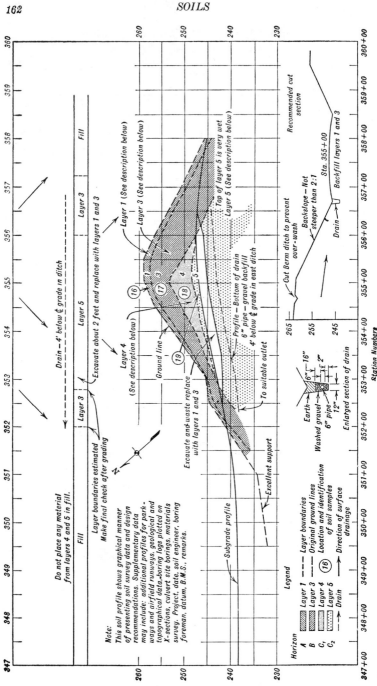

Results of Soil Tests

Mechanical Analysis

Identification Number	Layer	Per cent of particles having diameters smaller than					
		2 mm.	0.5 mm.	0.25 mm.	0.05 mm.	0.005 mm.	0.001 mm.
16	1	100	100	98	90	28	19
17	3	100	98	96	85	19	14
18	4	53	49	44	42	19	13
19	5	100	98	96	59	73	50

Physical properties of particles passing the 0.5 mm. sieve

Identification Number	Layer	Lower Liquid Limit	Plastic Index	Shrinkage		Moisture Equivalent		Group	Textural Class
				Limit	Ratio	Centrifuge	Field		
16	1	38	16	23	1.7	29	31	A – 4	Silt loam
17	3	27	18	18	1.7	36	22	A – 4	Silty clay loam
18	4	53	34	11	2.0	53*	33	A – 7	Clay and gravel
19	5	101	71	14	1.9	93*	58	A – 7	Plastic clay

*Waterlogged

Sample number 18 — Layer 4 contains coarse gravel. See description.

Description of Layers

Layer 1:
Reddish brown mellow silt loam. Friable when dry but of pasty consistency when wet.

Layer 3:
Grayish brown or mottled gray and rusty brown silty clay loam or silty clay of moderately compact structure. Compactness increases with depth. Friable when dry but plastic when wet. The compact nature of this layer does not seem to retard percolation to any degree.

Layer 4:
Similar to layer 5 but contains a very large quantity of gravel varying in size from 1/4" to 2" with the largest percentage between 3/4" and 1 1/2". The presence of gravel apparently does not affect the structure particles or their behaviour. On drying, shrinkage cracks develop and soil shrinks away from gravel. This layer also includes a brown or grayish brown compact clay which is a transition between layers 3 and 5, and shrinks considerably on drying.

Layer 5:
Mottled bluish gray and rusty brown plastic, sticky, and tenacious clay composed of angular structure particles which have a wet, shiny and slick surface. The particles are irregular in shape, easily crushed and when molded take on the appearance and consistency of putty. Upper 3 feet of layer is very wet. It blends gradually into a dense, plastic, cloddy structured bluish gray clay which retards the downward movement of water but does not stop it, since the water can penetrate between the cleavage planes which are well defined. White concretions, black, rusty brown and blood red stains are found throughout the layer. This material shrinks considerably on drying, leaving wide shrinkage cracks and on exposure the larger clods slake down to the smaller sized particles. This layer contains a high percentage of lime.

General Notes and Recommendations

Drainage is across the road from east to west.
Original ground gives excellent support for fill
Layers 1 and 3 are excellent subgrade materials
Construct drain as shown on plan, profile, and cross section
Cut and waste layer 5 material to a depth of about 2 feet below grade and backfill with layers 1 and 3. See plan, profile, and cross section
Cut Berm ditch as shown in cross section
Cut backslopes not steeper than 2:1
Waste all material excavated from layers 4 and 5
Pavement design should include longitudinal and transverse crack control

FIG. 95. Typical soil profile map as made for design and construction of road, runways, railroads, and canals. Adapted from *Surveying and Sampling Soils for Highway Subgrades*, A.S.T.M.

BORING LOG *

Location _____ Structure _____ Sheet No. __ of __

Boring No. __ Datum ____ Boring inspector ____ Date _____

Stratification			Description of Materials (Type, Color, and Consistency)	Casing		Sampler or Spoon		Sample No. †	Remarks ‡
Elevation	Depth	Legend		Blows	Penetration	Blows	Penetration		

† Write sample number at corresponding depth. Designate dry samples by *D*, wash samples by *W*, undisturbed samples by *U*, rock cores by *C*.

‡ When drilling cores in rock, record the percentage of recovery in each foot of penetration.

* From Caribbean Architect-Engineer.

BORING LOG (Continued)

Location of project _____

Location of boring _____

Coordinates _____ and _____

Drill No. _____

Boring foreman _____

Size and weight of casing _____ Depth _____

_____   _____

Length of hole _____ Earth _____ Rock _____

Type of rock drill used _____

Weight of hammer _____

Average fall of hammer _____

Elevation of ground water surface _____

Record of Work

Date							
Start							
Finish							
Hours							
Total Depth							
Weather							
Temperature							

Boring inspector _____

Remarks _____

_____

_____

_____

_____

Note. Mark samples with name of base, name of structure, hole number, sample number, depth, and material.

SOIL SHEAR TESTS

Purpose

To determine maximum shearing stresses, cohesion, and angle of internal friction from soil samples.

1. Unconfined Shear Test

Fig. 96. Portable machine for unconfined shear test, made of aluminum, by S. D. Teetor, Soils Engineer.

When Used. Field test for undisturbed soils. For clays and clayey mixtures. Cannot be used with materials without a binder.

Method. Specimen loaded to failure by weights added to loading arm. Approximately: Shear = $\frac{1}{2} \times$ p.s.i. on specimen.

Advantages. Test is easily performed; apparatus is portable and can be home-made. Test can be improvised. It could be made by adding water to a bucket resting on specimen. Unconfined shear test is more accurate for clays or mixtures of clay.

Note: Height of specimen = $2 \times$ width.

2. Angle-of-Repose Method

For sands and free draining deposits of sand. Determine angle of repose which equals lower limit of angle of friction.

3. Shear Box Method

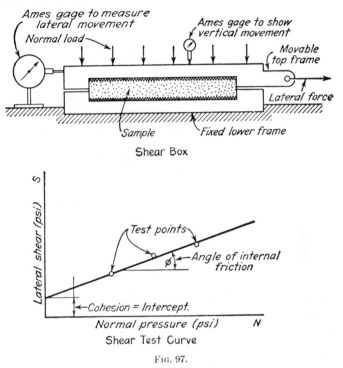

Fig. 97.

When Used. To simulate effect of material under pressure. Gives the maximum angle of internal friction. Use for free draining deposits, for soil mixtures above ground water, or for clays.

Method. With a constant normal load a lateral force is applied until slip occurs. A curve as shown in Fig. 97 may then be plotted. For clays and mixtures a quick shear test may be desirable.

Advantages. Apparatus is portable and easy to set up.

Note: Sample should be at field density.

4. Triaxial Shear Test

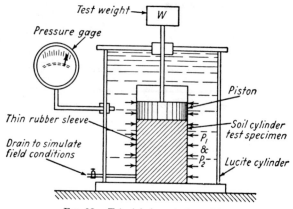

FIG. 98.   Triaxial shear test machine.

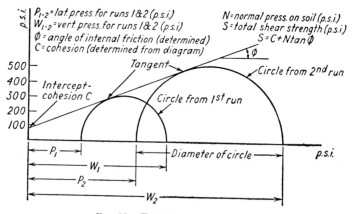

FIG. 99.   Triaxial shear test curve.

When Used. For silty soils generally. Can be used for all soils. Since apparatus is bulky, test is performed in a laboratory.

Method. Lateral hydrostatic pressure is maintained constant and a vertical force is applied until failure occurs. Test is repeated with increased lateral pressure, and results are plotted as shown, based on Mohr's circle theorem.

Advantages. Test can be used for all soils; gives most accurate results as actual drainage conditions are simulated.

IDENTIFICATION OF PRINCIPAL TYPES

TABLE 31. MAJOR DIVISIONS OF SOILS

Coarse-Grained (Granular)		Fine-Grained		Organic	
Gravel	Sand	Silt	Clay	Muck	Peat

IDENTIFICATION—VISUAL AND BY TEXTURE

GRAVEL

Rounded or water-worn pebbles or bulk rock grains. No cohesion. No plasticity. Gritty and granular. Crunchy under foot. As a soil, over $\frac{1}{10}$ in. in size. As an aggregate, over $\frac{1}{4}$ in. in size.

SAND

Granular, gritty, loose grains, passing No. 10 and retained on No. 270 sieve. Individual grains readily seen and felt. No plasticity or cohesion. When dry, a cast formed in the hands will fall apart. When moist, a cast will crumble when touched. The coarse grains are rounded; the fine grains are visible and angular. As an aggregate for construction sand consists of mineral grains between $\frac{1}{4}$ and $\frac{1}{200}$ in.

SILT

Fine, barely visible grains, passing No. 270 sieve and over 0.005 mm. in size. Little or no plasticity. No cohesion. A dried cast is easily crushed in the hands. Permeable; movement of water through voids occurs easily and is visible. When mixed with water the grains will settle in from 30 minutes to 1 hour. Feels gritty when bitten. Will not form a ribbon. Care must be used to distinguish fine sand from silt and fine silt from clay.

CLAY

Invisible particles under 0.005 mm. (or 0.002 mm. in M.I.T. scale) in size. Cohesive. Highly plastic when moist. When pinched between the fingers will form a long, thin, flexible ribbon. Can be rolled into a thread to a pin point. When bitten with the teeth will not feel gritty. Will form hard lumps or clods when dry, difficult or impossible to crush in hands. Impermeable; no movement of water apparent through voids. Will remain suspended in water from 3 hours to indefinitely.

MUCK AND ORGANIC SILT

Thoroughly decomposed organic material with considerable mineral soil material. Usually black, with a few fibrous remains. Odorous when dried and burnt. Found as deposits in swamps, peat bogs, and muskeg. Easily identified. May contain some sand or silt.

Peat

Partly decayed plant material. Mostly organic. Highly fibrous with visible plant remains.

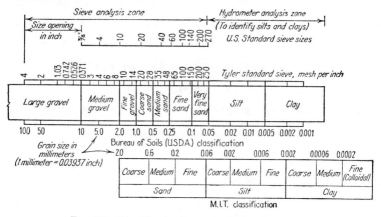

FIG. 100. Identification by mechanical grain size analyses.

Notes. Mechanical analysis is necessary to identify soils into the various divisions and into PRA and Casagrande systems. In general, the value of soils as a foundation for structures and as a material of construction is determined by the grain sizes and the gradation of the soil mixture. Other widely used grain-size classifications are International, M.I.T., Natl. Pk. Serv., A.S.T.M.

CLASSIFICATION OF SOILS BY HORIZONS

Soil Profile: A vertical cross section of the soil layers from the surface downwards

The upper layer, surface soil or top soil. The upper part is designated A_0 and is humus or organic debris. Indices are used for subdivision into transition zones as shown for A_1, A_2, etc. May range to 24 in. in depth.

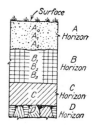

FIG. 101.

The heavier-textured underlayer or subsoil. May range from 6 in. to 8 ft. in depth. May be subdivided into transition zones B_1, B_2, etc., as shown. The products of the leaching or eluviation of the A horizon may be deposited in horizon B.

The unweathered or incompletely weathered parent material, C.

The underlying stratum such as hard rock, hard pan, sand, or clay, D.

Notes. Structures or pavements are not usually placed on A horizon soils. Also the organic content of these soils may adversely affect stabilization. In cuts the C horizon soil does not usually have as good bearing value as the more weathered B horizon. Foundations for heavy structures are preferably founded on the D horizon where it is bedrock or unyielding.

P.R.A. CLASSIFICATION

TABLE 32. CHARACTERISTICS FOR IDENTIFYING P.R.A. SOIL GROUPS *

Established by Public Roads Administration and Highway Research Board. Classification as shown is latest modification. Extensively used by engineers for highways, airfields, and dams.

Characteristics	A-1 Non-Plastic	A-1 Plastic	A-2 Non-Plastic	A-2 Plastic	A-3	A-4 and A-4-7†	A-5 and A-5-7†	A-6	A-7	A-8
Textural Class	Uniformly Graded Granular Coarse to Fine		Poorly Graded Granular, Coarse, and Fine		Clean Sand or Gravel	Silt or Silt Loam	Silt or Silt Loam	Plastic Clay	Plastic Clay Loam	Muck and Peat
Internal friction	High	High	High	High	High	Variable	Variable	Low	Low	Low
Cohesion	High	High	Low	High	None	Variable	Low	High	High	Low
Shrinkage	Not detrimental		Not significant	Detrimental if poorly graded	Not significant	Variable	Variable	Detrimental	Detrimental	Detrimental
Expansion	None		None	Some	Slight	Variable	High	High	Detrimental	Detrimental
Capillarity	None		None	Some		Detrimental	High	High	High	Detrimental
Elasticity	None		None	Some	None	Variable	Detrimental	None	High	Detrimental
Capillary rise	Low	High	36″ max.	Over 36″	6″ max.	High	High	High	High	Detrimental
Liquid limit	25 max.	35 max.	35 max.	40 max.	Non-plastic	40 max.	Over 40	35 min.	35 min.	35–400
Plasticity index	6 max.	4–9	Non-plastic	15 max.	Non-plastic	0–15	0–60	18 min.	12 min.	0–60
Shrinkage limit	14–20		15–25	25 max.	Not essential	20–30	30–120	6–14	10–30	30–120

Soil Constants — Atterberg Limits

Field moisture equivalent	Not essential		Not essential	Not essential	Not essential	30 max.	30-120	50 max.	30-100	30-400
Centrifuge moisture equivalent	15 max.		12-25	25 max.	12 max.	Not essential	Not essential	Not essential	Not essential	Not essential
Shrinkage ratio	1.7-1.9		1.7-1.9	1.7-1.9	Not essential	1.5-1.7	0.7-1.5	1.7-2.0	1.7-2.0	0.3-1.4
Volume change	0-10		0-6	0-6	None	0-16	0-16	17 min.	17 min.	4-200
Lineal shrinkage	0-3		0-2	0-4	None	0-4	0-4	5 min.	5 min.	1-30
% Sand	70-85		55-80	55-80	75-100	55 max.	55 max.	55 max.	55 max.	55 max.
% Silt	10-20		0-45	0-45		High	Medium	Medium	Medium	Not significant
% Clay	5-10		0-45	0-45		Low	Low	30 min.	30 min.	
% Passing No. 10	20-100	40-100								
% Passing No. 40	10-70	25-70								
% Passing No. 200	3-25	8-25	Less than 35	Less than 35	0-10					

Grading (Grain Size) brackets the rows % Sand through % Passing No. 200.

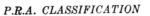

Subgrade or surface

Clay   Sand   Silt

Rapid succession of soil types, pockets of different soil types. Changes in soil profile or changes in soil structure.
Non-uniform support.

Group B-1

Subgrade or surface

Clay   Silt   Silt loam   Sand

Adjacent parts of fills constructed of widely differing soil types, as where hauling is from different sources.
Non-uniform support and settlement.

Group B-2

Original surface   Subgrade   Fill

A Horizon   B Horizon   C Horizon   D Horizon   Weak Zone

Cut through side hill creating subgrade on several types of natural soil and on fill within a short space.
Non-uniform support and settlement.

Group B-3

Fig. 102.   Classification of non-uniform subgrade soils.

* Adapted from Public Roads Administration and Highway Research Board Publications.
† A-4 or A-6 soil with A-7 characteristics.

TABLE 33. SOILS. USE AND TREATMENT. P.R.A. GROUPS. FOR ROADS AND AIRFIELDS

See also Table 36

A-1 Soils. Well-graded gravels and sand-clays, as Florida sand-clay or Georgia topsoil. Satisfactory treated surface. Good base with thin pavement. Excellent fill. Frost heave and break-up in North if plastic. Use subdrainage to lower water table. Stabilize: mechanically, chlorides or Portland cement.

A-2 Soils. Poorly graded sands and gravels, as South Carolina topsoil or bank run. Good base for moderate flexible or thin rigid pavement. Good fill. Frost heave, break-up if plastic. Softens when wet if plastic. Use base course when subgrade P.I. > 6. Subdrainage effective. Stabilize: with bitumen, chlorides, cement or admixture soils.

A-3 Soils. Clean sands and gravels, as Florida sand, glacial gravel, beach sand, wash gravel. Ideal base for moderate flexible or thin rigid pavement. Good fill. No frost heave or break-up. Subdrainage only through impervious shoulders. Stabilize: with soil binder, bituminous, or chemical admixtures.

A-4 Soils. Silty soils as New Hampshire silt or Minnesota silt. No good for surface. Poor base. Absorbs water. Unstable when wet. Bad frost heave and break-up. Use subdrainage and/or base and sub-base with flexible pavement. Use bituminous subgrade prime. Use thick concrete pavement (7 to 10 in.) with steel reinforcement and crack control.

A-5 Soils. Elastic silts as North Carolina micaceous silt or Maryland micaceous sandy loam. Use subdrainage and/or granular base and sub-base with bituminous subgrade prime. Use thick concrete pavement, reinforced with crack control.

A-6 Soils. Clays, as Missouri gumbo, Missouri collodial clay, sandy clays. Impermeable and stable when dry and undisturbed (hard clay). Plastic and absorbent if disturbed. Bad pumping into porous base, Macadam or pavement joints. Shrinks and cracks when dry. Use granular base and sub-base. Use subdrainage only when made pervious by cracks, root holes, and laminations. Frost heave slight when impermeable, bad when pervious. Use subgrade prime. Use thick, strong, dense, flexible pavement or reinforced crack controlled concrete.

A-7 Soils. Expansive, plastic clays, as adobe, Missouri clay, Illinois or Red River gumbo. Excessive volume change. Bad frost heave and break-up. Subdrainage not effective. Use thick, dense, flexible pavement with base and sub-base over subgrade prime or reinforced crack controlled concrete placed on impervious paper.

A-8 Soils. Muck and peat. No good for construction purposes. Excavate to solid stratum and replace with selected fill. Displacement by superimposed fill is doubtful. Displacement by explosive under superimposed fill is sometimes effective.

Use as Foundations for Structures. A-1 to A-3 soils: *best*. A-6 and A-7 soils: *next best*, when hard, undisturbed, and not plastic. A-5, A-8, plastic A-6 and A-7 soils: *require special treatment in each case*.

CLASSIFICATION

TABLE 34. CLASSIFICATION OF SOIL MIXTURES *

Class	Per Cent		
	Sand	Silt	Clay
Sand	80–100	0–20	0–20
Sandy Loam	50–80	0–50	0–20
Loam	30–50	30–50	0–20
Silt Loam	0–50	50–80	0–20
Sandy Clay Loam	50–80	0–30	20–30
Clay Loam	20–50	20–50	20–30
Silty Clay Loam	0–30	50–80	20–30
Sandy Clay	50–70	0–20	30–50
Clay	0–50	0–50	30–100
Silty Clay	0–20	50–70	30–45

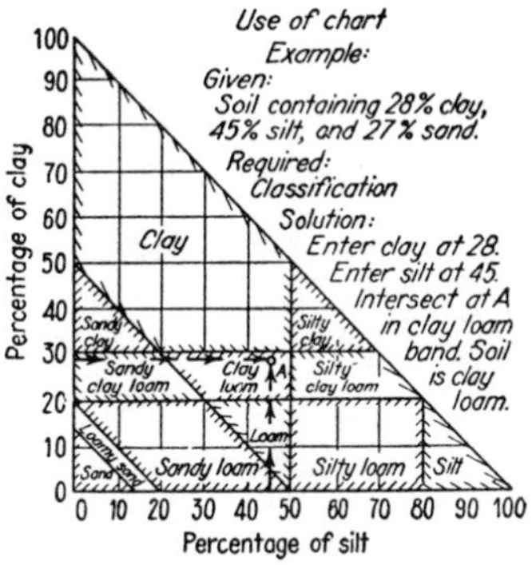

Use of chart
Example:
Given:
Soil containing 28% clay,
45% silt, and 27% sand.
Required:
Classification
Solution:
Enter clay at 28.
Enter silt at 45.
Intersect at A
in clay loam
band. Soil
is clay
loam.

FIG. 103. Right angle soil chart.

Note. Determine proportions of sand, silt and clay by sieve analysis or inspection.

(Natural soils seldom exist separately as gravel, sand, silt, clay, but are found as mixtures.)

TABLE 35. CLASSIFICATION OF SOILS BY ORIGIN

Residual:		Rock weathered in place—Wacke, laterite, podzols, residual sands, clays and gravels.
Cumulose		Organic accumulations—peat, muck, swamp soils, muskeg, humus, bog soils.
Transported	Glacial	Moraines, eskers, drumlins, kames—till, drift, boulder clay, glacial sands and gravels.
	Alluvial	Flood planes, deltas, bars—sedimentary clays and silts, alluvial sands and gravels.
	Aeolian	Wind-borne deposits—blow sands, dune sands, loess, adobe.
	Colluvial	Gravity deposits—cliff debris, talus, avalanches, masses of rock waste.
	Volcanic	Volcanic deposits—Dakota bentonite, volclay, volcanic ash, lava.
	Fill	Man-made deposits—may range from waste and rubbish to carefully built embankments.

Note. In general, residual or glacial deposits are preferable for heavy foundations. Important in soil surveys and engineering reports.

* *Adapted from Soil Cement Laboratory Handbook, Portland Cement Assoc.*

TABLE 36. CASAGRANDE CLASSIFICATION

Major Division		Soil Group Symbols	Soil Groups and Typical Names	General Identification			Observations and Tests Relating to Material in Place	Principal Classification Tests on Disturbed Samples
				Dry Strength	Other Pertinent Examinations			
Coarse-grained soils	Gravel and gravelly soils	GW	Well-graded gravel and gravel-sand mixtures; little or no fines	None	Gradation, grain shape		Dry unit weight or void ratio; degree of compaction; cementation; durability of grains; stratification and drainage characteristics; ground-water conditions; traffic tests; large-scale load tests; or California bearing tests	Sieve analysis
		GC	Well-graded gravel-sand-clay mixtures; excellent binder	Medium to high	Gradation, grain shape, binder examination, wet and dry			Sieve analysis, liquid and plastic limits on binder
		GP	Poorly graded gravel and gravel-sand mixtures; little or no fines	None	Gradation, grain shape			Sieve analysis
		GF	Gravel with fines, very silty gravel, clayey gravel, poorly graded gravel-sand-clay mixtures	Very slight to high	Gradation, grain shape, binder examination, wet and dry			Sieve analysis, liquid and plastic limits on binder if applicable
	Sands and sandy soils	SW	Well-graded sands and gravelly sands; little or no fines	None	Gradation, grain shape			Sieve analysis
		SC	Well-graded sand-clay mixtures; excellent binder	Medium to high	Gradation, grain shape, binder examination, wet and dry			Sieve analysis, liquid and plastic limits on binder
		SP	Poorly graded sands; little or no fines	None	Gradation, grain shape			Sieve analysis
		SF	Sand with fines, very silty sands, clayey sands, poorly graded sand-clay mixtures	Very slight to high	Gradation, grain shape, binder examination, wet and dry			Sieve analysis, liquid and plastic limits on binder if applicable

Major division	Symbol	Typical names			Field information	Laboratory tests
Fine-grained soils (containing little or no coarse-grained material) — Fine-grained soils having low to medium compressibility	ML	Silts (inorganic) and very fine sands, Mo, rock flour, silty or clayey fine sands with slight plasticity	Very slight to medium	Examination wet (shaking test and plasticity)	Dry unit weight, water content, and void ratio; consistency undisturbed and remolded; stratification; root holes; fissures, etc.; drainage and ground-water condition; traffic tests, large-scale load tests, California bearing tests, or compression tests	Sieve analysis, liquid and plastic limits, if applicable
	CL	Clays (inorganic) of low to medium plasticity, sandy clays, silty clays, lean clays	Medium to high	Examination in plastic range		Liquid and plastic limits
	OL	Organic silts and organic silt-clays of low plasticity	Slight to medium	Examination in plastic range, odor		Liquid and plastic limits from natural condition and after oven-drying
Fine-grained soils having high compressibility	MH	Micaceous or diatomaceous fine sandy and silty soils; elastic silts	Very slight to medium	Examination wet (shaking test and plasticity)		Sieve analysis, liquid and plastic limits if applicable
	CH	Clays (inorganic) of high plasticity; fat clays	High	Examination in plastic range		Liquid and plastic limits
	OH	Organic clays of medium to high plasticity	High	Examination in plastic range, odor		Liquid and plastic limits from natural condition and after oven-drying
Fibrous organic soils with very high compressibility	Pt	Peat and other highly organic swamp soils	Readily identified			Consistency, texture, and natural water content

C—clay, plastic-inorganic soil.
F—fines, material < 0.1 mm.
G—gravel, gravelly soil.
H—high compressibility.

L—relatively low to medium compressibility.
M—Mo, very fine sand, silt, rock flour
O—organic silt, silt clay or clay
P—poorly graded.

Pt—peat, highly organic fibrous.
S—sand, sandy soil.
W—well graded

TABLE 36. CASAGRANDE CLASSIFICATION, Continued

Major Division	Soil Group Symbols	Value as Foundation When Not Subject to Frost Action	Value as Wearing Surface *†		Potential Frost Action †	Shrinkage Expansion Elasticity	Drainage Characteristics	Compaction Characteristics and Equipment	Solids at Optimum Compaction, u, lb./cu. ft.‡ e, void ratio	California Bearing Ratio for Compacted and Soaked Specimen	Comparable Group in Public Roads Class (P.R.A.)
			With Dust Palliative	With Bit. Surf. Treat.							
Coarse-grained soils / Gravel and gravelly soils	GW	Excellent	Fair to poor	Excellent	None to very slight	Almost none	Excellent	Excellent; tractor	$u > 125$ $e < 0.35$	>50	A-3
	GC	Excellent	Excellent	Excellent	Medium	Very slight	Practically impervious	Excellent; tamping roller	$u > 130$ $e < 0.30$	>40	A-1
	GP	Good to excellent	Poor	Poor to fair	None to very slight	Almost none	Excellent	Good; tractor	$u > 115$ $e < 0.45$	25-60	A-3
	GF	Good to excellent	Poor to good	Fair to good	Slight to medium	Almost none to slight	Fair to practically impervious	Good; close control essential; rubber-tired roller, tractor	$u > 120$ $e < 0.40$	>20	A-2
Sands and sandy soils	SW	Excellent to good	Poor	Good	None to very slight	Almost none	Excellent	Excellent; tractor	$u > 120$ $e < 0.40$	20-60	A-3
	SC	Excellent to good	Excellent	Excellent	Medium	Very slight	Practically impervious	Excellent; tamping roller	$u > 125$ $e < 0.35$	20-60	A-1
	SP	Fair to good	Poor	Poor	None to very slight	Almost none	Excellent	Good; tractor	$u > 100$ $e < 0.70$	10-30	A-3
	SF	Fair to good	Poor to good	Poor to good	Slight to high	Almost none to medium	Fair to practically impervious	Good; close control essential; rubber-tired roller	$u > 105$ $e < 0.60$	8-30	A-2

Division	Symbol						Compaction characteristics	Unit weight ‡	CBR †	AASHO
Fine-grained soils (containing little or no coarse-grained material) — Fine-grained soils having low to medium compressibility	ML	Poor	Fair to poor	Poor	Medium to very high	Slight to medium	Fair to poor	Good to poor; close control essential; rubber-tired roller	$u > 100$ $e < 0.70$	6-25	A-4
	CL	Fair to poor	Fair to poor	Poor	Medium to high	Medium	Practically impervious	Fair to good; tamping roller	$u > 100$ $e < 0.70$	4-15	A-4 A-6 A-7
	OL	Poor	Poor	Very poor	Medium to high	Medium to high	Poor	Fair to poor; tamping roller	$u > 90$ $e < 0.90$	3-8	A-4 A-7
Fine-grained soils having high compressibility	MH	Poor	Very poor	Very poor	Medium to very high	High	Fair to poor	Poor to very poor	$u > 100$ $e < 0.70$	<7	A-5
	CH	Poor to very poor	Very poor	Very poor	Medium	High	Practically impervious	Fair to poor; tamping roller	$u > 90$ $e < 0.90$	<6	A-6 A-7
	OH	Very poor	Useless	Useless	Medium	High	Practically impervious	Poor to very poor	$u > 100$ $e < 0.70$	<4	A-7 A-8
Fibrous organic soils with very high compressibility	Pt	Extremely poor	Useless	Useless	Slight	Very high	Fair to poor	Compaction not practical; replace with compactible material			A-8

* Values are for subgrade and base courses, except for base courses directly under wearing surface.
† Values are for guidance only. Design should be based on test results.
‡ Unit weights apply only to soils with specific gravities ranging between 2.65 and 2.75.

TABLE 37. CONSTANTS FOR PAVEMENT DESIGN

SOILS – CONSTANTS FOR PAVEMENT DESIGN

Use: Identify soil and select bearing value for pavement design from diagram.
Example: Given: Plastic clay subgrade. Classification is CH or A-7 CBR would be 3 to 8.
"k" is 100 to 180 approx.
"C" is .900 to 1.000.
Bearing value is 10 to 19 lb per sq in.

SYMBOLS

C–Clay; F–Fines (<.1mm); G–Gravel, gr.soil.
H–High Compr.; L–Low, Med. Compr. M–Mo, very fine sand; O–Organic Silt; P–Poorly graded; Pt.–Peat; S–Sand, W–Well graded.

Warning: Value as foundation when not subject to Frost Action.

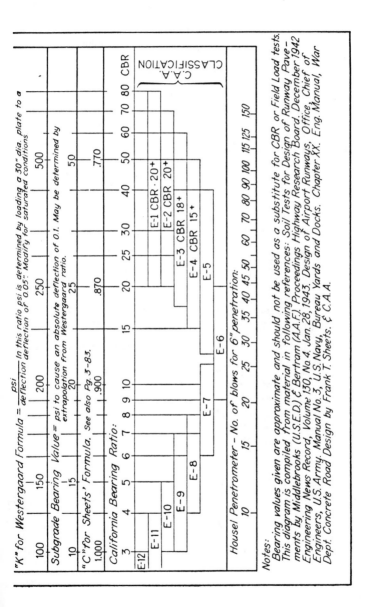

Notes:
Bearing values given are approximate and should not be used as a substitute for CBR or Field Load tests. This diagram is compiled from material in following references: Soil Tests for Design of Runway Pavements by Middlebrooks (U.S.E.D.) & Bertram (A.A.F.) Proceedings Highway Research Board. December 1942 Engineering News Record, Volume 130, No.4 Jan. 28, 1943. Design of Airport Runways, Office, Chief of Engineers, U.S. Army, Manual No.3, U.S. Navy, Bureau Yards and Docks. Chapter XX. Eng. Manual, War Dept. Concrete Road Design by Frank T. Sheets. & C.A.A.

ATTERBERG LIMIT TESTS

Purpose. 1. To classify soils into P.R.A., Casagrande, or C.A.A. groups. 2. To assign soils an approximate value as a foundation or construction material. 3. High values of liquid limit and plasticity index indicate high compressibility and low bearing capacity. 4. To determine soil suitability for road construction.

The liquid limit (L.L.) of a soil is the water content at which the groove formed in a soil sample with a standard grooving tool will just meet when the dish is held in one hand and tapped lightly 10 blows with the heel of the other hand. In the machine method the L.L. is the water-content when the soil sample flows together for $\frac{1}{2}''$ along the groove with 25 shakes of the machine at 2 drops per sec.

Diameter of brass cup or evaporating dish about $4\frac{1}{2}$ in.

Size of sample: By hand 30 grams; by machine 100 grams.

Several trials are made, the moisture content being gradually increased. Blows are plotted against water content and the liquid limit is picked off from the curve as shown, or

$$\text{L.L.} = \frac{\text{Weight of water}}{\text{Weight of oven-dried soil}} \times 100$$

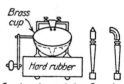

Crank and cam device Grooving
to produce 1 centimeter Tool
drop of cup.

Casagrande Liquid Limit Machine

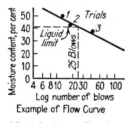

Example of Flow Curve

Adapted from Krynine, Soil Mechanics, McGraw-Hill Book Company.

Divided soil cake before test

Soil cake after test

Adapted from Hogentogler, Engineering Properties of Soil, McGraw-Hill Book Company.

FIG. 104. Liquid limit (L.L.), A.S.T.M. D423, A.A.S.H.O. T-89.

The plastic limit (P.L.) is the lowest water-content at which a thread of the soil can be just rolled to a diam. of $\frac{1}{8}$ in. without cracking, crumbling, or breaking into pieces.

$$\text{P.L.} = \frac{\text{Weight of water}}{\text{Wt. of oven-dried soil}} \times 100$$

Size of soil sample is 15 grams.

Soil which cannot be rolled into a thread is recorded as non-plastic (N.P.).

Soil thread above the plastic limit

Crumbling of soil thread below the plastic limit

FIG. 105. Plastic limit (P.L.), A.S.T.M. D424, A.A.S.H.O. T-90.

TABLE 38. LIMITING VALUES

Base Course	Subgrade	Sub-base	Stab. Surf.	Soil Cement	Cem. Treated Base
No Shrinkage L.L. = 25 P.I. = 6 max.	Lineal Shrinkage 3% to 5%	L.L. = 35 P.I. = 15 max.	P.I. = 4 to 9	L.L. = 40 P.I. = 18 max.	L.L. = 25 P.I. = 6 to 9

The water content or moisture content is expressed as a percentage of the oven-dried weight of the soil sample. These soil constants are determined from the soil fraction passing the No. 40 (420-micron) sieve.

Plasticity Index (*P.I.*): A.A.S.H.O. T-91. Numerical difference between liquid limit (L.L.) and plastic limit (P.L.) or P.I. = L.L. − P.L. Example: Given L.L. = 28, P.L. = 24, P.I. = 4. Cohesionless soils are reported as non-plastic (N.P.). When plastic limit is equal to or greater than liquid limit the P.I. is reported as 0, see Table 32.

Shrinkage Ratio (*R*): = bulk specific gravity of the dried soil pat used in obtaining shrinkage limit.

$$R = \frac{\text{Weight of oven-dried soil pat in grams}}{\text{Volume of oven-dried soil pat in cc.}} \quad \text{or} \quad \frac{W_0}{V_0}$$

Milk dish 1¾" dia. x ½" high
Wet soil
Before shrinkage
Dry soil
After shrinkage

Metal prong
Evaporating dish
Dry soil pat
Mercury
Top of glass dish ground surface
Mercury displaced by soil pat = volume of pat

Method of Obtaining Displaced Mercury

1"x1"x10" Mold
Soil
Shrinkage measure
Shrinkage
5"
Pins
Cigar shaped Soil
Shrinkage   Field Tests

Shrinkage Limit(s): A.A.S.H.O. T-92. Water content at which there is no further decrease in volume with additional drying of the soil but at which an increase in water content will cause an increase in volume.

$$S = \left(\frac{1}{\text{Shrinkage ratio}} - \frac{1}{\text{Spec. gravity}} \right) \times 100.$$

Size of sample 30 grams.

Lineal Shrinkage is the decrease in one dimension of the soil mass when the water content is reduced to the shrinkage limit or the % change in length occurring when a moist sample has dried out.

MOISTURE DETERMINATION

Purpose: 1. To determine moisture content for optimum moisture and maximum density relations. 2. To determine the amount of water in aggregates for concrete, bituminous, and other mixtures.

Gravelly soils: Use pycnometer method, Fig. 106, or heat method described below.

Sandy soils: Use heat method described below.

Silts and clays: Use heat method described below.

Heat Method: For total moisture content or surface moisture content.

1. Obtain a representative sample. If a metric scale is available the sample should not be smaller than 100 grams. If an avoirdupois scale graduated by ½ ounces is used, the sample should contain at least 50 ounces.

2. Weigh sample and record weight.

3. Place sample in pan and spread to permit uniform drying. Set pan in oven or on top of stove in a second pan to prevent burning of soil.

4. Dry to constant weight when total moisture is to be found; dry until surface moisture disappears when surface moisture content is desired. Temperature should not exceed 105° C. (221° F.). Stir constantly to prevent burning.

5. After the sample has been dried to constant weight, remove from oven and allow to cool sufficiently to permit absorption of hygroscopic moisture. Weigh dried sample and record weight.

6. Compute the moisture content as follows:

$$\text{Per cent moisture} = \frac{\text{weight of wet soil} - \text{weight of dry soil}}{\text{weight of dry soil}} \times 100$$

MAXIMUM DENSITY, OPTIMUM MOISTURE, PROCTOR NEEDLE PLASTICITY TEST

Purpose of maximum density-optimum moisture test is to determine the percentage of moisture at which the maximum density can be obtained when soil is compacted in fill, earth dams, embankments, etc.

After the maximum density curve has been obtained, these samples may be subjected to the Proctor needle for resistance to penetration. Then subjecting soil at the site to the Proctor needle, the amount of compaction of soil at the site may be obtained. See Fig. 109.

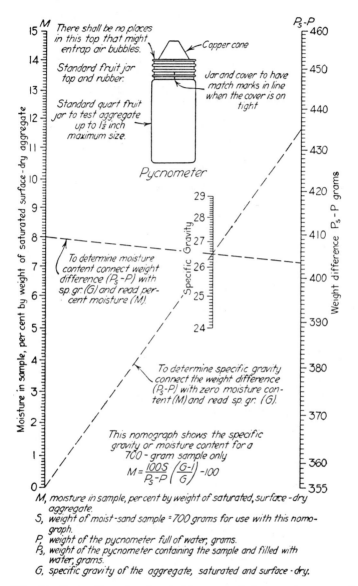

M, moisture in sample, per cent by weight of saturated, surface-dry aggregate.
S, weight of moist-sand sample = 700 grams for use with this nomograph.
P, weight of the pycnometer full of water, grams.
P₅, weight of the pycnometer containing the sample and filled with water, grams.
G, specific gravity of the aggregate, saturated and surface-dry.

IG. 106. Specific gravity and surface moisture content of aggregate, pycnometer method.

Maximum Density, Optimum Moisture, A.S.T.M. D-698; A.A.S.H.O. T-99.

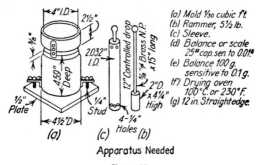

(o) Mold ⅟₃₀ cubic ft.
(b) Rammer, 5½ lb.
(c) Sleeve.
(d) Balance or scale 25# cap. sen. to 0.01#
(e) Balance 100 g. sensitive to 0.1 g.
(f) Drying oven 100°C. or 230°F.
(g) 12 in. Straightedge.

Apparatus Needed

FIG. 107.

Testing Procedure. 6 lb. ± (3000 grams) of air-dried soil slightly damp and passing the No. 4 sieve is mixed thoroughly, then compacted in the mold in 3 equal layers, each layer receiving 25 blows from the rammer with a controlled drop of 1 ft. The collar is removed, the soil struck off level, and the mold weighed.

(Wt. of soil plus mold − wt. of mold) × 30

= wet weight per cubic foot or wet density

A 100-g. sample from the center of the mold is weighed, then dried at 230° F., and the moisture content is determined.

Repeat test with additional sample of same material, adding about 1% water. Repeat until soil becomes saturated (about 5 times) using a fresh sample for each test. Plot wet-density curve. See Fig. 108. Compute dry density by formula and plot curve:

$$\text{Dry density} = \frac{\text{Wet wt., lb. per cu. ft.}}{\% \text{ moisture} + 100} \times 100$$

In Fig. 108 enter at top of dry density curve and read optimum moisture and maximum weight of soil 20.2% and 103.5 lb.

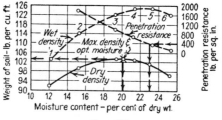

FIG. 108.

Modified A.A.S.H.O. Method.*

Same as above except:

1. Rammer to weigh 10 lb.
2. Rammer to have controlled drop of 18 in.
3. Soil compacted in mold in 5 equal layers, 25 blows to each layer.

The highest dry density is recorded as laboratory unit weight.

Note. Modern air field compaction equipment can secure greater densities than can be obtained by the standard Proctor or A.A.S.H.O. Test. If field compaction or vibration will give greater densities on any job than the test, the higher density should be used to control compaction.

Proctor Needle Plasticity Test †

Five pounds of dry soil passing a No. 10 sieve is mixed thoroughly with just enough water to make it slightly damp, then compacted in the mold in 3 layers. Each layer is given 25 blows with the rammer dropped 1 ft. The soil is then struck off level with the cylinder, weighed, and the stability determined with the plasticity needle by measuring the force required to press it into the soil at the rate of $\frac{1}{2}$ in. per sec. A small portion of the soil is oven-dried to determine the moisture content. This procedure is repeated 3 to 6 or more times, each time adding about 1% more water until the soil becomes very wet. The density and plasticity needle readings are plotted against moisture content. See Fig. 108. Thus in Fig. 108 a needle reading of 400 gives a moisture content of 23%.

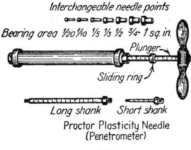

Interchangeable needle points

Bearing area $\frac{1}{20}$ $\frac{1}{10}$ $\frac{1}{5}$ $\frac{1}{3}$ $\frac{1}{2}$ $\frac{3}{4}$ 1 sq. in.

Plunger

Sliding ring

Long shank   Short shank

Proctor Plasticity Needle
(Penetrometer)

Fig. 109.

* *Engineering Manual, O.C.E., War Dept.*
† *Engineering News-Record, Aug. 31 to Sept. 28, 1933, R. R. Proctor.*

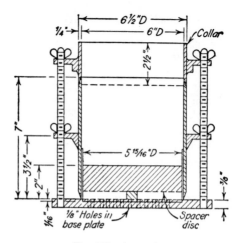

Fig. 110. Apparatus.

California Bearing Ratio

Purpose is to obtain relative resistance of a soil in place or soil to be placed and compacted to a specified degree to a standard broken stone layer. The resistance of the standard layer is given in the last column of the report form for California Bearing Ratio on p. 200.

For a soil in place apply a 3 sq. in. end area piston at a constant rate of penetration of 0.05 in. per minute to a total penetration of 0.5 in. The penetration force required per square inch at the values in the left-hand column of the report form for California Bearing Ratio on p. 200 is recorded and stated as a ratio of the corresponding values in the right-hand column of the report; usually the values for 0.1-in. deflection are used.

Laboratory determination is made by remolding the samples of the soil until it has the specified density using the A.S.T.M. or A.A.S.H.O. methods given above, except that 55 blows of the rammer are used instead of 25 and material is passed through a ¾-in. sieve instead of a No. 4 sieve. These samples are then loaded by means of the same piston and recorded as given above for the field test.

For the purpose of determining the effect of saturating conditions on the soil, tests may be made on soaked samples.

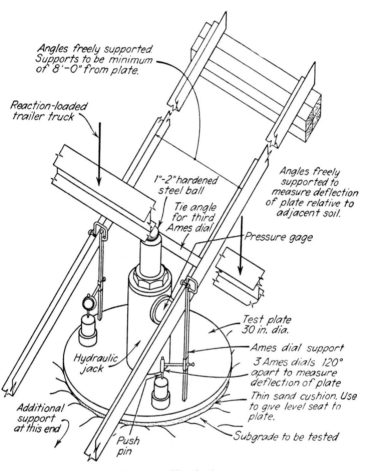

Fig. 111. Plate load test.

Plate Load Test to Obtain Modulus of Subgrade Reaction

The thickness of a concrete pavement depends upon the strength of the subgrade. This strength may be measured by the Plate Load Test.

Load is applied to a steel or aluminum plate 30 in. in diameter and 2 in. thick, and the settlement is measured by three Ames deflection gages symmetrically placed on the plate. A convenient method of loading is by means of a jack reacting against a steel beam bearing against the bottom of two loaded trucks. The load in p.s.i. between the plate and the soil that causes a settlement of 0.05 in. is the value k used in the Westergaard formulas for determining concrete slab thickness. (See *Data Book*, Vol. I, for these formulas and their application.)

The equipment should be capable of applying a load up to 50,000 lb. The plate must be set level with 100% bearing; plaster of paris bedding should be used when testing clay soils or old pavements.

Weather is not an important factor in making this test.

In *sandy soils*, no correction for moisture need be made.

In *plastic soil* a correction must be made if the moisture content at the time of the test is less than the maximum which may be anticipated in operation. In such a case, determine k for the soil as it is. Also make unconfined compression tests on samples of the soil having both the same moisture content as present in the field at the time of the Plate Load Test and also the maximum anticipated moisture content. Should the latter figure be unknown use the saturated moisture content. The ratio between these unconfined compression tests applied to the k value found by the plate test will give the corrected k value to use in the Westergaard formulas.

An Army method of correction is to use the ratio of the deformation of samples at the two moisture contents under a unit load of 10 p.s.i. in a consolidation test.

Check List for Inspector for Plate Test

Plate level and properly bedded.

Angle-iron supports for dials anchored and bearing on ground at least 8 ft. away from the test plate or truck wheels.

Dials vertical and mounted on heavy angle-iron frame.

Dials set so that they have ½-in. travel.

Settlement and load readings made at such intervals and loads as specified.

Check density and moisture of subgrade adjacent to area being tested.

Steel ball between jack and I beams to insure vertical loading.

Dials protected from rain.

Angle-iron supports and all testing equipment protected from wind.

Area graded to drain away from test location.

FIELD DENSITY (UNIT WEIGHT) TEST *

Purpose. 1. To obtain the natural density of soil in place (*a*) as an indication of its stability or bearing value as foundation, (*b*) to compute

* *Adapted from Public Roads, Vol. 22, No. 12 by Harold Allen, Public Roads Administration.*

the shrinkage or swell when the soil is removed and placed in embankment at a higher or lower density. 2. To determine the per cent of compaction being obtained to check against requirements of specifications.

Method of Determining Weight per Cubic Foot of Soil in Place. Calibrated Sand Method

The density of a soil layer may be determined by finding the weight of a disturbed sample and measuring the volume of the space occupied by the sample prior to removal. This volume may be measured by filling the space with a weighed quantity of a medium of predetermined weight per unit volume. Sand, heavy lubricating oil, or water in a thin rubber sack may be used.

1. Determine the weight per cubic foot of the dry sand by filling a measure of known volume. The height and diameter of the measure should be approximately equal, and its volume should be not less than 0.1 cu. ft. The sand should be deposited in the measure by pouring through a funnel or from a measure with a funnel spout from a fixed height. The measure is filled until the sand overflows and the excess is struck off with a straightedge. The weight of the sand in the measure is determined, and the weight per cubic foot computed and recorded.

2. Remove all loose soil from an area large enough to place a box similar to the one shown in Fig. 113 and cut a plane surface for bedding the box firmly. A dish pan with a circular hole in the bottom may be used.

3. With a soil auger or other cutting tools bore a hole the full depth of the compacted lift.

4. Place in pans all soil removed, including any spillage caught in the box. Remove all loose particles from the hole with a small can or spoon. Extreme care should be taken not to lose any soil.

5. Weigh all soil taken from the hole, and record weight.

6. Mix sample thoroughly, and take sample for water determination.

7. Weigh a volume of sand in excess of that required to fill the test hole, and record weight.

8. Deposit sand in test hole by means of a funnel or from a measure as illustrated in Fig. 113 by exactly the same procedure as was used in the determination of unit weight of sand until the hole is filled almost flush with original ground surface. Bring the sand to the level of the base course by adding the last increments with a small can or trowel and testing with a straightedge.

9. Weigh remaining sand, and record weight.

10. Determine the moisture content of soil samples in percentage of dry weight of sample.

11. Compute dry density from the following formulas:

$$\text{Vol. soil} = \frac{\text{Wt. of sand to replace soil}}{\text{Wt. per cu. ft. of sand}}$$

$$\%\text{ moisture} = \frac{\text{Wt. of moist soil} - \text{Wt. of dry soil}}{\text{Wt. of dry soil}} \times 100$$

$$\text{Moist density} = \frac{\text{Weight of soil}}{\text{Volume of soil}}$$

$$\text{Dry density} = \frac{\text{Moist density}}{1 + \dfrac{\%\text{ of moisture}}{100}}$$

$$\%\text{ compaction} = \frac{\text{Dry density}}{\text{Maximum density}} \times 100$$

EXAMPLE. *Given:*

Wt. per cubic foot of sand = 100 lb.
Wt. of moist soil from hole = 5.7 lb.
Moisture content of soil = 15%
Wt. of sand to fill hole = 4.5 lb.

Required: Density and per cent compaction.

Solution: Vol. soil $= \dfrac{4.5}{100} = 0.045$ cu. ft.

$$\text{Moist density} = \frac{5.7}{0.045} = 126.7 \text{ lb.}$$

$$\text{Dry density} = \frac{126.7}{1 + 15/100} = 110.0 \text{ lb.}$$

Given maximum density = 115 lb. (from density test).

$$\%\text{ compaction} = \frac{110}{115} \times 100 = 95.7\%$$

Note. In gravel soils material over $\frac{1}{4}$ in. is screened out and correction made.

Chunk Sample Method. 1. Cut sample 4″–5″ in diameter full depth of layer. 2. Determine per cent moisture. 3. Trim sample and weigh to $\frac{1}{2}$ oz. 4. Immerse sample in hot paraffin, remove, cool, and weigh again. 5. Compute volume of paraffin using 55 lb. per cu. ft. 6. Compute volume of sample by weighing in water (correcting for volume of paraffin). 7. Compute density data by formulas above.

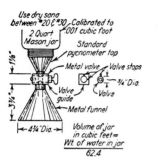

FIG. 112. Field density determination apparatus, dry sand method

FIG. 113. Field density test.

FIG. 114. Rubber sack inflated to fill hole with known volume of water.

FIG. 115. Pump and jar to fill hole with known volume of oil. S.A.E.-40.

TABLE 39. BEARING VALUES AND PER CENT COMPACTION REQUIRED

Max. Dry Density	Soil Rating	Recommended Compaction
90 lb. and less	No good	
90 lb.–100 lb.	Very poor	95–100%
100–110 lb.	Poor to very poor	95–100%
110–120 lb.	Poor to fair	90–95%
120–130 lb.	Good	90–95%
130 lb. and over	Excellent	90–95%

Note. Density or $\dfrac{\text{Wt.}}{\text{Vol.}}$ may be expressed as pound per cubic foot or grams per cubic centimeter. Density in grams per cubic centimeter = bulk specific gravity.

MECHANICAL ANALYSIS (GRAIN SIZE)

Purpose. 1. To identify homogeneous soils in the major divisions See pp. 169 and 170. 2. To classify soil mixtures occurring in a natural state, Table 34 and Fig. 101. 3. To classify soil into the P.R.A. or Casagrande groups. See pp. 172 and 173, also Table 36. 4. To design or control stabilized soil mixtures. 5. To determine frost heaving potentialities. 6. To determine effective size (D_{10}) and uniformity coefficient (Cu) for the design and control of filters and subdrainage backfill.

Sieve Analysis

Size of sample to be 400 to 750 grams—the coarser the material the larger the sample required.

Take sample by quartering or with sample splitter.

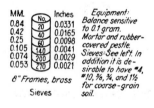

MM.	No.	Inches
0.84	20	0.0331
0.42	40	0.0165
0.25	60	0.0098
0.105	140	0.0041
0.074	200	0.0029
0.053	270	0.0021

8" Frames, brass

Sieves

Equipment:
Balance sensitive to 0.1 gram.
Mortar and rubber-covered pestle.
Sieves: See left. In addition it is desirable to have #4, #10, 3/8, 3/4, and 1 1/2 for coarse-grain soil.

Dry surface moisture by heating the quartered sample at less than 212° F., or boiling point of water at high altitudes, in open pan until surface water disappears and sample is apparently dry and will not lose more weight with additional heating.

Break up cakes with mortar and pestle.

Record dry weight of sample.

Proceed to pass material through screens by placing sample in a stack of sieves, largest size on top, and shake vigorously with horizontal rotating motion balancing on bumper or pad until no more material will pass through each screen.

Weigh amount retained on each sieve, compute per cent of total weight of sample, and plot curve.

Washing is recommended for No. 200 sieves and smaller.

Partly immerse the largest sieve in a pan of water and agitate.

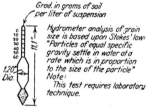

Grad. in grams of soil per liter of suspension

120 Dia.

Hydrometer analysis of grain size is based upon Stokes' law: "Particles of equal specific gravity settle in water at a rate which is in proportion to the size of the particle"
Note:
This test requires laboratory technique.

Hydrometer Test

Fig. 116. Mechanical analysis of soils.

Take material and water from pan and repeat for next smaller size sieve. Agitate smallest sieve in several water baths until water remains clear. Air-dry portions retained in sieves, weigh and plot curve.

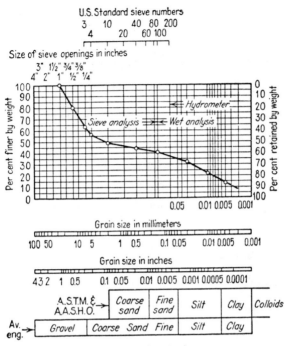

Fig. 117. Typical grain size curve.

Effective size (D_{10}) of a soil is the particle size that is coarser than 10% (by weight) of the soil; that is, 10% of the soil consists of particles smaller than the effective size (D_{10}) and 90% consists of larger particles. *Example.* In Fig. 118, effective size (D_{10}) is 0.02 mm.

Uniformity coefficient (Cu) is computed by first determining the size that is coarser than 60% of the soil and dividing that size by the effective size (D_{10}), i.e., $Cu = \dfrac{60\% \text{ size}}{10\% \text{ size}}$.

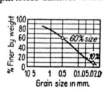

Fig. 118. Effective size (D_{10}) and uniformity efficient (Cu).

Example. In chart, $Cu = \dfrac{0.5}{0.02} = 25$.

Note. The Cu of filter backfill should not be over 20. The D_{10} of non-frost heaving uniform soil is 0.02 mm. minimum.

Engineer

OPTIMUM MOISTURE—MAXIMUM DENSITY

LABORATORY TEST

Location _____   Soil sampler _____

Date _____   Soil tester _____

Control soil #

Item	Run 1	Run 2	Run 3	Run 4	Run 5	Run 6
Weight of cylinder + wet soil						
Weight of cylinder						
Weight of wet soil						
Weight of wet sample + pan						
Weight of pan						
Weight of wet sample						
Weight of dry sample + pan						
Weight of pan						
Weight of dry sample						
Weight of moisture						
% of moisture						
Wet density						
Dry density						

Optimum moisture =   Maximum density =

SEELYE, STEVENSON, VALUE & KNECHT

Consulting Engineers

101 Park Avenue

New York 17, N. Y.

REPORT OF COMPACTION CONTROL TESTS

Test No. _____ By _____ Date _____
Location _____ Elev. _____

1. Weight of container and sample _____ gm.
2. Weight of container _____ gm.
3. Weight of sample _____ gm.

4.   Mass Specific Gravity of Test Sand _____

5. Weight of container and sand _____ gm.
6. Weight of container and sand remaining _____ gm.
7. Weight of sand in hole and funnel _____ gm.
8. Weight of sand in funnel _____ gm.
9. Weight of sand in hole _____ gm.
10. Volume of hole: $\underline{\text{Item 9}} \div \underline{\text{Item 4}}$ equals _____ cc.
11. Wet specific gravity of sample: $\underline{\text{Item 3}} \div \underline{\text{Item 10}}$ equals _____
12. Wet density of sample: $\underline{\text{Item 11}} \times 62.4$ equals _____ lb./cu. ft.
13. Weight of container and lab. sample (moist) _____ gm.
14. Weight of container and lab. sample (dry) _____ gm.   Note: When stone con-
15. Weight of container _____ gm.   tent is more than 15%
16. Weight of water lost: $13 - 14$ _____ gm.   dry whole sample
17. Weight of dry sample: $14 - 15$ _____ gm.
18. % Moisture: $\underline{\text{Item 16}} \div \underline{\text{Item 17}}$ equals _____%
19. Dry density of sample: $\underline{\text{Item 12}} \div \underline{1 + \text{Item 18}}$ equals _____ lb./cu. ft.
20. Total weight of sample _____ gm.
21. Weight retained on No. 4 sieve _____ gm.
22. % Retained on No. 4 sieve _____%

Correction to field density for stone content of 15–40% retained on No. 4:

23. $\underline{\text{Item 21}} \div \underline{\text{sp. gr. stone}}$ equals _____ c.c. of stone
24. $\underline{\text{Item 20}} - \underline{\text{Item 21}}$ equals _____ gm. of dry fines
25. $\underline{\text{Item 10}} - \underline{\text{Item 23}}$ equals _____ c.c. of dry fines
26. $\underline{\text{Item 24}}$ gm. of dry fines $\div \underline{\text{Item 25}}$ cc. of dry fines equals _____ mass sp. gr. dry fines
27. Dry density of fines equals $\underline{\text{Item 26}} \times 62.4$ equals _____ lb./cu. ft.
28. Proctor # $\underline{\text{(Test No.)}}$: _____ lb./cu. ft. @ _____% moisture
29. Correction to Proctor density for stone content of over 40% retained on No. 4:

$$\left[\frac{(100\% - \text{Item 22})}{\% \text{ Fines}} \times \frac{\text{Item 28}}{\text{Max. density fines}} \right]$$
$$+ \left[\frac{\text{Item 22}}{\% \text{ Stone}} \times 0.9 \times 62.4 \times \frac{\text{sp. gr.}}{\text{stone}} \right] \text{ equals } \_\_\_\_\_$$

30. Per cent compaction: _____

     Item 19 ÷ Item 28   For less than 15% stone
     Item 19 ÷ Item 29   For more than 40% stone
     Item 27 ÷ Item 28   For 15 to 40% stone

Engineer

SOIL STUDIES *

Report for _____ Date _____
Material _____ Report No. _____
Project _____

Sample identification							Location of Samples
Classification							
Hygroscopic moisture							

Gradation							
Pass.　Ret.							
1"　　3¾"							
¾"　　½"							
½"　　4 mesh							
4　　10							
10　　20							
20　　40							
40　　60							
60　　100							
100　　200							
200 (wash)							

Hydrometer test							Remarks:
% Finer #10							
#200							
0.005 mm.							
0.001 mm.							
Liquid limit							
Plastic limit							
Plastic Index							
Specific gravity							

Absorption and Stability Tests of Materials Used	Opt. Mois.	Density Wt./cu. ft.	% Blended Soil	% Water	% Binder	7-day Absorp. %	7-day Stab. (lb.) Bottom ½-in.

Inspector

* From *Haller Testing Laboratories, Inc.*

Engineer

SOILS CLASSIFICATION *

Client			Date	
			Report No.	
Site				
Sample No.				
Location				

SOIL TYPE

Size (mm.)	%	%	%	
Gravel 2.0 +				
Sand 2.0 − 0.05				
Silt 0.05 − 0.005				
Clay 0.005 −				

SIEVE ANALYSIS

Sieve Size	Diameter (mm.)	% Passing	% Passing	% Passing	% Passing
2″	50.80				
1½″					
1″	26.67				
¾″	18.85				
⅜″	9.423				
No. 4	4.699				
No. 10	1.981				
No. 40	0.417				
No. 60	0.246				
No. 100	0.147				
No. 200	0.074				

HYDROMETER ANALYSIS

Size of Particle	% Smaller than	% Smaller than	% Smaller than	% Smaller than
0.05 mm.				
0.005 mm.				
0.001 mm.				

Inspector

* *From Halter Testing Laboratories, Inc.*

Engineer

CALIFORNIA BEARING RATIO *

Client _____ Date _____

Report No. _____

Site

Sample No.				
Location				

MAXIMUM DENSITY, OPTIMUM MOISTURE

Optimum water content (percentage of dry weight)				
Maximum density (pounds per cu. ft.)				

CALIFORNIA BEARING TEST DATA

Condition of Sample							
Penetration (inches)	Lb. per Sq. In.	C/B Ratio	Lb. per Sq. In.	C/B Ratio	Lb. per Sq. In.	C/B Ratio	Standard
0.025							
0.050							
0.075							
0.10							*1000*
0.20							*1500*
0.30							
0.40							
0.50							
Unit dry weight (pounds per cu. ft.)							
Expansion %							

WATER CONTENTS—PERCENTAGE OF DRY WEIGHT

Unsoaked				
Soaked—Top 1 in.				
Soaked—Total				

Inspector

* *From Haller Testing Laboratories, Inc.*

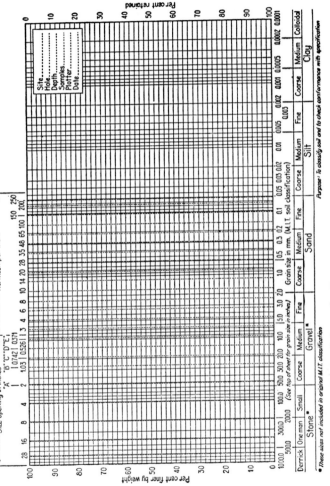

From the *Haller Testing Laboratories, Inc.*

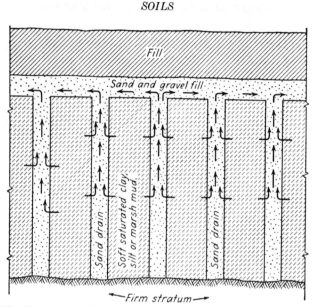

Fig. 119. Section of area being drained and consolidated by sand drain method.

CONSOLIDATION OF COMPRESSIBLE SOILS

Sand-Drain Method. Excess pore water is bled off from compressible subsoils by the construction therein of vertical columns of porous sand, called sand drains or sand piles, placed at intervals through the material to be drained. The weight created by a gradually added cover of fill causes the pore water to flow from the holding subsoil into the sand drains and upward.

As there is little shearing strength in soil oversaturated with pore water, the weight of fill added must not exceed the load that the soil can bear at any stage of consolidation. Otherwise the shearing strength will be exceeded and a mud wave will result. Check on the pore-water pressure must be made as the fill progresses to keep within the safe bearing capacity of the soil.

The effectiveness of sand drains depends upon the type of soil. In granular soils they are not necessary, because pore water can be drained away in a short time when given a chance to escape.

Clayey soils are usually so impervious to the flow of any contained water that sand drains would have to be too close together for economy and the length of required time for the water to reach them would be excessive.

Sand drains are most applicable in highly compressible soils such as silt and marsh muck.

Sand drains were successfully used in building the New Jersey Turnpike through salt meadows, which consist of black muck (organic silt) to depths of over 100 ft.

Laboratory control is necessary to determine the feasible use of sand drains, as well as their spacing, the probable time required for subdrainage, and the maximum rate at which fill should be added.

Rip rap as distinguished from manually placed slope paving is rough stone of various sizes deposited usually by mechanical means for slope protection to prevent erosion or undermining. Unless such stone is obtainable as spoil from mass rock excavations, it would ordinarily be ordered from a commercial crushed stone quarry. In this case stone which passes the initial crusher and is screened thereafter through a grizzly should be well graded from a minimum of 2½ in. to a maximum of 18 in. The next-larger-size stone would vary from zero to 1-ton pieces. The largest stone would be 2- to 8-ton pieces on the quarry floor.

SOIL MECHANICS

SOIL BEHAVIOR

Shearing Resistance vs. Stability of Soils

All stability in soils is derived from shearing strength.

Corollaries

(a) The ability of soils to support spread footings is directly proportional to their shearing strength.

(b) The steepness of the safe slope of embankments depends on the shearing strength of the soil.

(c) The safety of the supporting soil under the embankment from mud wave action depends on the shearing strength of the soil.

Shearing Resistance—Granular vs. Cohesive Soils

As granular soils have internal friction resistance, their shearing strength increases in relation to the normal pressure to which they are subjected.

Granular soils have no cohesion.

Clays are cohesive soils and have no angle of internal friction. They have a resistance to shearing due to their cohesive or molecular strength.

Hence their shearing strength is the same regardless of the normal pressure to which they are subjected.

Soil mixtures of clay and sand partake of both cohesive and internal friction resistance. Thus, most sands have some cohesion; most clays have some internal friction.

Corollaries

(a) The safe slope of a granular bank does not decrease as the height increases because its shearing strength increases to resist the increasing shearing stresses as the bank becomes higher.

(b) The safe slope of a clay bank becomes flatter as the bank becomes higher because its shearing strength does not increase to resist the corresponding increasing shearing stress.

Field Test for Shearing Strength

(a) The so-called unconfined compression test is very simple and may be improvised in the field.

(b) For granular soils, the angle of repose is approximately equal to the angle of internal friction.

(c) Load tests are of little value except on cohesive soils.

Judging Strength of Soils

(a) Classification is by grain size such as clay, silt, sand, and gravel. It is generally used in connection with empirical tables to give strength for building foundations.

(b) Refinement of this classification is the measurement of resistance to a standard driving spoon which is tied into the empirical strength table.

(c) Load tests on soils and piles, see pp. 71 and 133.

(d) For determining the wheel bearing capacity of composite sections of flexible pavements and bases, there is the California Bearing Ratio, which is the resistance of a standard plunger to being forced into a layer of the base or subbase compared to its resistance to being forced into a standard layer of broken stone. Empirical curves are available.

(e) For rigid pavements, a plate loaded according to a standard procedure gives a value known as the modulus of subgrade reaction, K, which is used in connection with established formulas to estimate the bearing capacity of concrete pavements.

(f) In general, the heavier the unit weight of soil, the greater the strength.

(g) In general, the less voids, the greater the strength.

Worked-over Clays

Clays are subject to remolding; that is, an apparently stiff clay if worked will give off water and become soft. This is because a certain amount of water is fixed or attracted to the submicroscopic particles of clay, and the adhered water is broken off by working. If the clay is allowed to rest, the water will readhere and the clay will regain some of its stiffness.

Corollaries

(a) Piles driven into clay, if left overnight, will set up and be difficult to start.

(b) Freezing and thawing works the clay and weakens it, causing settlement or slides.

Settlement of Footings on Clays

Clay settlement varies directly with water content and inversely with the cohesive strength.

For instance, clay sample No. 1 under pressure of 2 tons to the square foot might have an expected total settlement of 2 in. After this settlement has taken place and the pressure is increased to 3 tons, there would be a certain additional settlement of, let us say, ¾ in.

There are also definite periods of time for these settlements to become complete.

The settlement is caused by the squeezing out of water.

Corollaries

(*a*) The larger the footing, the greater the settlement, because the area of the bulb of pressure will increase in proportion to the sizes of the footings. See Fig. 120.

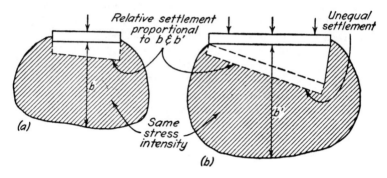

Fig. 120. Effect of increased footing size on cohesive soils.

(*b*) Although increase in size of footing increases the amount of settlement in the relation of *b′* to *b*, it does not increase the danger of failure in cohesive soil because the intensity of shearing stresses in (*a*) are no greater than in (*b*).

(*c*) However, unequal settlement as shown in (*b*) might produce critical shearing stresses, or a vicious cycle of rotation.

Shrinkage

Clays tend to hold free water in addition to their adhered water. They do not drain nor do they dry out rapidly.

Granular soils do not hold water readily.

Corollaries

(*a*) Clays are subject to a large amount of shrinkage, but the loss of water that causes this shrinkage is slow.

(*b*) These shrinkages might amount to as much as 20% in volume.

(*c*) Granular soils do not shrink much when drying, and they shrink more rapidly.

Settlement of Footings on Sands

Compressibility of sands and silts varies inversely with density.

Pore Water Pressure

When granular soils are saturated with water and the water is trapped, the footing may be partially supported on hydraulic pressure. This is called pore water pressure. Examples are boils or soft spots in a subgrade, or a ground with quakes under the passage of trucks.

Corollaries

(a) Soils under pore water pressure are without shearing strength.

(b) The seepage out of pore water under pressure will cause settlement.

(c) Drainage to relieve the water is indicated.

Compaction-Optimum Moisture

For most soils there is a percentage of moisture at which the soil will compact to its greatest density.

This optimum moisture may run from 8% for sands and 15% for silts to 15–20% for clays.

Corollaries

(a) Water content should be controlled in making fills.

(b) Reduction of water in clays is very difficult. Suggestions for doing this are as follows:

> Spread out in thin layers to dry.
> Choose a borrow drier than optimum moisture.
> Modern construction equipment supplemented by rollers is recommended for consolidation of fills.

(c) Deep trenches with clay backfill may require puddling, which will result in considerable shrinkage, or power tampers may be used to break up the lumps.

(d) Sand backfill may be inundated in compacting around a foundation wall, which will leave the sand loose and subject to settlement, but it may be of benefit as it will improve the drainage around the basement walls.

(e) For pavements and floor fills, optimum moisture and heavy rolling and tamping are the only answers.

(f) A well-graded gravel may be compacted with a bulldozer in 12-in. layers.

Soil Investigations

(a) Borings giving the material and the resistance to driving with a standard spoon.

(b) Undisturbed samples which can be tested for density, shearing resistance, and settlement.

(c) Atterberg limit tests, which are tests of fluidity and sensitivity to vibration. Their use includes determination of resistance of soils to frost, soil classification, compressibility, and shrinkage.

Capillary Action

Soils possess capillary action similar to a dry cloth with one end immersed in water.

Corollaries

(a) Coarse gravel; no capillary action.

(b) Coarse sand: up to 12 in.

(c) Fine sands and silts: up to 3 ft., low in clay.

(d) Pure clay: low value.

Frost Action

Most damage from frost action occurs at the time of thawing.

Corollaries

(a) With granular soils and mixtures, especially silt, frost aided by capillary action may suck up water and cause actual ice lenses. When this material thaws, the soil is over-irrigated and incapable of supporting a load. A pore water condition may result.

(b) With a clay soil, the heave may be less because the clay soil does not readily suck up water from below on account of its imperviousness, but when it thaws the frozen clay will lose its strength as when it is worked.

(c) Frost boils in pavement occur from similar causes.

Prevention

All structural foundations to extend below frost. Subgrades of roads to have a gravel capillary cutoff.

Quicksand

Quicksand is a condition in which a current of water passing upward through a soil is of sufficient velocity to cause a flotation or boiling up of the particles.

Corollaries

(a) The bearing value of the material is destroyed.

(b) Material may be carried off from under a structure.

(c) Correct by lowering head of water or otherwise cutting off the flow.

Sand Filters

Water passing out of a fine soil into a coarse aggregate may tend to carry material with it and clog the drainage.

Corollary

(a) The grains around the drain should be graded so that there is no abrupt change in size.

Vibration

Vibration may cause damage to structures by:

(a) Pile driving, which

1. May endanger tender adjoining masonry.
2. May start overloaded skin friction piles down.
3. May cause cracks in plaster or masonry.

(b) Effect of resonance: The natural period of the earth crust may vary between 1000 and 2000 cycles per minute, and machinery, for instance, vibrating within these limits might cause trouble.

(c) Truck vibrations may act similarly to pile driving.

(d) Blasts are not continuous, and a large part of the effect of blast is an air wave. However, blasts are a distinct danger.

Precautions

(a) Shoring of weak masonry and foundations.

(b) Insulation of foundations of vibrating machinery.

(c) Open windows during blasting.

(d) Line hole rock supporting adjacent buildings.

(e) Shore exposed rock faces where dangerous shear planes or faults occur in rock face.

(f) Protect client by preconstruction survey of adjoining properties together with photographs or sketches properly identified for court evidence.

Effect of Water on Soils

Water in soil is the all-round enemy of the engineer and constructor. Some of the troubles caused by water are:

(a) Frost failures.

(b) Settlement due to drying out of water in soil.

(c) Reduction of the bearing power of soil due to weakening of the shearing strength.

(d) Removal of soil under foundations due to pumping operations.

(e) Cavitation and piping in dams.

(f) Construction difficulties.

Water is fought by:

(a) Gravity drainage.

(b) Ordinary pumping.

(c) Pumping with well points in soils that are somewhat granular.

(d) Sealing of the casing of well points to produce better vacuum.

Advantages of water:

(a) Protection of timber below water line.

(b) Prevention of shrinkage in soils due to drying out.

Flow through Soils

Flow through soils is dependent on the hydraulic gradient and the fineness of the material, and can be estimated.

(a) The uplift on the base of a dam should be estimated.

(b) Piping which is a channelized outflow of soil carried away by this seepage must be controlled.

GRADING

CHECK LIST FOR INSPECTORS

Inspectors' Equipment

    Complete set of approved plans and specifications.

    Surveying instruments if required.

    100-ft. tape and 6-ft. rule.

    Line level and line.

    Equipment for sampling and testing soils as required.

Procedure in Inspection

    Preparation of Site.   Check against specifications for:

        Stripping.

        Storage of topsoil.

        Removal of obstructions.

        Clearing and grubbing.

        Protection of trees.

        Removal of peat, muck, humus, sod.

        Removal or resetting of poles.

        Resetting or installation of culverts.

        Drains, sewers, water pipes, utilities.

        Cavities and trenches to be backfilled and tamped.

        Stake grades and slopes.

        Cross-section borrow pits.

        Cross-section rock as exposed before excavating.

        Approval of material for backfill.

   Selection of Material.   Follow specifications in selecting material such as placing granular material under paved areas.

    Broken rocks on slopes and in marshy foundations.

    Wasting peat, muck, frozen clods, organic matter.

   Soil Compaction.   Check specification requirements such as:

Weight of equipment and number of passes. Eight to twelve passes with sheepsfoot roller are customary. Three-wheel roller, 8 to 12 tons for final rolling of each layer and on the subgrade beneath base course. Caterpillar tractors may be used for granular soils when sheepsfoot or three-wheel rollers are not effective.

    Thicknesses of layers rolled (usually 4 in. to 12 in.).

    Harrows, rotary tillers, reduction of moisture and soil mixture.

    Provision of water distribution in dry weather.

    Provision of uniform travel for construction equipment.

    Do not permit end dumping over face of high fills.

    Stable slopes may be obtained by filling beyond final grade and subsequently excavating to that grade.

Protection of pipes from injury by equipment during construction.

Field Checks. Make test for compaction.

Establish percentage of moisture required for optimum density by Proctor test.

Determine moisture in fill.

Sprinkle with water or air-dry filling material to attain optimum moisture content.

Consolidate fill in layers as per specifications.

No frozen filling material to be used.

Do not place fill on frozen ground.

Check consolidated material for percentage of optimum density required by specifications.

BITUMINOUS PAVING—GENERAL

CHECK LIST FOR INSPECTORS

Inspectors' Equipment

Complete set of latest approved plans and specifications.

Penetrometer with extra needles and 3-oz. tins (optional; needed only when asphalt penetration is checked on job).

Supply of report forms, sample tags, cartons, cans, and sacks for shipping samples.

Metal dipper, pans, shovels, pails, etc., for sampling.

Armored thermometers of specified temperature range for both plant and field.

Marshall Testing Machine, molds, tamper, and sample extractor.

Centrifuge.

Set of screens or sieves of specified aggregate sizes.

Wire brush for cleaning sieves.

1 balance of 500-gram capacity.

1 scale or balance of 10- to 25-lb. capacity.

Supply of carbon tetrachloride or other solvent such as benzol, carbon disulfide, chloroform, or gasoline.

Putty knife for checking pavement depth.

6-ft. folding rule and 50-ft. steel tape.

10-ft. straightedge, 3-ft. straightedge, and template cut to required crown.

Grade line and string level.

Field books, pencils, keel or crayon.

Fruit jar or hot plate and pan for moisture content (not necessary for mixes with hot, dry aggregates).

Procedure in Inspection

Bituminous Treatments

Prime Coat. Applied to receptive surfaces; should soak in.

Subgrade or Base. Compacted to specified density; should not shove, creep, or weave under a moving road roller.

Width, elevation, and cross section.

Condition to receive prime; excess loose material removed but surface not so tightly bound as to be impervious; slightly moist surface better for cutbacks and tars than dry and dusty; surface may be quite damp for asphalt emulsions.

Application. Bituminous material tested, approved, and of specified type.

Distributor truck calibrated and volume of material in load determined.

Distance each load should cover, at the width spread and at the gallonage per square yard specified, measured off and marked conspicuously. Amount of bitumen used is usually 0.20 to 0.45 gal. per sq. yd. for tight surfaces and 0.4 to 0.6 for open surfaces. Blot excess material with sand.

Distributor checked for specified requirements, usually: mechanical circulator, dual tires, pressure gage, range of application rates, positive shut-off, thermometer, spray bar width, measuring stick, tachometer, application pressure, wheel load or tire pressure, clean apertures or jets, load calibration and capacity.

Specified temperature of application adhered to.

Net gallonage computed by applying temperature conversion factor to gallonage measured at application temperature, see Table 50.

Provision to prevent overlap at beginning and end of application strip; usually building paper is laid down to insure a clear-cut joint.

Cover Material. May or may not be specified. If not specified, a light cover in spots may be necessary to prevent migration of bitumen on steep grades and banked curves.

If specified, check following: gradation, type, moisture content, rate and uniformity of application, dragging, rolling, brooming and sweeping.

Curing Period. As specified, should elapse before subsequent applications or pavement courses.

Tack Coat. Usually applied to hard, dense impervious surfaces, without soaking in.

Surface. Cleaned or swept, dry but not dusty, patched, brought to line, grade and cross section as specified.

Application. Same as for prime coat except for following precautions:

As application is very light (0.08 to 0.15 gal. per sq. yd.) distributor must travel at very high speed; tachometer is a necessity.

All distributor bar apertures or jets must be open and functioning.

Uniformity can be obtained by use of burlap drag behind distributor.

Great care must be exercised to prevent overlapping at sides and ends of strips; resulting fat spots will seriously affect pavement.

Surface must be tacky or sticky when pavement is laid and must not be allowed to be covered with dust or dirt; traffic must be kept off.

Seal Coat. Surface. Prepared as per specifications.

Application. Same as for prime coat with same precautions as for tack coat except bitumen is usually immediately covered with aggregate. Leave an 8-in. strip of bitumen uncovered for lapping adjacent strips.

Cover Material. May or may not be specified. If specified, check type, gradation, moisture content, rate and uniformity of application.

Applied at once after bitumen is spread so particles can be embedded. Material should be spread out ahead in piles or windrows or spreader trucks should be on job before bitumen is applied.

Specified method of uniformly distributing cover material followed.

Rolling, if specified, began at once and continued until aggregate is embedded. Excessive rolling, causing crushing of particles, avoided.

Broom or wire mesh dragging carried on simultaneously with rolling unless otherwise specified.

Excess cover material swept off after rolling if specified.

Back spotting of bleeding areas with cover material for several days.

Mix-in-Place (Road Mix)

Subgrade or Base. Compacted to specified requirements and shaped to correct width, grade and cross section.

Prime Coat. May or may not be specified. Same as for bituminous treatments.

Aggregates for Mix. Source approved and laboratory testing verified. Gradation checked before use and continuously during operations.

Aggregate may be bank run or artificially mixed as specified; in either case the aggregate, before mixing with bitumen, should conform to specified gradation.

Continuous check on any special requirements such as liquid limit, plasticity index, percentage of silt and clay, either by sending samples to laboratory or by field testing as directed by superiors.

Preparation of Aggregate. Loose aggregate spread flat or in windrows in such volume and to such depth as to produce specified thickness when compacted.

Coarse or fine material mixed into aggregates to produce specified gradation if necessary.

Mixed aggregate brought to specified moisture content by pulverizing and aeration if wet or by sprinkling if dry. If not specified otherwise, usually maximum 2% moisture for cutback asphalts and tars, and 4 to 5% moisture for emulsions. Sprinkling necessary only when aggregate is very dry and dusty.

As contractor will demand quick moisture readings, use of fruit jar pycnometer is recommended; see p. 185.

Application of Bituminous Material. (a) By Set Quantity per Square Yard. Same as for bituminous treatments, prime coat. Follow job specification for increments and sequence of application. If not specified, best practice is to apply in increments of 0.5 to 0.6 gal. per sq. yd. with partial mixing between increments. For dense graded mixes, 0.5 to 0.6 gal. per sq. yd. per inch of depth of finished mix should suffice.

(b) Quantity Varied per Aggregate Gradation. Inspector must make continual screen analysis and compute required quantity of bitumen by formula or method as specified or directed. Screen analysis made either at pit, plant, or on the site, preferably on the site. Bitumen usually 4 to 7% by weight.

Mixing. (a) By Blade Graders. Graders to cut clear down to base (but not to cut into or tear up the base) and make complete turnover. Mixture to roll over in front of grader blade. Mixing to begin at once behind bituminous application to prevent migration of bitumen. Graders to manipulate mixture back and forth across entire width of road or strip being placed. Mix in as long strips as possible keeping turnarounds to minimum. Mixing to continue until all aggregate particles are coated; usually 12 to 15 complete turnovers are necessary.

Areas deficient in bitumen, i.e., dry, brownish color, powdery, no cohesion, large particles uncoated, should receive additional bitumen and remixing.

Areas with excess bitumen, i.e., greasy, fat, sloppy, unstable, free bitumen in evidence, corrected by adding more aggregate and remixing.

(b) By Rotary Tillers (Pulvi-Mixers, Roto-Tillers, etc.). Same general procedure as for blade graders except:

Aggregate is usually spread flat and mixed flat.

Aggregate is not manipulated back and forth.

Bitumen applied in 0.4 to 0.6 gal. per sq. yd. increments with partial mixing between applications is best practice.

Watch for balling up of aggregate, i.e., lumps of uncoated aggregates.

If road or area is wide enough, transverse, diagonal or figure-8 travel of the Rotary-Tiller is recommended.

Mixing continued till all aggregates are coated for full depth.

Note. Rotary tillers and blade graders are sometimes operated in combination. Blade grader throws up windrow directly in front of rotary tiller, which mixes and spreads out flat; 10 to 12 repetitions of this process will usually produce uniform mixture.

(c) By Travel Plant Methods. Check calibration of measuring devices on machine.

Control of moisture content of aggregates by constant checking.

Gradation of material in windrows; continual screen analysis.

Accurate windrowing of aggregates ahead of travel plant to produce required finished thickness and width of pavement.

Mixed material as it leaves plant to have all aggregates coated, well mixed, and uniform in gradation and bitumen content.

Bituminous material introduced within specified temperature range.

Mixture may be spread with blade graders or paving machine; follow job specifications.

Curing. As specified.

Rolling. Equipment and methods as specified, to continue until mix is compacted to specified density, is smooth, and shaped to specified cross section and elevations.

Seal Coat. Same as for Bituminous Treatments.

Penetration Macadam

Subgrade or Base. Compacted to specified requirements and shaped to correct width, grade, and cross section.

Aggregates. Coarse stone, choke stone, and chips tested and approved for gradation and quality before use.

Inspection of gradation primarily visual, but screen analysis should be made once a day.

Avoid an excess of stone under 1¼-in. size, dust, and screenings, which will form mats that bitumen cannot penetrate.

Placing Aggregates. May be spread by hand, spreader boxes, machines, or blade graders.

Avoid segregation of coarse and fine stone.

Spread in layers as specified; 3½ in. to 4 in. is about the maximum thickness one layer can be built.

Depressions removed by working coarse stone into low areas; do not fill depressions with fine stone.

Pockets or areas of fine stone or choked with dust removed and replaced with properly graded stone.

Surface true, "spotted" to grade and cross section and without areas of excess fine or coarse stone before rolling begins.

Initial Rolling. Begin at sides and progress to center, overlapping shoulder and each previous wheel mark.

Rolling to continue until all stone keyed together.

Depressions developing during rolling corrected.

Roll in as long strips as possible to avoid reversing roller.

Rollers to operate in straight, not wavy, lines, and reverse motion smoothly, not in jerks.

Bituminous Application. Do not begin until surface is dry (except for emulsions), not dusty or excessively choked, and uniformly compacted.

Application is same as for prime coat, Bituminous Treatments.

Choke Stone (applied after bituminous material). Spread uniformly, just sufficient to fill voids in stone.

Rolled and broom dragged simultaneously until surface is thoroughly consolidated and free from large voids.

In hot weather or with asphalt emulsions this rolling and brooming may be postponed until day following bituminous application.

Continue rolling and broom dragging until all roller creases and marks are removed and surface does not creep or shove under roller wheels. Additional small amounts of keystone may be added during this process.

Note. Follow job specifications for quantity of bitumen and increments of application. Practice varies from applying bitumen in one heavy application with one choking and rolling to applying bitumen in two or three increments with choking and rolling after each.

Seal Coat. Same as for seal coat, Bituminous Treatments.

Pay Items. Accurate record of all pay items in contract.

Gallons of bituminous material placed (corrected for temperature).

Tons, square yards, or cubic yards of aggregates or completed pavement as specified.

Extra applications of bitumen and aggregates.

CHECK LIST FOR INSPECTORS

PLANT-MIX BITUMINOUS PAVING

Procedure in Inspection

Plant Inspection

Follow design mix requirements furnished by laboratory.

Tested and Approved Materials. Bituminous material, aggregates, and fillers tested and approved before use.

Samples of aggregates, bitumen, and mixture shipped to laboratory at least once a week.

Daily screen analysis of aggregates and completed mixture.

Storage and Handling of Materials. Aggregates stock piled to avoid segregation and intermingling.

Mineral filler stored in dry place.

Plant. Plant equipment to meet specifications.

Weighing devices to work properly. Check scales with standard weights.

Tare weight of asphalt bucket checked twice daily. Tare weight is weight of empty bucket including residue and adhering bitumen.

Bucket kept clean or correction made for adhering bitumen.

Weigh box large enough to prevent spilling, with tight gates and in good condition.

No segregation or intermingling of aggregates before mixing.

Screens of specified size to completely separate various sizes required.

Asphalt thermometers checked for correct reading.

Weighing facilities for mineral fillers.

Correction of aggregate grading if variation occurs.

Scales for aggregate and bitumen set to produce specified mixture.

No change in basic mix proportions without approval from engineer.

Mixing Operations. Specified moisture content of aggregates adhered to for cold aggregate mixes.

All aggregates coated with bitumen and mix of uniform color and consistency.

Bitumen bucket completely emptied or drained.

Mixing time as specified and sufficient to coat aggregate thoroughly.

On sheet-asphalt jobs sand gradation checked hourly.

Weekly check of aggregate scales or more often if variation occurs.

Net weight of truck loads to equal total batch weights; check once a week.

Aggregates and bitumen heated to specified or approved temperatures; keep daily record.

Aggregates or bitumen never to be heated above the specified limits.

Mixture leaves plant at specified or approved temperature.

Proportions of mixture checked daily by dissolving the bitumen of a representative sample and making screen analysis of aggregates.

Transporting Mixture. All trucks covered with canvas or tarpaulin.

Trucks cleaned and sprayed with light oil or soap emulsion before mixture is placed therein; avoid excess.

Insulated truck bodies preferable if available.

No loads sent out if weather will hinder proper laying; cooperate with field inspector and contractor in this respect.

Field Inspection

Subgrade or Base. Compacted and shaped according to plans and specifications.

Prime or tack coats, if specified, properly applied and curing time elapsed.

Holes and depressions repaired and rolled in advance of paving.

Base dry before mix is placed.

Note. Proper compaction and contour of base and subgrade are essential to a smooth and satisfactory pavement.

Forms. If specified, must be rigidly supported and accurately set to line and grade.

Placing. Paving machines and rollers inspected and approved before use for conformance with specified requirements.

Screeds on paver checked for crown ordinates. See p. 315 for crown offsets.

Screeds cleaned at noon and night shutdowns with fuel oil and scrapers.

Contact surfaces of paving equipment lightly oiled.

Avoid excessive hand raking behind paver. Paver should be so adjusted that only occasional touching behind will be necessary by hand.

Notify plant to shut down if rain begins. Loads in transit are customarily allowed to be placed if they are covered and temperature is sufficiently high. Surface course joints should not be over base course joints.

Mixture delivered at proper temperature and not too rich or too lean.

Note. Excessive bitumen in mix will flush to surface during rolling and mix will be fat, greasy, and soupy. Deficient bitumen is indicated by cracking under roller, pushing into lumps, and dull, lusterless appearance. Either condition must be reported immediately to plant inspector.

Check temperature frequently by use of immersion armored thermometer of Weston type or equal.

An overheated or burnt-up batch will usually give off a cloud of acrid, white smoke when dumped.

If bitumen drains off or migrates to bottom of truck and aggregate on top is uncoated, the plant inspector should be notified immediately.

Check thickness of course as follows: (1) Compute number of square yards a load will cover, and make a mark on the base to which a load should spread. (2) After initial rolling make small hole with putty knife in mixture and check depth with rule. (3) Check square yards laid against tons hauled at noon and at end of day. For dense bituminous concrete mixes, the yield should be about 1 sq. yd. for 1 in. in depth for every 110 lb. of mix.

Mixture spread to a loose depth that will produce specified finished thickness; loose depth must be determined by experiment.

Hand Spreading. Each shovelful turned over as placed and load so dumped that entire batch is shoveled into place.

Workmen not to walk in loose mixture.

Avoid excessive raking that pulls coarse stone to surface.

Control depth with spreading blocks of correct height.

Shovels, rakes, and tampers kept hot and clean.

Rolling. Rollers of type and weight specified, and equipped with water spray and scrapers on wheels.

Begin rolling as soon after spreading as mixture will bear the roller without shoving or hair cracking.

When specified, check square yards rolled per hour per roller.

Begin rolling at sides and proceed toward center, overlapping one-half width of roller on successive passes.

If not specified otherwise, use tandem rollers for initial rolling and keep 3-wheel rollers off until mix is somewhat cooled.

Rollers to reverse motion smoothly, not in jerks.

Length of roller passes to be staggered.

Surface checked immediately after initial rolling with straightedge and template. This must be done before mix cools so corrections can be made. Tolerance usually $\frac{1}{8}$ in. to $\frac{1}{4}$ in. in 10 ft. Try to correct surface before mix hardens to avoid unsightly skin patches later.

Rollers and trucks not to park on pavement while it is still plastic.

Excessive rolling avoided; it will cause crushing of aggregate and displacement of mix.

Rolling diagonally and at right angles very desirable if width of street or road is sufficient.

Rolling continued until all roller creases are removed and specified density is attained.

Joints. At shutdowns and end of day's work, transverse joints are formed by rolling over edge and then cutting back a vertical joint at full depth.

All cold joints painted with liquid bitumen and fresh mixture rolled firmly against the joint face.

Seal Coat. If specified, check gallons per square yard, temperature, and type of material.

Final Inspection. Depressions and bumps over specified tolerance corrected by concentrated rolling or skin patches.

Oil spots and fat spots cut out and refilled and tamped.

Disintegrated spots where mixture is raveling cut out to full depth with vertical faces and refilled with fresh mixture thoroughly tamped and rolled.

Opening to Traffic. Edges protected from traffic runover before opening pavement, usually after final rolling when mix has cooled off and hardened or from 4 to 12 hr. after placing.

Pay Items. Accurate record kept of all contract pay items, such as:

Tons, square yards, or cubic yards (as specified) of mixture laid.

Volume of embedded structures if deducted from unit price.

Gallons or square yards of any prime, tack, or seal coats applied.

Record of batches condemned or wasted.

Any other contract pay items.

Outline of Operations

Formula for Developing Mix. The engineer sends to the testing laboratory samples of the materials proposed for use on the job.

From a series of tests made on these samples the laboratory furnishes a formula for the bituminous mix and test requirements for stability and flow which are established as job standards.

Field and Plant Operations. The construction force receives deliveries of materials, mixes them in accordance with the laboratory formula, and lays and finishes the pavement.

As variations in the materials may develop during the progress of the work, samples are taken at the mixing plant and laboratory tests are made to maintain continuous conformity with the requirements. (See pp. 221 and 222.) Tests may be made by the main laboratory or by a field laboratory.

Cores are cut from the finished pavement, checked for density, and tested in the Marshall machine as a final proof of performance (see p. 223 and Fig. 122).

Placing of pavement is checked for variations from grade, "bird baths," and other details.

BITUMINOUS PAVING

TABLE 40. FIELD SAMPLING

Material and Method	When Sampled	Size of Sample	Instructions
Asphalt, cement, crude asphalt, refined asphalt, bituminous materials, A.S.T.M. D-140-41T	From each source in advance of work and from each carrier as delivered	1 qt. min. Asphalt emulsion or cut-back 2 qt. min.	Draw sample from top, bottom, and middle of tank by lowering bottle or can fitted with a stopper or lid lifted by attached wire, or sample may be taken from drain cock after initial draining. Solid or semi-solid asphalt sampled with clean hatchet or putty knife. Place liquids in small-mouth cans with cork-lined screw top. Place semi-solid material in friction lid cans. Ship crated or boxed. Mark cans.
Asphalt, A.S.T.M. D-290-39	Daily, for penetration test	3 oz. min.	Draw sample into can from valve over asphalt bucket on plant. Mix and pour into tin or glass container.
Asphalt sand, screenings, crushed stone and gravel, mineral fillers, A.S.T.M. D-75-42T	Each source First shipment and if any change for laboratory tests Daily from piles or bins for plant tests	Fine aggregates 5 lb. min.; coarse aggregates 20 lb.	Quarter samples to size required. Sample from pits by channeling open face or from test hole. Sample from stock piles in various places avoiding base of pile. From cars, sample from top, middle, and bottom. Ship in strong, tight bags or boxes. At plant, sample separate sizes and composite mixture for daily sieve tests.
Heated and dried aggregates, A.S.T.M. D-290-39	Daily from bins	Fine, 5 lb.; coarse, 20 lb.	Pass shovel or pan quickly through stream of hot material as it flows from bin for daily sieve tests.
Bituminous mixtures (sheet asphalt, bituminous concrete, road mix, sand asphalt, plant mixes), A.A.S.H.O. T-41-35, A.S.T.M. D-290-39	Daily, or as specified or directed	Sheet asphalt, 1 lb. min.; bituminous concrete, 5 lb. min.; cold mixes, 15 lb. min.; compressed mixture, 6 to 12 in. sq. by full depth	At plant, take small portions from a number of batches during day, mix, and quarter to size. At paving site, compose sample from top, bottom, front, and back of load. Road mixes, shovel from course full depth, mix, and quarter. Ship samples in clean, tight box, carton, or friction lid can. Compressed samples, select location where mix is representative, before seal coat and after final rolling. Cut exact square to full depth of course.

MARKING SAMPLES—ALL MATERIALS

General. Same as for concrete field sampling, p. 46.

Bituminous Material. Railroad car number, refinery, type, grade, proposed use.

Aggregates. Kind, source, where sampled, separated size or combined mixture.

Bituminous Mixtures. Type, plant, date, specified mix, station or location placed.

FIELD OR PLANT TESTS

May be used when full-scale laboratory tests are not practicable

Penetration of Asphalt (A.S.T.M. D-5) is the distance, measured in units of $\frac{1}{10}$ mm., that a standard blunt-point needle will penetrate a sample of asphalt at 77° F. when the needle is loaded with 100 grams applied for 5 seconds. Sample selected per Table 40, melted, stirred, and poured into container, 2.17 in. diameter by 1.38 in. Place in water for 1 hour at 77° F. to a depth of 4 in. and 2 in. off bottom of vessel. Sample is penetrated in at least 3 places, and average penetration is reported.

Notes. Sample must be maintained at 77° F. during the test by placing in a transfer dish filled with water and by returning the sample to the water bath after each test. The needle must be wiped after each test. Metal "ointment box" of above dimensions may be obtained at drug store. The inspector should have orders as to action to take if penetration is not as specified. Test apparatus is called a penetrometer.

NORMAL PENETRATION LIMITS

(77° F., 100 g., 5 sec.)

25–30	50–60	85–100
30–40	60–70	100–120
40–50	70–85	120–150
		150–200

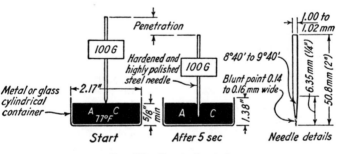

FIG. 121. Penetration test.

Pat Test of Sheet Asphalt. Select small sample of hot mix and note the temperature; it should be between 275° and 325° F. Place at once upon a sheet of unglazed manila paper resting upon a flat board. Fold the paper over the sample, and press heavily with the flat of a wood paddle 6 in. long by 4 in. wide. Strike the paper a sharp blow with the paddle; open the paper and remove the sample. If the stain is medium dark, bitumen content is about right provided the texture appears normal. If it is very dark or sloppy, bitumen is excessive or mix is too low in fines.

If it is light and dry, bitumen is insufficient or mix has excess of fines. If only the imprint of single sand grains appears, the amount of filler is deficient. If the space between sand grains is filled in, aggregate grading is good. A small magnifying glass will help in examining gradation.

Cigar-Box Test of Asphaltic Concrete. Select sample of hot mix and note the temperature; it should be between 275° and 325° F. Place sample in cigar box, with white paper over it, filling the box ½ to ⅔ full. Holding lid on tightly, pound end of box 6 times squarely on firm base. Open box, and examine stain on lid. If stain is medium dark, bitumen content is about right. If it is very dark or sloppy, bitumen is excessive or mix is too low in fines. If it is light and dry, bitumen is insufficient or mix has excess of fines. The texture of the compressed sample indicates the appearance of the mixture after rolling. Oily spots indicate improper mixing, which may be due either to mechanical trouble or to variation in temperature of the aggregates.

Percentage of Bitumen and Mechanical Analysis of Mixtures. The following method is for emergency checking where A.A.S.H.O. Tests T-58 * and T-30 † are not practicable. Dissolve and wash all the bitumen from a weighed sample of the mix with carbon tetrachloride, gasoline, or other solvent such as benzol, xylene, or chloroform, and weigh the recovered aggregate.

% of bitumen

$$= \frac{\text{weight of original sample} - \text{weight of recovered aggregate}}{\text{weight of original sample}} \times 100$$

Note. Wash aggregate clean. Avoid loss of any aggregate. If the percentage of bitumen varies from that specified, check the plant scales and the weighing operation. For sieve analysis of dried recovered aggregate (A.S.T.M. C-136 and C-117), see pp. 45 and 48, Aggregate Field Testing.

Field Density of Compressed Mixture. Immerse the weighed sample in hot paraffin, remove, cool, and weigh again. Weight gain is weight of paraffin. Volume of paraffin coat is calculated using 55 lb. per cu. ft. as weight of paraffin. Weigh the coated sample in water, record weight, and calculate the volume of the sample or measure the volume of the displaced water by an overflow device (weight water = 62.4 lb. per cu. ft.). Deduct the volume of the paraffin coat. Field density (lb. per cu. ft.) = net weight of sample in pounds ÷ volume of sample in cubic feet. The percentage of compaction = field density ÷ theoretical maximum density (from laboratory).

$$\% \text{ of voids} = \frac{\text{theoretical maximum density} - \text{field density}}{\text{theoretical maximum density}} \times 100$$

* Centrifuge method for bitumen.
† Sieve analysis of aggregate after bitumen extraction.

Note. Compaction to 94–96% of maximum density is usually specified.

Marshall Test. A test for stability and flow of asphaltic concrete using a Marshall testing machine illustrated in Fig. 122.

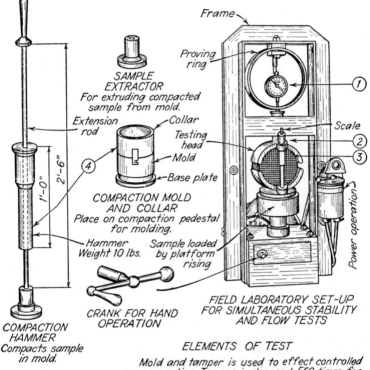

ELEMENTS OF TEST

Mold and tamper is used to effect controlled compaction. Tamper is dropped [50 times for road work, 75 times for airfield] from a height of 18" onto sample in the compaction mold. Sample molded at 225° F.

Compression gage registering total pounds pressure at sample failure indicates stability. Flow meter registering total deformation in hundreths of an inch indicates flow. Sample, 4" diam. 2½" thick, formed by controlled compaction. Tested at 140° F.

FIG. 122. Marshall test.

TYPICAL EXAMPLE OF PLANT OR FIELD CONTROL TESTS OF ASPHALTIC CONCRETE

Job mix formula furnished by testing laboratory.   Given:

1. Aggregate gradation requirements:

Sieve, square openings	3/4″	1/2″	3/8″	No. 4	No. 10	No. 20	No. 40	No. 80	No. 200
Percentage passing	100	93	84	66	49	40	29	13	4.6

Check gradation by sieve analysis.   Tolerances: 1.5% on material passing No. 200 sieve, 4% on all others.

2. Bitumen: grade 85–100 penetration.

Check at source before sealing and shipping to job; or take field samples and check by penetrometer.   See p. 221.

3. Bitumen in mixture: by weight 5.3%, tolerance 0.3%.

Check: Take sample of paving mixture and extract bitumen.   See **Percentage of Bitumen and Mechanical Analysis of Mixture,** p. 222.

4. Use only if Marshall machine was used in mix design.

Stability 1000 lb. plus—tolerance 10%.

Flow value 16 points or lower—tolerance 2 points.

Check: Take samples of paving mixture from trucks at plant, tamp into molds, and test on Marshall machine.   See Fig. 122.

5. Density of Marshall sample taken at plant: 160.2 lb. per cu. ft.; tolerance 1.5 lb.   Per cent voids 3–5.

Check: Take sample from Marshall mold and follow as outlined under **Field Density of Compressed Mixture,** p. 222.

6. Per cent voids filled with bitumen: 75–82.

This is computed from information obtained in test Nos. 3 and 5.

Note that percentage of bitumen by weight must be converted to percentage by volume.

% Bitumen by volume

$$= \frac{\% \text{ Bitumen by weight} \times \text{Density of Marshall sample}}{\text{Specific gravity of bitumen} \times 62.4}$$

% Voids filled with bitumen

$$= \frac{\% \text{ Bitumen by volume}}{\% \text{ Bitumen by volume} + \% \text{ Voids of mixture}}$$

7. Field density: 96–98% of density of Marshall test sample.

Obtain sample by core or other means, and determine field density.

Date __June 10__ 194 _0_

Engineer

Report No. __12__

S. P. No. __2006-05__

F. A. P. No. __174__

DAILY BITUMINOUS REPORT (CONSTRUCTION) *

FOR MACADAM, BITUM. TREATMENTS, MIX-IN-PLACE

T. H. No. __56__ From __West Concord__ To __Jct. T.H. 14__

Inspector __A. C. Johnson__ Contractor __Pioneer Co.__

Course	Station		Aggregate Lb. per Sq. Yd.	Bituminous Material		Application Temperature	Width of Material
	From	To		Gal. per Sq. Yd.	Per cent of Mix		
Base prime	326 + 00	356 + 00		0.22		110°	26'
Wearing	256 + 00	308 + 00	150	0.88	4.6	200°	24'
Wearing	308 + 00	326 + 00	100	0.60	4.7	200°	24'
Tack	308 + 00	326 + 00		0.07		200°	24'
Seal	0 + 00	56 + 00	18	0.30		220°	24'

AGGREGATE GRADING

Course	Pit No. or Station	Total Per Cent Passing								
		¾	⅝	⅜	10	20	40	100	200	
Wearing	506	100	98	80	47		18		3.1	
Wearing	506	100	97	80	51		19		3.4	
Seal	Doe Gravel Co.			100	5.0	2.0				

BITUMINOUS MATERIALS USED

Course	Kind and Grade	Gallons Applied		
		Today	Prev.	To Date
Base prime	MC – 1	1,700	10,100	11,800
Tack	MC – 3	350	1,200	1,550
Wearing	MC – 3	25,300	101,000	126,300
Seal	RT – 9	4,450	15,000	19,450

Weather A.M. _clear_ P.M. _clear_

Temperature 8 A.M. 67 12 N. 85 5 P.M. 78

EQUIPMENT

Description	No. Units	Hours Worked
Traveling plant	1	15
Distributor	1	8
Trucks	5	40
Blade graders	3	24
Roller	1	10
Chip spreader	1	3
Sweeper	1	5
Drag broom	1	3
Transfer tank	1	15

OPERATING TIME AND DELAYS

Time start __4:00__ A.M. Time stop __7:30 P.M.__ Gross time __15½__ Time delayed __½__ Net operating time __15__

Delays due to	Lunch ½ hr.

Remarks: _Aggregate mixed and coated very well._

Signed by __J. M. Smith__

Project Engineer

* From Minnesota State Highway Department.

Date _____
          Engineer
          Report No. _____

Type of
pavement _____
    PLANT-MIX BITUMINOUS INSPECTION
    S. P. No. _____

Width _____
      DAILY PAVING REPORT *
    F. A. P. No. _____

T. H. No. _____ From _____ To _____ Length _____

Engineer _____ Contractor _____

Street inspector _____ Plant inspector _____

Course	Station		Tons Mixture Placed	Area in Sq. Yd.	Yield Lb. per Sq. Yd.	Temperature When Laid
	From	To				
Total						

EQUIPMENT

Spreading Machine			Total Hours Worked
Make and type	No.	Width	
Rollers			
Make and type	No.	Wt.	
Other Equipment			
Description		No.	

PLACING AND FINISHING

	Good	Fair	Poor	Remarks
Workability___				
Temperature___				
Spreading___				
Shoveling___				
Raking___				
Rolling___				
Finished surface ___				

HAUL DATA

Average round-trip time _____ Min.
Average length of haul _____ Miles

WEATHER AND TEMPERATURE

Weather: A.M. _____ P.M. _____
Temperature: 7:00 A.M. _____ 10:00 A.M. _____
           2:00 P.M. _____ 5:00 P.M. _____

TIME DISTRIBUTION AND DELAYS

Time start _____ Time stop _____ Gross time _____ Time delayed _____ Net paving time _____

Moving	Weather	Non wk. days	Rock	Sand	Filler	Bit. cement	Chips
Paver	Plant	Switching	Haul road	Trucks	Rollers	Base	

Remarks _____

_____

Signed by _____
          Project Engineer

* *From Minnesota State Highway Department.*

Engineer

Date _____

Type of
Pavement _____

Report No. _____
S. P. No. _____
F. A. P. No. _____

PLANT-MIX BITUMINOUS INSPECTION

DAILY PLANT REPORT *

T. H. No. _____ From _____ To _____ Length _____

Contractor _____ Location of plant _____ Type of plant _____

Engineer _____ Plant inspector _____ Street inspector _____

BATCH PROPORTIONS

	Wearing Course				Binder Course			
	Lb.	Per Cent	Lb.	Per Cent	Lb.	Per Cent	Lb.	Per Cent
Sand _____								
Filler _____								
Bit. cem. _____								
Batch totals _____								
Tons Mixed _____								

AGGREGATE GRADINGS

	Separate Bins					Composite Grading		
				Sand	Filler	Wearing C.		Binder C.	

Total % Passing
" 1½
" 1¼
" 1
" ¾
" ⅝
" ⅜
" 4
" 10
" 20
" 40
" 80
" 100
" 200

MATERIALS USED

	Source	Tons Today	Tons Prev.	Tons to Date
Coarse agg.				
Fine agg. _				
Filler _____				
Bit. cement				
Totals				

TEMPERATURES

	Max.	Min.	Aver.
C. Agg. _____			
Fine agg. _____			
Bit. cement _____			
Wear. course			
Binder course			

MIXING TIME

Mixing time _____ Sec.
Spec. req. min. _____ Sec.

Time start _____ Time stop _____ Gross time _____ Time delayed _____ Net operating time _____

Remarks: _____

Signed by _____

Project engineer

* From Minnesota State Highway Department.

Engineer

BITUMINOUS PAVING ANALYSES *

————————194

Client ——————————————————————————————

Project ——————————————————————————————

Material ——————————————————————————————

Report No.			
Job sample No.			
Date laid			
Sampled by			
Taken at			

SCREENS AND SIEVES USED	ANALYSES	REQUIRED Min. %	Max. %
Passing 200 Mesh Filler			
Bitumen content			

Remarks:

————————————————
Inspector

* *From Haller Testing Laboratories, Inc.*

_____
Engineer

ASPHALT REPORT *

Report for _____

Material _____

Project _____

Producer _____

Contractor _____ Quantity _____

Shipped via _____ Report No. _____ Date ____

Specific gravity @ 25° C.				
Flash point degrees F. (open cup)				
Asphalt content @ 100 pen.				
Furol viscosity @ ° C.				
Penetration 25° C. 100 g. 5 sec.				
Ductility, centimeters @ 25° C.				
Loss on evaporation 163° C. 5 hr.				
Penetration of residue from evaporation 25° C. 100 g. 5 sec.				
Total distillate % by volume to 320° F. (160° C.)				
Total distillate % by volume to 374° F. (190° C.)				
Total distillate % by volume to 437° F. (225° C.)				
Total distillate % by volume to 600° F. (315° C.)				
Total distillate % by volume to 680° F. (360° C.)				
Penetration residue from Distillation 25° C. 100 g. 5 sec.				
Ductility residue from Distillation cm. @ 25° C.				
Total bitumen (soluble in CS_2)				

Remarks:

The above fulfills the specification requirements.

_____
Inspector

* *From Haller Testing Laboratories, Inc.*

TABLE 41. USE OF BITUMINOUS MATERIALS

Material	Road Tar (RT) / Cutback Road Tar (RTCB)														Road Oil, Slow Curing Asphaltic (SC)						
Commercial grade	RT-1	RT-2	RT-3	RT-4	RT-5	RT-6	RT-7	RT-8	RT-9	RT-10	RT-11	RT-12	RTCB-5	RTCB-6	SC-0	SC-1	SC-2	SC-3	SC-4	SC-5	SC-6
Application temperature, °F.	125/60	125/90	150/80	150/80	150/80	150/80	235/150	225/150	225/150	260/175	250/175	250/175	120/60	120/60	120/50	175/125	225/120	250/160	275/175	350/225	400/300
Penetration macadam, hot				x							x*	x								x	x
Penetration macadam, cold						x					x*									x	x
Plant-mix, open-graded mixes					x*	x*	x*	x*	x*	x*	x*									x†	x
Mix-in-place, open graded (crushed stone)											x									x	x
Plant-mix, dense-graded mixes				x*			x†	x†	x†									x	x	x	x
Mix-in-place, dense-graded mixes (gravel mulch)							x†	x†	x†								x	x	x	x	x
Bituminous concrete							x	x	x	x	x						x	x	x	x	x
Surface treatment		x	x		x	x	x	x	x	x	x	x	x	x			x	x	x	x	x
Seal coats (fog, flush, or sand cover)					x	x							x	x		x	x	x			
Seal coats (carpet, stone, or armor coat)					x	x	x	x	x				x	x			x	x		x	x
Tack coats	x*	x		x	x	x								x		x	x				
Prime coats (dense or tight bases)													x	x	x	x					
Prime coats (porous or open bases)	x	x*	x	x											x	x					
Soil stabilization	x	x	x	x†	x										x	x					
Dust palliative	x	x	x												x	x	x				
Patching mixtures (cold patch)																					
Sand tar				x*	x*	x	x†	x†									x*		x†	x†	
plant mix						x	x	x†	x												
road mix																					
Crack and joint filler													x	x			x				

* Cool weather work (spring and fall).   † Hot weather work (summer).

Note. For both tars and road oils, the higher the number the more viscous the material. Thus an RT-1, RT-2, and SC-0 are non-viscous liquids at ordinary temperatures suited for soaking into a tightly bound base like clay-gravel. RT-4 to RT-7 and SC-2 to SC-4 will remain semi-liquid and are suited for mixing in place. RT-9 to RT-12 and SC-5 or SC-6 become solid or semi-solid at air temperatures and are suited for hot plant mixes, sealing, and macadam.

References. P. R. A., Koppers Co., Barrett Co., A.S.C.E., Barber-Greene Co.

TABLE 41. USE OF BITUMINOUS MATERIALS, Continued

Material	Medium Curing Cut-back Asphalt (MC)						Rapid-curing Asphalt (RC)						Asphalt Emulsion (AE)				Asphalt Cement (AC) Penetration				
Commercial grade	MC-0	MC-1	MC-2	MC-3	MC-4	MC-5	RC-0	RC-1	RC-2	RC-3	RC-4	RC-5	SS-1 & 2	MS-1	MS-2 & 3	RS-1	25 to 45	45 to 70	70 to 85	85 to 150	150 to 200
Application Temperature, °F.	150/60	175/80	200/130	225/175	250/175	275/200	125/60	150/90	175/125	200/150	225/175	250/200	70/50	120/60	120/60	135/80	350/275	350/250	350/250	350/250	350/250
Penetration macadam — hot																			x†	x	x*
cold												x								x	x
Plant-mix, open-graded (crushed stone)				x	x	x		x	x	x	x	x		x	x						x
Mix-in-place, open-graded				x	x	x		x	x	x	x	x		x	x						
Plant-mix, dense-graded			x	x	x*															x	x
Mix-in-place, dense-graded (mulch)			x	x	x†								x		x						x*
Bituminous (asphaltic) concrete																		x	x	x	
Sheet asphalt																		x	x		
Sand asphalt plant-mix			x	x	x	x		x	x	x				x					x	x	
Sand asphalt mix-in-place (road mix)		x	x	x			x	x	x				x	x					x		
Surface treatment			x	x	x	x		x	x	x	x	x		x		x		x	x	x	
Seal coats (fog, flush, or sand cover)	x	x					x		x				x					x			
Seal coats (carpet, stone, or armor coat)			x	x		x			x	x	x	x		x		x		x	x	x	
Tack coats	x* x†	x					x						x			x					
Prime coats, dense or tight bases	x		x																		
Prime coats, open or porous bases								x	x	x	x		x								
Soil stabilization	x	x					x						x								
Dust palliative	x		x				x						x								
Patching mixtures (cold patch)		x		x	x	x									x		x				
Joint filler, brick																		x	x		
Crack and joint filler																	x	x	x	x	

* Cool weather work (spring and fall).   † Hot weather work (summer).

Note. For cutback asphalts, the higher the number the more viscous the material. For asphalt cements, the higher the penetration number the softer the asphalt; thus for heavy traffic in hot weather a stiff grade such as 45 to 70 or 70 to 85 is used and for cool weather or light traffic softer grades are used. For asphalt emulsions: SS = slow setting; MS = medium setting; RS = rapid setting.

References. P.R.A., Asphalt Institute, A.S.C.E., Barber-Greene Co., Texas Co.

TABLE 42. GALLONS ASPHALTIC MATERIALS REQUIRED AT VARIOUS RATES OF APPLICATION *

GALLONS PER 100 LINEAR FEET

Width, ft.	9	12	15	16	20
Gal. per Sq. Yd.					
0.10	10.	13.3	16.7	17.8	22.2
0.15	15.	20.0	25.0	26.7	33.3
0.20	20.	26.7	33.3	35.6	44.4
0.25	25.	33.3	41.7	44.5	55.5
0.30	30.	40.0	50.0	53.4	66.6
0.35	35.	46.7	58.3	62.3	77.7
0.40	40.	53.3	66.7	71.2	88.8
0.45	45.	60.0	75.0	80.1	99.9
0.50	50.	66.7	83.4	89.0	111.1
1.25	125.	166.3	208.4	222.3	277.7
2.00	200.	266.7	333.4	355.6	444.4

GALLONS PER MILE

Width, ft.	9	12	15	16	20
Gal. per Sq. Yd.					
0.10	530	700	880	940	1,170
0.15	790	1,050	1,320	1,410	1,760
0.20	1,050	1,410	1,760	1,880	2,350
0.25	1,320	1,760	2,200	2,350	2,930
0.30	1,580	2,110	2,640	2,820	3,520
0.35	1,840	2,460	3,080	3,290	4,110
0.40	2,110	2,820	3,520	3,750	4,690
0.45	2,330	3,170	3,960	4,220	5,280
0.50	2,640	3,520	4,400	4,690	5,870
1.25	6,600	8,800	11,000	11,730	14,670
2.00	10,560	14,080	17,600	18,770	23,470

* *From Asphalt Handbook, Asphalt Institute.*

TABLE 43. TONS MINERAL AGGREGATE REQUIRED AT VARIOUS RATES OF APPLICATION *

Tons per 100 Linear Feet

Width, ft.	9	12	15	16	20
Lb. per Sq. Yd.					
10	.5	.67	.84	.89	1.11
15	.75	1.0	1.25	1.33	1.67
20	1.0	1.33	1.67	1.77	2.22
25	1.25	1.67	2.08	2.22	2.78
30	1.5	2.0	2.50	2.67	3.33
35	1.75	2.33	2.92	3.11	3.89
40	2.0	2.67	3.33	3.56	4.44
45	2.25	3.0	3.75	4.0	5.0
50	2.5	3.33	4.16	4.44	5.55

Tons per Mile

Width, ft.	9	12	15	16	20
Lb. per Sq. Yd.					
10	27	35	44	47	59
15	40	53	66	71	88
20	53	71	88	94	117
25	66	88	110	118	147
30	80	106	133	141	176
35	93	124	155	165	205
40	106	141	177	188	234
45	119	159	199	212	264
50	133	177	221	236	293

* *From Asphalt Handbook, Asphalt Institute.*

TABLE 44. CUBIC YARDS OF AGGREGATE REQUIRED PER 100 LINEAR FEET AND PER MILE FOR VARIOUS LOOSE DEPTHS ON ROADS OF VARIOUS WIDTHS *

Width of Road	Area Per	Sq. Yards	1″	1½″	2″	2½″	3″	3½″	4″	5″	6″
6′	100′	66.6	1.9	2.8	3.7	4.6	5.6	6.5	7.4	9.3	11.1
	Mile	3520.0	97.8	146.7	195.6	244.4	293.3	342.2	391.1	488.9	586.7
7′	100′	77.7	2.2	3.2	4.3	5.4	6.5	7.6	8.6	10.8	13.0
	Mile	4106.6	114.1	171.1	228.1	285.2	342.2	399.3	456.3	570.4	684.4
8′	100′	88.8	2.5	3.7	4.9	6.2	7.4	8.6	9.9	12.3	14.8
	Mile	4693.3	130.4	195.6	260.7	325.9	391.1	456.3	521.5	651.9	782.2
9′	100′	100.0	2.8	4.2	5.6	6.9	8.3	9.7	11.1	13.9	16.7
	Mile	5280.0	146.7	220.0	293.3	366.7	440.0	513.3	586.7	733.3	880.0
10′	100′	111.1	3.1	4.6	6.2	7.7	9.3	10.8	12.3	15.4	18.5
	Mile	5866.6	163.0	244.4	325.9	407.4	488.9	570.4	651.9	814.8	977.8
12′	100′	133.3	3.7	5.6	7.4	9.3	11.1	13.0	14.8	18.5	22.2
	Mile	7040.0	195.6	293.3	391.1	488.9	586.7	684.4	782.2	977.8	1173.3
14′	100′	155.5	4.3	6.5	8.6	10.8	13.0	15.1	17.3	21.6	25.9
	Mile	8213.3	228.1	342.2	456.3	570.4	684.4	798.5	912.6	1140.7	1368.9
16′	100′	177.7	4.9	7.4	9.9	12.3	14.8	17.3	19.8	24.7	29.6
	Mile	9386.6	260.7	391.1	521.5	651.9	782.2	912.6	1043.0	1303.7	1564.4
18′	100′	200.0	5.6	8.3	11.1	13.9	16.7	19.4	22.2	27.8	33.3
	Mile	10560.0	293.3	440.0	586.7	733.3	880.0	1026.7	1173.3	1466.7	1760.0
20′	100′	222.2	6.2	9.3	12.3	15.4	18.5	21.6	24.7	30.9	37.0
	Mile	11733.3	325.9	488.9	651.9	814.8	977.8	1140.7	1303.7	1629.6	1955.6
21′	100′	233.3	6.5	9.7	13.0	16.2	19.4	22.7	25.9	32.4	38.9
	Mile	12320.0	342.2	513.3	684.4	855.6	1026.7	1197.8	1368.9	1711.1	2053.3
23′	100′	255.5	7.1	10.6	14.2	17.7	21.3	24.8	28.4	35.5	42.6
	Mile	13493.3	374.8	562.2	749.6	937.0	1124.4	1311.9	1499.3	1874.1	2248.9
24′	100′	266.6	7.4	11.1	14.8	18.5	22.2	25.9	29.6	37.0	44.4
	Mile	14080.0	391.1	586.7	782.2	977.8	1173.3	1368.9	1564.4	1955.6	2346.7

Rolling compacts crushed aggregate base course approximately 20% and wearing course approximately 25% Ordinary bank gravel compacts approximately 33⅓%.

For road 5′ wide take half of 10′ quantity.

For road 22′ wide add quantities for 10′ and 12′ widths.

For road 26′ wide add quantities for 20′ and 6′ widths.

For road 28′ wide take twice quantity for 14′ width.

For road 30′ wide take three times quantity for 10′ width.

* *From Tarmac Handbook, Koppers Co.*

TABLE 45. AREAS OF PAVEMENT SURFACES *

Width in Feet	Square Feet per Mile	Square Yards per Mile	Square Yards per Linear Foot
1	5,280	587	0.1111
8	42,240	4,693	0.8888
9	47,520	5,280	1.0000
10	52,800	5,867	1.1111
11	58,080	6,453	1.2222
12	63,360	7,040	1.3333
15	79,200	8,800	1.6667
16	84,480	9,387	1.7778
18	95,040	10,560	2.0000
20	105,600	11,733	2.2222
22	116,160	12,906	2.4444
24	126,720	14,080	2.6667
26	137,280	15,253	2.8888
28	147,840	16,426	3.1110
30	158,400	17,600	3.3333
32	168,960	18,773	3.5555
36	190,080	21,120	4.0000
40	211,200	23,467	4.4444
50	264,000	29,333	5.5556

* From Bitumuls Handbook, American Bitumuls Co.

TABLE 46. LINEAR FEET COVERED BY 1 TON OF AGGREGATE AT VARIOUS RATES OF APPLICATION *

Width, ft.	9	12	15	16	20
Lb. per Sq. Yd.					
10	200	150	120	113	90
15	133	100	80	75	60
20	100	75	60	56	45
25	80	60	48	45	36
30	67	50	40	38	30
35	57	43	34	32	26
40	50	38	30	28	23
45	44	33	27	25	20
50	40	30	24	23	18

* From Asphalt Handbook, Asphalt Institute.

TABLE 47. WEIGHT AND VOLUME RELATIONS MINERAL AGGREGATES *

BROKEN STONE

Pounds per Cubic Yard

Kind	Sp. Gr.	Loose Spread 45% Voids	Compacted 30% Voids
Trap	2.8	2590	3300
	2.9	2680	3420
	3.0	2770	3540
	3.1	2870	3650
Granite	2.6	2400	3060
	2.7	2500	3180
	2.8	2590	3300
Limestone	2.6	2400	3060
	2.7	2500	3180
	2.8	2590	3300
Sandstone	2.4	2220	2830
	2.5	2310	2940
	2.6	2400	3060
	2.7	2500	3180

GRAVEL AND SAND

Approximate Number of Pounds per Cubic Yard

Voids	Weight	Voids	Weight
50%	2240	35%	2910
45%	2460	30%	3130
40%	2680	25%	3350

* *From Asphalt Handbook, Asphalt Institute.*

TABLE 48. WEIGHT AND VOLUME RELATIONS OF ASPHALTIC MATERIALS AT 60° F. *

Specific Gravity	Pounds per Gallon	Gallons per Ton	Specific Gravity	Pounds per Gallon	Gallons per Ton	Specific Gravity	Pounds per Gallon	Gallons per Ton
0.930	7.745	258.2	0.980	8.162	245.0	1.030	8.578	233.2
0.935	7.786	256.8	0.985	8.203	243.8	1.035	8.620	232.0
0.940	7.828	255.6	0.990	8.245	242.6	1.040	8.662	230.8
0.945	7.870	254.2	0.995	8.287	241.4	1.045	8.704	229.8
0.950	7.911	252.8	1.000	8.328	240.2	1.050	8.745	228.6
0.955	7.953	251.4	1.005	8.370	239.0	1.055	8.787	227.6
0.960	7.995	250.2	1.010	8.412	237.8	1.10	9.161	218.3
0.965	8.036	248.8	1.015	8.453	236.6	1.20	9.994	200.1
0.970	8.078	217.6	1.020	8.495	235.4	1.30	10.826	184.8
0.975	8.120	246.4	1.025	8.537	234.2	1.40	11.659	171.6

* *From Principles of Highway Construction, Public Roads Administration.*

TABLE 49. DISTANCE IN LINEAL FEET COVERED BY A 1000-GALLON DISTRIBUTOR TANK LOAD *

Application Rate, gallons per square yard	Width of Spread, feet									
	2	3	4	5	6	7	8	9	10	11
0.1	45,000	30,000	22,500	18,000	15,000	12,857	11,250	10,000	9000	8182
0.15	30,000	20,000	15,000	12,000	10,000	8,571	7,500	6,667	6000	5455
0.2	22,500	15,000	11,250	9,000	7,500	6,429	5,625	5,000	4500	4091
0.25	18,000	12,000	9,000	7,200	6,000	5,143	4,500	4,000	3000	3273
0.3	15,000	10,000	7,500	6,000	5,000	4,286	3,750	3,333	3000	2727
0.333	13,500	9,000	6,750	5,400	4,500	3,857	3,375	3,000	2700	2455
0.35	12,857	8,571	6,429	5,143	4,286	3,673	3,214	2,857	2571	2338
0.4	11,250	7,500	5,625	4,500	3,750	3,214	2,813	2,000	2250	2045
0.45	10,000	6,667	5,000	4,000	3,333	2,857	2,500	2,222	2000	1818
0.5	9,000	6,000	4,500	3,600	3,000	2,571	2,250	2,000	1800	1636
0.6	7,500	5,000	3,750	3,000	2,500	2,143	1,875	1,667	1500	1364
0.667	6,750	4,500	3,375	2,700	2,250	1,929	1,688	1,500	1350	1227
0.7	6,429	4,286	3,214	2,571	2,143	1,837	1,607	1,429	1286	1169
0.75	6,000	4,000	3,000	2,400	2,000	1,714	1,500	1,333	1200	1091
0.8	5,625	3,750	2,813	2,250	1,875	1,607	1,406	1,250	1125	1023
0.9	5,000	3,333	2,500	2,000	1,667	1,429	1,250	1,111	1000	909
1.0	4,500	3,000	2,250	1,800	1,500	1,286	1,125	1,000	900	818
1.25	3,600	2,400	1,800	1,440	1,200	1,029	900	800	720	655
1.5	3,000	2,000	1,500	1,200	1,000	857	750	667	600	545
1.75	2,571	1,714	1,286	1,029	857	735	643	571	514	468
2.0	2,250	1,500	1,125	900	750	643	563	500	450	409
2.25	2,000	1,333	1,000	800	667	571	500	444	400	364
2.5	1,800	1,200	900	720	600	514	450	400	360	327

Application Rate, gallons per square yard	Width of Spread, feet								
	12	13	14	15	16	17	18	19	20
0.1	7500	6923	6429	6000	5625	5294	5000	4737	4500
0.15	5000	4615	4286	4000	3750	3529	3333	3158	3000
0.2	3750	3462	3214	3000	2813	2647	2500	2368	2250
0.25	3000	2769	2571	2400	2250	2118	2000	1895	1800
0.3	2500	2308	2143	2000	1875	1765	1667	1579	1500
0.333	2250	2077	1929	1800	1688	1588	1500	1421	1350
0.35	2143	1978	1837	1714	1607	1513	1429	1353	1286
0.4	1875	1731	1607	1500	1406	1324	1250	1184	1125
0.45	1667	1538	1429	1333	1250	1176	1111	1053	1000
0.5	1500	1385	1286	1200	1125	1059	1000	947	900
0.6	1250	1154	1071	1000	938	882	833	789	750
0.667	1125	1038	964	900	844	794	750	711	675
0.7	1071	989	918	857	804	756	714	677	643
0.75	1000	923	857	800	750	706	667	632	600
0.8	938	865	804	750	703	662	625	592	563
0.9	833	769	714	667	625	588	556	526	500
1.0	750	692	643	600	563	529	500	474	450
1.25	600	554	514	480	450	424	400	379	360
1.5	500	462	429	400	375	353	333	316	300
1.75	429	396	367	343	321	303	286	271	257
2.0	375	346	321	300	281	265	250	237	225
2.25	333	308	286	267	250	235	222	211	200
2.5	300	277	257	240	225	212	200	189	180

* *From Principles of Highway Construction Public Roads Administration.*

TABLE 50.   STANDARD ABRIDGED VOLUME CORRECTION TABLE FOR BITUMINOUS MATERIALS *

[Volume at 60° F. occupied by unit volume at indicated temperature; t = observed temperature °F.; M = multiplier to reduce volume to 60° F.]

GROUP 0.   SPECIFIC GRAVITY AT 60° F., ABOVE 0.966

t	M	t	M	t	M	t	M
60	1.0000	145	0.9707	230	0.9425	315	0.9154
65	.9982	150	.9691	235	.9409	320	.9138
70	.9965	155	.9674	240	.9392	325	.9123
75	.9948	160	.9657	245	.9376	330	.9107
80	.9931	165	.9640	250	.9360	335	.9092
85	.9914	170	.9623	255	.9344	340	.9076
90	.9896	175	.9606	260	.9328	345	.9061
95	.9879	180	.9590	265	.9312	350	.9045
100	.9862	185	.9573	270	.9296	355	.9030
105	.9844	190	.9556	275	.9280	360	.9014
110	.9827	195	.9539	280	.9264	365	.8999
115	.9809	200	.9523	285	.9248	370	.8984
120	.9792	205	.9507	290	.9233	375	.8969
125	.9775	210	.9490	295	.9217	380	.8953
130	.9758	215	.9474	300	.9201	385	.8938
135	.9741	220	.9458	305	.9185	390	.8923
140	.9724	225	.9441	310	.9169	395	.8908
						400	.8893

GROUP 1.   SPECIFIC GRAVITY AT 60° F., 0.850 TO 0.966

t	M	t	M	t	M	t	M
60	1.0000	145	0.9667	230	0.9345	315	0.9034
65	.9980	150	.9647	235	.9326	320	.9016
70	.9960	155	.9628	240	.9307	325	.8998
75	.9940	160	.9608	245	.9289	330	.8980
80	.9921	165	.9590	250	.9270	335	.8962
85	.9901	170	.9570	255	.9252	340	.8945
90	.9881	175	.9551	260	.9234	345	.8927
95	.9861	180	.9532	265	.9215	350	.8909
100	.9841	185	.9513	270	.9197	355	.8892
105	.9822	190	.9494	275	.9179	360	.8874
110	.9803	195	.9476	280	.9160	365	.8856
115	.9783	200	.9457	285	.9142	370	.8839
120	.9763	205	.9438	290	.9124	375	.8821
125	.9744	210	.9419	295	.9106	380	.8804
130	.9724	215	.9401	300	.9088	385	.8786
135	.9705	220	.9382	305	.9070	390	.8769
140	.9686	225	.9363	310	.9052	395	.8752
						400	.8734

GROUP 00.   TAR PRODUCTS, A.A.S.H.O.

GRADES RT-5, RT-6, RT-7, RT-8, RT-9, RT-10, RT-11, RT-12, RTCB-5, RTCB-6

t	M	t	M	t	M	t	M
60	1.0000	105	0.9867	155	0.9723	205	0.9583
65	.9985	110	.9852	160	.9709	210	.9569
70	.9970	115	.9838	165	.9695	215	.9556
75	.9955	120	.9823	170	.9681	220	.9542
80	.9940	125	.9809	175	.9667	225	.9528
85	.9926	130	.9794	180	.9653	230	.9515
90	.9911	135	.9780	185	.9639	235	.9501
95	.9896	140	.9766	190	.9625	240	.9488
100	.9881	145	.9751	195	.9611	245	.9474
		150	.9737	200	.9597	250	.9461

* From *Principles of Highway Construction, Public Roads Administration.*

TABLE 51. AMOUNTS OF MATERIAL PER SQUARE YARD FOR A TYPICAL PENETRATION MACADAM SURFACE *

	BASE		SURFACE	
	Size	Amount	Size	Amount
Coarse stone	3 to 2 in.	285 lb.	2½ to 1½ in.	270 lb.
Bitumen		1.85 gal.		1.5 gal.
Medium stone	1 to ¾ in.	30 lb.	¾ in. to No. 4	30 lb.
Bitumen		0.3 gal.		0.5 gal.
Fine stone	1 to ¾ in.	25 lb.	¾ in. to No. 4	25 lb.
Bitumen				0.3 gal.
Stone chips	½ in. to No. 4	10 lb.	⅜ in. to No. 8	15 lb.
Do			⅜ in. to No. 8	10 lb.
Total aggregate		350 lb.		350 lb.
Total bitumen		2.15 gal.		2.3 gal.

* *From Principles of Highway Construction, Public Roads Administration.*

SEWERS, DRAINS, WATER, GAS, ETC.
POLLUTION CONTROL WORKS

Inspector's Equipment

Complete contract or working drawings and specifications of all structures, copy of contractors' working schedule, plotted boring logs; also approved detailed shop drawings of reinforcement, structural, architectural, mechanical, electrical, and equipment installation work.

Suitable field office with heat, lockers, telephone, water, toilet, safe place for instruments and records where needed, and watchman service.

Procedure in Inspection

Be sure that excavation operations do not endanger or undermine other structures, public or private, including water mains, sewers, drains, gas, pole lines, etc.

When excavating in running or boiling sands watch for expected loss of ground and lowering of ground surface; use of well points or other methods for lowering of ground water is indicated.

Insist on adequate sheeting, bracing, and drainage.

Adequately support underground structures such as pipe lines and conduits.

Report the necessity of underpinning foundations of buildings, walls, bridge piers or abutments, and similar existing structures which are close to influence slope lines of earth rupture.

In deep excavations check that adequate bracing and shoring of sheeting is being done in accordance with approved plans indicating structural stability.

Check equipment for grit channels, collectors or drags, link chains, sprockets, grit elevators and washing devices, bins for receiving grit, and lifting cranes; test their operation and bin discharge gates, also means of agitating or aiding gravity discharge of grit.

Check List for Inspectors

Check permits, code requirements, and ordinances.

Accurate placing of reinforcement in concrete masonry.

Follow concrete design mix.

Check keys and water stops as per plans.

Check lines and grades.

Check setting of anchor and other bolts.

Check wall castings for valves and sluice gates, and wall sleeves for other apparatus (avoid temporary openings in concrete masonry).

Check sluice gates and valves with operating mechanisms for compliance with specifications.

Protective coatings to be applied to concrete and other surfaces as specified.

Check structural steel framing, sizes, riveting, welding, or bolting.

Check welding.

Masonry, brick, and tile work finished as called for.

Window sash, frames, doors, and hardware as specified.

Roofing and other details as specified.

Lighting, communication, and alarm systems as per specifications.

Shop and/or certified tests on pumps, motors, engines, and similar operating mechanisms requiring performance proof as specified.

Require all equipment to be covered and properly protected to prevent injury from the elements, water, dirt, and construction operations.

Bolt settings to be done by equipment manufacturer's templates.

The bottom of equipment or its base and foundation should be cleaned before setting and grouting. Grout, usually 1 to 1, must completely fill the space between the equipment base and the foundation, as a rule about 1 inch. Remove the leveling wedges only after grout has set; then fill the voids with grout. Trowel exposed grout to a neat finish.

See that electrical appliances that have been subjected to injury by water are thoroughly dried out and put through a special dielectric test or replaced.

See that metal ladders are safely installed, also that manhole steps are not too far apart (12 to 14 in.).

Floor gratings and manhole covers should be machined for even seating.

All pipe lines to be cleaned before use after pressure or other tests. Open ends to be temporarily plugged while work is suspended to prevent entrance of foreign matter.

Check on proper facing and machining, proper bearing, and seals for gates.

Reject unsound timber and lumber or any that has been seriously injured by cuts. Kind and quality to be as specified.

Reject treated lumber that has been injured to the extent of impairing structural or preservative properties.

See that cuts, chamfers, holes, and minor injuries to treated lumber are thoroughly saturated with hot creosote and covered by hot asphaltum unless otherwise specified. The same applies to treated timber piles also.

Check on proper means of fastening timbers.

In painting metal, machinery, and equipment see that rust, loose scale, oil, grease, and dirt are removed by washing with benzine or other solvent before applying approved paints.

Equipment having a standard machinery finish to be shop or field painted the specified color. For all other painting follow specifications and manufacturer's instructions.

See that all finished surfaces and equipment are protected by drop cloths to prevent spattering with paint.

Check specification for required galvanized work, rejecting any that has been damaged.

Check on adjustments and calibration of venturi metering and other indicating recording and metering devices.

Check on landscaping and grass seeding, topsoiling, and preparation and fertilization of soil.

Excavation and Foundations

Require safe fences or guard rails around excavations.

Check blasting operations for observation of safety protection regulations and competent licensed blasters.

Check on materials and pile driving or other type of foundation operations.

If subsurface conditions are different from those anticipated and provided for, and seem to require a change, report them to your superior.

If work is in a street or road, check that provisions are made for maintaining or rerouting traffic, and also for access to individual properties.

Require proper street drainage, and prohibit pumping muddy or silty water into catch basins or sewers without first pumping into a settling box.

Avoid excessive surface loadings close to excavations such as storage of excavated material which will create a surcharge exerting high pressures against sheeting.

Limit the width of trench excavations for pipe lines to avoid excessive loads on the pipes; maximum width at top of pipe should not exceed its outside diameter plus 2.5 to 3.0 ft.

Check on subgrade conditions; poor soil, if shallow, calls for consolidation with broken stone or gravel; if deep, perhaps bearing piles are required. In wet clayey bottom a layer of straw will often improve conditions.

Completely backfill trenches or other excavations as per specifications. The use of large stone should be prohibited, and the nesting of small ones avoided.

Restore backfilled or disturbed surface to original condition, or place temporary or permanent pavement in streets or roads; see specifications.

Structures

Check with specifications and plans for requirements on such items as heating and ventilation, plumbing, and power- and blowerhouse equipment; also other details such as workshop, laboratory, chemical building, storage and feed equipment, grit washing and storage, and sludge handling and/or drying equipment.

Check gas-utilization equipment to be certain that drainage traps are correctly installed and that valves are tightly packed to prevent gas escape.

Make sure that overflow weirs in tanks are level.

Where stone and/or sand filters are included check size, grading, kind, and cleanness of broken stone and sands actually tested by laboratory for effective size and uniformity coefficient before delivery to job, and repeat tests often enough to be sure that specifications are being complied with, or make tests in the field; see p. 194 for procedure.

Mechanical

All piping should be of the kind and size shown on plans and should be firmly supported in the locations shown on plans.

Pipe railings to be welded or connected to support stanchions by standard fittings as specified. Base flanges to be rigidly held by expansion bolts to masonry or framing, or stanchions may rest in pipe sleeves embedded in masonry.

Conduits, saddles, boxes, cabinets, sleeves, inserts, foundation bolts, anchors, and similar work in walls, floors, or elsewhere should be provided for as specified.

Nuts for bolts should be American standard, heavy.

Pumps, engine generators, blowers, and all other operating equipment and apparatus should be installed according to approved shop and working drawings.

Float and limit switches and other mechanically actuated switches enclosed in cast-metal boxes should be installed in suitable locations for rigid conduit connections.

Check on scheme of control sequence and timing of operations as required by plans.

Suitable tools and appliances needed to adjust, maintain, or repair each piece of equipment should be installed in a convenient, accessible location.

Necessary or specified spare parts should be provided for equipment as specified.

Be sure that the valves and gates are of the types indicated with rising or non-rising stems as specified.

Mechanical—Painting

Check against specification.

Mechanical—Piping

See that all drilling of bolt holes matches companion equipment for center-line or staggered drilling.

Check on lining or coatings of pipe lines where such are specified. Insist on careful handling of coated pipe to prevent damage thereto.

See that piping conforms to the required class or schedule.

Check castings for shape, dimensions, and thicknesses specified.

DISCUSSION

Mechanical

Check that operating stems and hydraulic operating cylinders meet specifications.

Check that metals of specified composition, quality, and strength, such as cast iron, cast steel, steel, galvanized steel, and bronzes, are as called for.

Avoid low-grade castings where subject to rugged use; use cast steels alloy-treated for sub-zero temperatures and shock conditions.

Give particular attention to composition of bronzes both forged and cast, considering workability and durability, and avoidance of dezincification.

Metals in contact should be of the types that "get along" together.

See that each piece of equipment for which pressure, duty, capacity, rating, efficiency, performance, function, or special requirement is specified is tested in the shop of the maker in a manner to prove that its required characteristics are fully complied with and certified to, if required.

Shafts of direct driven equipment, such as pumps, sluice gates, and similar mechanisms, to be aligned accurately by experienced mechanics.

Sewage regulating and diversion chamber equipment and tide gates sometimes required on combined sewerage systems should meet specifications.

Specialized equipment for emplacement in such structures as pumping stations, grit chambers, flocculators, settling tanks, and digestors should be installed under the supervision of the manufacturer and should be field-tested.

Where aluminum is called for, alloys and properties should conform to specifications. Check against any local conditions unfavorable to its use, such as salt water or electrolysis.

Motor-operated valves and sluice gates to be equipped as required for the service specified.

In sluice gates look for seating or unseating types, position of wedges, and conservative spacing of bolts between frame and wall castings; check for good details on attachment of facing strips, e.g., dovetailed and securely fastened by heavy bronze machine screws.

Where sluice gates are to be hydraulically operated, check details of operating cylinder, check minimum water supply pressure, and see that adequate support (preferably independent of the gates guides) is provided; also, where opening and closing are automatic from flowing water float control levels, see that proper water supply connection is made to the cylinder through a damping system to coils, valves, strainers, etc., to reduce hunting effect.

Red Light. Long water lines from a main will cause a serious reduction in water pressure at the hydraulically operated mechanisms.

Check equipment such as trash racks, bar screens, grinders, conveyors for flocculators, settling tanks, trickling filters, aeration tanks, sludge digestors, chlorination facilities, screening and sludge handling, dewatering and disposal, pumping facilities, engine generators, and gas handling and storage with their accessories.

Check the specifications and plans for requirements on such items as heating and ventilation, plumbing, power- and blowerhouse equipment and apparatus, workshop, laboratory, chemical building, storage and feed equipment, and sludge handling and/or drying equipment.

Mechanical—Piping

See that check valves are provided with counterbalancing devices to prevent slamming.

Electrical

For electrical installations the inspector should be provided with approved drawings which give dimensioned outlines showing provision for and location of conduit connections, also detailed and schematic wiring diagrams, installation layouts, material schedules, and test and descriptive data.

Follow the plans for proper location of all electrical device controls and other apparatus.

Electrical equipment with operating devices should be completely wired, and all appliances and conduits should be integral therewith; connections should terminate in a junction box of ample size. A special electrical inspector may be required.

Check that electrical equipment vendors or manufacturers maintain service stations or spare parts stocks where replacements for equipment furnished by them may be obtained on short notice.

See that electric motors are of the types called for and have the required horsepower rating and other specified characteristics.

Check that cranes, hoists, elevators, and other equipment are provided with complete control equipment, accessories, and safety devices for safe and efficient operation. Check for spare parts.

Lumber

Lumber should be of form, grade, and dressing specified and should have preservative treatment where subjected to alternate wetting and drying as specified; otherwise creosote treatment should be by the empty cell process (A.W.P.A.-35) except that timber and lumber used in tide-water structures should be done by the full cell process.

Minimum retention of preservative in the empty cell process is 12 lb. per cu. ft. of wood; in the full cell process it is 16 lb. per cu. ft. for timbers 5 in. thick and less, and 12 lb. per cu. ft. for larger timbers.

Landscaping

Mid-August through September is usually the most favorable for seeding; the next best is early spring.

Unless specified, usual good perennial grasses are Kentucky Blue Grass, creeping fescues, and bents, depending on location; red top and rye grass are used for nurse grass.

For tree planting check with local conditions as to most favorable season. Proper preparation of soil is important in tree planting. Adequate watering for tree and grass seeding is important; see specifications.

PRECAST CONCRETE PIPE MANUFACTURE

CHECK LIST FOR INSPECTORS

See that:

Plans and specifications are followed for kind of pipe required.

Forms are clean, true, and oiled, and that bolted joints match up.

Bell and spigot ring castings are true, machined to true circle if rubber joint gaskets are called for; suitable means are provided for maintaining accurate positions of inner and outer forms for uniform wall thickness.

Welded steel wire fabric has required area per foot of pipe in ring wires, which are usually spaced 2 or 3 in. whereas tie wires are usually spaced 8 in. Cutting of wires for any purpose is prohibited.

Laps in circular wires are ample, and placed so that they will be at the eighth points of circle when pipe is cast; wire is tied or welded on both sides after rust is removed by wire brushing.

Tied or welded laps are staggered where two sections of wire fabric are used.

Wire fabric is shaped to proper curve on steel rolls; use of intermediate or other grade of steel is prohibited.

Metal chairs or spacers are used for proper position of reinforcement.

Indented letters are cast in outer surface of pipe to indicate its *top* in final laid position, and small hole on this line in wall is centered for lifting eye bolt if permitted.

Cement (not air-entraining), sand, stone (or gravel), and water comply with specifications, also proportions and means for measuring quantities for each batch; slump down to 3 in. except machine cast pipe zero to 1 in. Coarse aggregate ¾ in. graded stone or gravel, except grits used for machine compacted pipe.

Internal vibration during continuous pour but excessive use in any place avoided.

Spigot ring properly placed at end of barrel; place concrete without delay and work it in, smoothing off with hand trowel.

Pipe is marked with project identification, consecutive number, producer's name, diameter, date cast, inspector's approval mark.

Proper protection, steam curing, and time elapse before removal of forms is observed, especially in cold weather.

Test cylinders for 28-day compression tests of concrete conform to proper handling, storing, and curing.

In making 3-edge load bearing tests on 28-day-old pipe, A.S.T.M. is followed and the 0.01 in. gage leaf is not forced into crack.

Rubber gaskets, if used, comply with specifications, also cemented on pipe spigot one day in advance of laying and protected by a cover from hot sun; rubber cement is not smeared on concrete surface outside of gasket preventing bonding of cement mortar.

SANITARY AND DRAINAGE CONSTRUCTION
PIPE LAYING
CHECK LIST FOR INSPECTORS

Inspector's Equipment

Complete set of final or contract plans, specifications, and approved shop drawings.

6-ft. rule, 50-ft. tape, and mason's level.

Plumb bob and line.

Calipers.

Shop or Plant Inspection

See p. 246 for Precast Concrete Pipe Manufacture. Waive shop inspection and require certified mill reports except where specifications require load bearing or other tests for cast-iron pipe, vitrified-clay sewer pipe, asbestos-cement pipe, and corrugated metal pipe.

Procedure in Job Inspection

Check all pipe delivered to job for conformity with requirements such as diameter, length, wall thickness, joint detail, and general uniformity of shape. Compare with tables pp. 251 to 256.

Check all pipe and fittings for cracks or other serious defects or damaged material.

Interior of precast concrete pipe should be reasonably smooth except for shallow air pits. Exterior roughness is common and usually not important.

When precast reinforced-concrete pipe is delivered to the job without having a representative load test made to check its compliance with requirements, one pipe in every 200 or fraction thereof in each kind and diameter shall be broken with sledges to check the size, spacing, and position of the reinforcing and the quality of the concrete.

Accept no elliptically reinforced-concrete pipe unless the top (or bottom) is clearly indicated by an indentation in the exterior surface of the concrete such as by the letter "T" or a 2-in. hole at mid-length for a lifting eyebolt. When pipe is being installed in the trench, require these points to line up on the center line of the work.

The spacing of cross bracing in sheeted excavations is dependent on the clearance required for pipes which are to be lowered therein or other similar use.

Report any unsatisfactory subgrade condition not anticipated or provided for which may require special correctives such as removal of unsound material and its replacement with compacted stone or gravel for consolidation thereof, concrete cradle, or pile supports.

Insist on dry trench during laying and jointing of pipes.

Check proper bedding for pipes; see specifications for concrete cradle or shaped subgrade for even bearing of pipe.

Permit no change from the type of pipe bedding or support called for; such change affects load bearing strength of pipe.

See that grade boards are set firmly not more than 25 ft. apart, and that grade string is drawn taut and kept so.

If contractor is responsible for establishing working lines and grades, verify accuracy of such working basis and require care in preserving all points against damage.

Be sure that joints of pipe are clean before each pipe length is laid, and require spigots and tongues to be inserted to full depth of bell or socket, and to be so centered and supported that interior surfaces will match, with uniform annular spaces.

Check each length as laid for size, position, type, line, and grade.

Do not allow trench backfill to be placed until leakage test has been made unless specifications permit an alternative infiltration test, in which case sufficient lapse of time should be allowed for return of ground water to normal level.

Require backfill to be compactly placed in manner specified with overfill at ground surface. No stone to be in contact with pipes.

Common practice requires laying of bell-and-spigot pipe with bells headed upstream, but ordinarily it is unimportant whether they are upstream or downstream.

Red Light

See that no sudden flooding condition can arise to cause flotation of pipe such as in making a leakage test with ends closed with bulkheads.

Water, Gas, Ducts, Etc.

See preceding notes for applicable checks.

Keep dirt and foreign matter out of pipe lines during process of laying.

Where pipe bends or tees occur, also in back of hydrants, require proper concrete backing to prevent the opening of joints from internal water pressure.

Where pipe lines are laid in filled or disturbed ground require that suitable supports be furnished such as piers or piles resting on firm material.

Test pipe lines for air or watertightness in the manner required.

In making leakage tests with air pressure, use a soapy solution for painting joints. Bubbles will reveal leaks.

Sterilize water mains and water storage tanks with chlorine before placing in service. See specifications.

Joints

Precast concrete pipes: Portland cement mortar (preferably poured using an exterior band of strong fabric 12 in. wide to prevent escape of mortar for large tongue-and-groove pipe); hand-packed mortar with gaskets in bell-and-spigot pipes; rubber gaskets for hand-cast pipes (machined end forms) with or without mortar packing. Use a "come-along" for bringing pipes home where joints are made with rubber gaskets. *Allow no mortaring of joints while laying of pipe; otherwise fresh joints will be damaged by shifting of pipes.*

Clay sewer pipe: preferably hot poured bituminous compound using a snake after solidly rammed twisted hemp or oakum gasket is placed.

Cast-iron pipe, bell and spigot: calked lead following a twisted gasket, except where Leadite is permitted.

Cast-iron pipe, flanged: carefully bolted, full dimension bolts with suitable gaskets for service required.

Steel pipe, threaded and coupled, screw flanged and bolted (with gaskets) or welded. Joints to be properly shaped and cleaned before welding.

Asbestos-cement pipe, sleeve type with rubber seals supplied by manufacturer.

Corrugated metal pipe, connecting or coupling bands.

REPORT ON PIPE LAYING

Engineer

REPORT ON CLAY AND CONCRETE PIPE
SHOP INSPECTION

Material _____

Project _____

Producer _____

Contractor _____ Date _____

					Reported to
Sample taken from					
Quantity represented					
Marks on sample					
Sampled by					
Date pipe was made					
Date received at lab.					
Job Sample No.					
Laboratory report No.					

				Required Min.	Required Max.
Type or class of pipe, ASTM Des.					
Internal diameter of pipe, in.					
External diameter of pipe, in.					
Min. and max. wall					
Internal diameter of bell, in.					
External diameter of bell, in.					
Thickness of bell, in.					
Length of spigot					
Length of bell					
Total length of pipe, ft.					
Total length of bearing, ft.					
Load applied at first crack, lb.					
Load per lineal foot at first crack, lb.					
Load at 0.01 crack					
Load per lineal foot at 0.01 crack					
Load required per lineal foot at 0.01 crack					
Maximum load applied, lb.					
Maximum load required (ultimate)					
Maximum load per lineal foot, lb.					

ABSORPTION TEST

Weight after immersion, grams					
Weight after drying, grams					
Loss of weight					
Absorption, %					

REINFORCEMENT

Spacing of circular wires, welded fabric					
Area of circular reinforcing per ft. of pipe, sq. in.					
Gages of circular and longitudinal wires					
Spacing longitudinal wires, in.					
Total area of longitudinals, sq. in.					

Remarks:

The above tests $\begin{smallmatrix}do\\do\ not\end{smallmatrix}$ fulfill ASTM Spec. _____

_____

Inspector

TABLE 52. CLAY PIPE, STANDARD STRENGTH, A.S.T.M. C-13

Size, in.	Laying Length		Maximum Difference in Length of Two Opposite Sides, in.	Outside Diameter of Barrel, in.		Inside Diameter of Socket at 1/2 in. above Base, in.		Depth of Socket, in.		Thickness of Barrel, in.		Thickness of Socket at 1/2 in. from Outer End, in.		Average Strength Requirements, min., lb per lin. ft.		Weight per foot of Pipe *
	Nominal, ft.	Limit of Minus Variation,[a] in. per ft. of length		Min.	Max.	Min.	Max.	Nominal	Min.	Nominal	Min.	Nominal	Min.	Three-Edge-Bearing Method	Sand-Bearing Method	
4	2, 2 1/2, 3	1/4	5/16	4 7/8	5 1/8	5 3/4	6 1/8	1 3/4	1 1/2	1/2	7/16	7/16	3/8	1000	1430	9.0
6	2, 2 1/2, 3	1/4	3/8	7 1/16	7 7/16	8 3/16	8 5/8	2 1/4	2	5/8	9/16	1/2	7/16	1000	1430	15.5
8	2, 2 1/2, 3	1/4	7/16	9 1/4	9 3/4	10 1/2	11	2 1/2	2 1/4	3/4	11/16	9/16	1/2	1000	1430	23.8
10	2, 2 1/2, 3	1/4	7/16	11 1/2	12	12 3/4	13 1/4	2 5/8	2 3/8	7/8	13/16	5/8	9/16	1100	1570	33.8
12	2, 2 1/2, 3	1/4	7/16	13 3/4	14 5/16	15 1/8	15 3/4	2 3/4	2 1/2	1	15/16	3/4	11/16	1200	1710	46.8
15	3, 4	1/4	1/2	17 3/16	17 13/16	18 5/8	19 1/4	2 7/8	2 5/8	1 1/4	1 1/8	15/16	7/8	1400	2000	67.7
18	3, 4	1/4	1/2	20 5/8	21 7/16	22 1/4	23	3	2 3/4	1 1/2	1 3/8	1 1/8	1 1/16	1700	2430	97.7
21	3, 4	1/4	9/16	24 1/8	25	25 7/8	26 3/4	3 1/4	3	1 3/4	1 5/8	1 5/16	1 3/16	2000	2860	139
24	3, 4	3/8	9/16	27 1/2	28 1/2	29 3/8	30 3/8	3 3/8	3 1/8	2	1 7/8	1 1/2	1 3/8	2400	3430	180
27	3, 4	3/8	9/16	31	32 1/8	33	34 1/8	3 1/2	3 1/4	2 1/4	2 1/8	1 11/16	1 9/16	2750	3930	
30	3, 4	3/8	5/8	34 3/8	35 5/8	36 1/2	37 3/4	3 5/8	3 3/8	2 1/2	2 3/8	1 7/8	1 3/4	3200	4570	277
33	3, 4	3/8	5/8	37 5/8	38 15/16	39 7/8	41 1/4	3 3/4	3 1/2	2 5/8	2 1/2	2	1 13/16	3500	5000	
36	3, 4	3/8	1 1/16	40 3/4	42 1/4	43 1/4	44 3/4	4	3 3/4	2 3/4	2 5/8	2 1/16	1 7/8	3900	5570	392

[a] There is no limit for plus variation.

* *From Robinson Clay Products Co.*

TABLE 53. CLAY PIPE, EXTRA STRENGTH, A.S.T.M. SPEC. C-200

Nominal Size, in.	Laying Length: Nominal, ft.	Laying Length: Limit of Minus Variation,[a] in. per ft. of length	Maximum Difference in Length of Two Opposite Sides, in.	Outside Diameter of Barrel,[b] in.: Min.	Outside Diameter of Barrel,[b] in.: Max.	Inside Diameter of Socket at 1/2 in. above Base, in.: Min.	Inside Diameter of Socket at 1/2 in. above Base, in.: Max.	Depth of Socket, in.: Nominal	Depth of Socket, in.: Min.	Thickness of Barrel, in.: Nominal	Thickness of Barrel, in.: Min.	Thickness of Socket at 1/2 in. from Outer End, in.: Nominal	Thickness of Socket at 1/2 in. from Outer End, in.: Min.	Average Strength Requirements, min., lb. per lin. ft.: Three-Edge-Bearing Method	Average Strength Requirements, min., lb. per lin. ft.: Sand-Bearing Method	Weight per Foot of Pipe *
6	2, 2½, 3	¼	⅜	7 1/16	7 7/16	8 3/16	8⅝	2¼	2	1 1/16	9/16	½	7/16	2000	2850	16
8	3	¼	7/16	9¼	9¾	10½	11	2½	2¼	⅞	¾	9/16	½	2000	2850	26
10	3	¼	7/16	11½	12	12¾	13¼	2⅝	2⅜	1	⅞	⅝	9/16	2000	2850	38
12	3	¼	7/16	13¾	14 5/16	15⅛	15¾	2¾	2½	1 3/16	1 1/16	¾	1 1/16	2250	3200	54
15	3, 4	¼	½	17 3/16	17 13/16	18⅝	19¼	2⅞	2⅝	1½	1⅜	15/16	⅞	2750	3925	78
18	3, 4	¼	½	20⅝	21 7/16	22¼	23	3	2¾	1⅞	1¾	1⅛	1 1/16	3300	4700	124
21	3, 4	¼	9/16	24⅛	25	25⅞	26¾	3¼	3	2¼	2	1 5/16	1 3/16	3850	5500	170
24	3, 4	⅜	9/16	27½	28½	29⅜	30⅜	3⅜	3⅛	2½	2¼	1½	1⅜	4400	6300	214
30	3, 4	⅜	⅝	34⅜	35⅝	36½	37¾	3⅝	3⅜	3	2¾	1⅞	1¾	5000	7100	323
36	3, 4	⅜	1 1/16	40¾	42¼	43¼	44¾	4	3¾	3½	3¼	2 1/16	1⅞	6000	8575	444

[a] There is no limit for plus variation.

[b] The bells of reduced-diameter pipe are not large enough to receive the spigots of full-diameter pipe, so check to see that job deliveries are either one or the other exclusively. For full-diameter pipe see Federal Spec. SS-P-361a.

* *From Robinson Clay Products Co.*

TABLE 54. NON-REINFORCED-CONCRETE SEWER PIPE, A.S.T.M. C-14

Internal Diameter, D, in.	Laying Length, L, ft.	Inside Diameter at Mouth of Socket, D_s, in.[a]	Depth of Socket, L_s, in.	Minimum Taper of Socket, H	Thickness of Barrel, T, in.	Thickness of Socket, T_s	Average Strength, lb. per lin. ft.		Maximum Absorption, %	Limits of Permissible Variation in:				
							Three-Edge-Bearing Method	Sand-Bearing Method		Length, in. per ft. (−)[b]	Internal Diameter, in. Spigot (±)[b]	Internal Diameter, in. Socket, (±)[b]	Depth of Socket, in. (−)[b]	Thickness of Barrel, in. (−)[b]
4	2, 2½, 3	6	1½	1:20	9/16	The thickness of the socket ¼ in. from its outer end shall be not less than ¾ of the thickness of the barrel of the pipe.	1000	1500	8	¼	⅛	⅛	⅛	1/16
6	2, 2½, 3	8¼	2	1:20	⅝		1100	1650	8	¼	3/16	3/16	¼	1/16
8	2, 2½, 3, 4	10¾	2¼	1:20	¾		1300	1950	8	¼	¼	¼	¼	1/16
10	2, 2½, 3, 4	13	2½	1:20	⅞		1400	2100	8	¼	¼	¼	¼	1/16
12	2, 2½, 3, 4	15¼	2½	1:20	1		1500	2250	8	¼	¼	¼	¼	1/16
15	2, 2½, 3, 4	18¾	2½	1:20	1¼		1750	2620	8	¼	¼	¼	¼	3/32
18	2, 2½, 3, 4	22¼	2¾	1:20	1½		2000	3000	8	¼	¼	¼	¼	3/32
21	2, 2½, 3, 4	26	2¾	1:20	1¾		2200	3300	8	¼	5/16	5/16	¼	⅛
24	2, 2½, 3, 4	29½	3	1:20	2⅛		2400	3600	8	⅜	5/16	5/16	¼	⅛

[a] When pipes are furnished having an increase in thickness over that given in last column, then the diameter of socket shall be increased by an amount equal to twice the increase of thickness of barrel.

[b] The minus sign (−) alone indicates that the plus variation is not limited; the plus-and-minus sign (±) indicates variation in both excess and deficiency in dimension.

TABLE 55. REINFORCED-CONCRETE SEWER PIPE, A.S.T.M. C-75

Internal Diameter, in.	Strength Test Requirements, lb. per lin. ft.				Minimum Design Requirements [a]					
	Three-Edge-Bearing Method		Sand-Bearing Method		Concrete, 3000 psi.		Concrete, 3500 psi.		Concrete, 4000 psi.	
	Load to Produce a 0.01-in. Crack	Ultimate Load	Load to Produce a 0.01-in. Crack	Ultimate Load	Shell Thickness, in.	Total Steel Area, sq. in. per lin. ft.	Shell Thickness, in.	Total Steel Area, sq. in. per lin. ft.	Shell Thickness, in.	Total Steel Area, sq. in. per lin. ft.
12	1,800	2,700	2,700	4,050	2	1 line 0.06	1¾	1 line 0.07	....	....
15	2,000	3,000	3,000	4,500	2¼	1 line 0.06	2	1 line 0.07	....	....
18	2,200	3,300	3,300	4,950	2½	1 line 0.06	....	....	2	1 line 0.07
21	2,400	3,600	3,600	5,400	2¾	1 line 0.06	....	....	2¼	1 line 0.07
24	2,400	3,600	3,600	5,400	3	1 line 0.06	2⅝	1 line 0.08	2½	1 line 0.09
27	2,550	3,800	3,800	5,700	3	1 line 0.07	2¾	1 line 0.10	2⅝	1 line 0.12
30	2,700	4,050	4,050	6,100	3½	1 line 0.09	3	1 line 0.12	2¾	1 line 0.14
33	2,850	4,300	4,300	6,400	3¾	1 line 0.11	3¼	1 line 0.14	2¾	1 line 0.17
36	3,000	4,500	4,500	6,750	4	2 lines b totalling 0.14	3⅜	2 lines b totalling 0.20	3	2 lines b totalling 0.23
42	3,200	4,800	4,800	7,200	4½	2 lines b totalling 0.16	3¾	2 lines b totalling 0.23	3⅜	2 lines b totalling 0.27
48	3,400	5,100	5,100	7,650	5	2 lines b totalling 0.21	4¼	2 lines b totalling 0.27	3¾	2 lines b totalling 0.32
54	3,700	5,550	5,550	8,300	5½	2 lines b totalling 0.25	4⅝	2 lines b totalling 0.32	4¼	2 lines b totalling 0.38
60	4,000	6,000	6,000	9,000	6	2 lines b totalling 0.29	5	2 lines b totalling 0.38	4½	2 lines b totalling 0.44
66	4,250	6,350	6,350	9,550	6½	2 lines b totalling 0.32	5⅜	2 lines b totalling 0.44	4¾	2 lines b totalling 0.47
72	4,500	6,750	6,750	10,100	7	2 lines b totalling 0.36	5¾	2 lines b totalling 0.47	5	2 lines b totalling 0.55
78	....	....	....	....	7½	2 lines b totalling 0.40	....	....	....	....
84	....	....	....	....	8	2 lines b totalling 0.43	....	....	....	....
90	....	....	....	....	8	2 lines b totalling 0.49	....	....	....	....
96	....	....	....	....	8½	2 lines b totalling 0.57	....	....	....	....
108	....	....	....	....	9	2 lines b totalling 0.67	....	....	....	....

a The distance from the center line of the reinforcement to the nearest surface of the concrete has been assumed in the design tables as 1 in.
b Where two lines of steel are specified, a single line placed elliptically may be used, and the area of this shall be at least 50% of the total steel area specified in the design table.

TABLE 56. STANDARD STRENGTH REINFORCED-CONCRETE CULVERT PIPE, A.S.T.M. C-76

Internal Diameter of Pipe, in.	Concrete, 3,500 psi.				Concrete, 4,500 psi.				Strength Test Requirements, lb. per lin. ft. of pipe	
	Minimum Shell Thickness, in.	Minimum Reinforcement,[a] sq. in. per lin. ft. of pipe barrel [b]		Weight per Lin. Ft., lb.[c]	Minimum Shell Thickness, in.	Minimum Reinforcement,[a] sq. in. per lin. ft. of pipe barrel [b]		Weight per Lin. Ft., lb.[c]	Three-Edge-Bearing Method [d]	
		Circular Reinforcement in Circular Pipe	Elliptical Reinforcement in Circular Pipe and Circular Reinforcement in Elliptical Pipe			Circular Reinforcement in Circular Pipe	Elliptical Reinforcement in Circular Pipe and Circular Reinforcement in Elliptical Pipe		Load to Produce a 0.01-in. Crack	Ultimate Load
12	2	1 line 0.07	....	88	1¾	1 line 0.08	....	75	2,250	3,500
15	2¼	1 line 0.09	....	125	2	1 line 0.11	....	110	2,625	4,065
18	2½	1 line 0.12	1 line 0.10	160	2	1 line 0.14	....	140	3,000	4,500
24	3	1 line 0.17	1 line 0.13	260	2½	1 line 0.20	1 line 0.17	225	3,000	5,000
30	3½	1 line 0.22	1 line 0.17	370	3	1 line 0.28	1 line 0.21	315	3,375	5,750
36	4	2 lines, each 0.18	1 line 0.18	520	3⅜	2 lines, each 0.22	1 line 0.22	450	4,050	6,600
42	4½	2 lines, each 0.21	1 line 0.21	680	3¾	2 lines, each 0.25	1 line 0.25	560	4,725	7,350
48	5	2 lines, each 0.25	1 line 0.25	850	4¼	2 lines, each 0.31	1 line 0.31	720	5,400	8,000
54	5½	2 lines, each 0.30	1 line 0.30	1,050	4⅝	2 lines, each 0.37	1 line 0.37	880	5,850	9,000
60	6	2 lines, each 0.33	1 line 0.33	1,280	5	2 lines, each 0.41	1 line 0.41	1,060	6,000	10,000
66	6½	2 lines, each 0.37	1 line 0.37	1,480	5½	2 lines, each 0.45	1 line 0.45	.....	6,300	11,000
72	7	2 lines, each 0.40	1 line 0.40	1,835	6	2 lines, each 0.48	1 line 0.48	.....	6,600	12,000
78	7½	2 lines, each 0.43	1 line 0.43	2,150	6½	2 lines, each 0.51	1 line 0.51	.....	.....	......
84	8	2 lines, each 0.46	1 line 0.46	2,300	7	2 lines, each 0.54	1 line 0.54	.....	.....	......
90	8	2 lines, each 0.56	1 line 0.56	2,600	...	....	....	.....	.....	......
96	8½	2 lines, each 0.60	1 line 0.60	2,750	...	....	....	.....	.....	......
102	8½	2 lines, each 0.72	1 line 0.72	3,050	...	....	....	.....	.....	......
108	9	2 lines, each 0.78	1 line 0.78	3,450	...	....	....	.....	.....	......

[a] The distance from the center line of the reinforcement to the nearest surface of the concrete has been assumed in the design tables as 1¼ in. for pipe with a shell 2½ in. or more in thickness.

[b] For 2 lines or elliptical reinf., provide 1-in. cover.

[c] *From Universal Concrete Pipe Co. for tongue and groove pipe.*

[d] Test loads for sand-bearing tests shall be 1½ times those specified in this table for the three-edge-bearing tests.

TABLE 57. EXTRA-STRENGTH REINFORCED-CONCRETE CULVERT PIPE, A.S.T.M. C-76

| Internal Diameter of Pipe, in. | Minimum Shell Thickness, in. | Concrete, 4500 psi. | | Strength Test Requirements, lb. per lin. ft. of pipe | | Weight per Lin. Ft. in Lb.[d] |
| | | Minimum Reinforcement,[a] sq. in. per lin. ft. of pipe barrel [b] | | Three-Edge-Bearing Method [c] | | |
		Circular Reinforcement in Circular Pipe	Elliptical Reinforcement in Circular Pipe and Circular Reinforcement in Elliptical Pipe	Load to Produce a 0.01-in. Crack	Ultimate Load	
24	3	1 line 0.26	1 line 0.20	4,000	6,000	260
30	3½	1 line 0.31	1 line 0.24	5,000	7,500	370
36	4	2 lines, each 0.28	1 line 0.28	6,000	9,000	520
42	4½	2 lines, each 0.33	1 line 0.33	7,000	10,500	680
48	5	2 lines, each 0.38	1 line 0.38	8,000	12,000	850
54	5½	2 lines, each 0.44	1 line 0.44	9,000	13,500	1,050
60	6	2 lines, each 0.50	1 line 0.50	9,000	15,000	1,280
66	6½	2 lines, each 0.56	1 line 0.56	9,500	16,500	1,480
72	7	2 lines, each 0.60	1 line 0.60	9,900	18,000	1,835
78	7½	2 lines, each 0.65	1 line 0.65	.....	......	2,150
84	8	2 lines, each 0.72	1 line 0.72	.....	......	2,300
90	8	2 lines, each 0.84	1 line 0.84	.....	......	2,600
96	8½	2 lines, each 0.90	1 line 0.90	.....	......	2,750
102	8½	2 lines, each 1.08	1 line 1.08	.....	......	3,050
108	9	2 lines, each 1.17	1 line 1.17	.....	......	3,450

[a] The distance from the center line of the reinforcement to the nearest surface of the concrete has been assumed in the design tables as 1¼ in. for pipe with a shell 2½ in. or more in thickness.

[b] For 2 lines or elliptical reinforcement provide 1-in. cover.

[c] Test loads for sand-bearing tests shall be 1½ times those specified in this table for the three-edge-bearing tests.

[d] *From Universal Concrete Pipe Co. for tongue and groove pipe.*

TABLE 58. CONCRETE CRADLE FOR CONCRETE PIPE
DIMENSIONS, CRADLE AREAS AND TRENCH WIDTHS

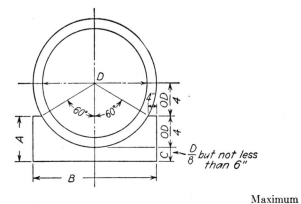

Pipe I.D., in.	Pipe O.D., in.	Cradle Dimensions, in.			Cradle Area, sq. ft.	Maximum Trench Width = O.D. + 3' 0"	
		A	B	C		ft.	in.
12	16	10	21⅞	6.00	1.25	4	4
15	19½	10⅞	24⅞	6.00	1.47	4	7½
18	23	11¾	27⅞	6.00	1.71	4	11
21	26½	12⅝	31	6.00	1.97	5	2½
24	30	13½	33	6.00	2.23	5	6
27	33½	14⅜	37	6.00	2.50	5	9½
30	37	15¼	40	6.00	2.78	6	1
36	44	17	46⅛	6.00	3.38	6	8
42	51	18¾	52½	6.00	4.02	7	3
48	58	20½	58¼	6.00	4.70	7	10
54	65	23	64⁵⁄₁₆	6.75	5.77	8	5
60	72	25½	70⅜	7.50	6.94	9	0
66	79	28	76⁷⁄₁₆	8.25	8.20	9	7
72	86	30½	82½	9.00	9.58	10	2
78	93	33	88½	9.75	11.07	10	9
84	100	35½	94⅝	10.50	12.66	11	4
90	107	38	100⅝	11.25	14.36	11	11
96	114	40½	106¾	12.00	16.16	12	6
102	121	43	112¹³⁄₁₆	12.75	18.07	13	1
108	128	45½	118⅞	13.50	20.08	13	8
114	135	48	124⅞	14.25	22.21	14	3
120	142	50½	131	15.00	24.43	14	10

TABLE 59.* STANDARD THICKNESSES AND WEIGHTS OF CAST IRON PIT CAST PIPE

Note. These weights are for pipe laid without blocks, on flat bottom trench, with tamped backfill, under 5 ft. of cover.

Size, in.	Class 50 50 Lb. Pressure 115 Ft. Head			Class 100 100 Lb. Pressure 231 Ft. Head		
	Thickness, in.	Wt. Based on 12-Ft. Length †		Thickness, in.	Wt. Based on 12-Ft. Length †	
		Avg. per Foot	Per Length		Avg. per Foot	Per Length
3	0.37	14.2	170	0.37	14.2	170
4	0.40	19.2	230	0.40	19.2	230
6	0.43	30.0	360	0.43	30.0	360
8	0.46	42.9	515	0.46	42.9	515
10	0.50	57.1	685	0.50	57.1	685
12	0.54	73.3	880	0.54	73.3	880
14	0.54	85.4	1025	0.58	91.3	1095
16	0.58	105.4	1265	0.63	113.3	1360
18	0.63	127.9	1535	0.68	136.7	1640
20	0.66	148.8	1785	0.71	158.8	1905
24	0.74	198.8	2385	0.80	212.9	2555
30	0.87	288.3	3460	0.94	311.3	3735
36	0.97	384.2	4610	1.13	449.2	5390
42	1.07	497.5	5970	1.16	542.1	6505
48	1.18	625.8	7510	1.37	726.3	8715
54	1.30	777.1	9325	1.51	906.3	10875
60	1.39	922.5	11070	1.62	1077.1	12925

Size, in.	Class 150 150 Lb. Pressure 346 Ft. Head			Class 200 200 Lb. Pressure 462 Ft. Head		
	Thickness, in.	Wt. Based on 12-Ft. Length †		Thickness, in.	Wt. Based on 12-Ft. Length †	
		Avg. per Foot	Per Length		Avg. per Foot	Per Length
3	0.37	14.2	170	0.37	14.2	170
4	0.40	19.2	230	0.40	19.2	230
6	0.43	30.0	360	0.43	30.0	360
8	0.46	42.9	515	0.46	42.9	515
10	0.54	60.8	730	0.58	64.6	775
12	0.58	77.9	935	0.63	83.8	1005
14	0.63	100.8	1210	0.68	107.9	1295
16	0.68	125.0	1500	0.79	142.5	1710
18	0.79	161.3	1935	0.85	172.1	2065
20	0.83	188.8	2265	0.90	202.5	2430
24	0.93	252.5	3030	1.00	269.2	3230
30	1.10	367.1	4405	1.19	402.9	4835
36	1.22	491.3	5895	1.43	578.3	6940
42	1.35	637.9	7655	1.58	749.2	8990
48	1.48	799.6	9595	1.73	940.0	11280
54	1.63	997.1	11965	1.90	1168.8	14025
60	1.89	1270.9	15250	2.20	1488.3	17860

* Table reproduced from A.S.A. Spec. A21.2-1953 by permission of the American Water Works Association.
† Including bell-and-spigot bead. Calculated weight of pipe rounded off to nearest 5 lb.

TABLE 60.* STANDARD THICKNESSES AND WEIGHTS OF CAST IRON PIT CAST PIPE (Continued)

Note. These weights are for pipe laid without blocks, on flat bottom trench, with tamped backfill, under 5 ft. of cover.

Size, in.	Class 250 250 Lb. Pressure 577 Ft. Head			Class 300 300 Lb. Pressure 693 Ft. Head		
	Thick-ness, in.	Wt. Based on 12-Ft. Length †		Thick-ness, in.	Wt. Based on 12-Ft. Length †	
		Avg. per Foot	Per Length		Avg. per Foot	Per Length
3	0.37	14.2	170	0.37	14.2	170
4	0.40	19.2	230	0.40	19.2	230
6	0.43	30.0	360	0.46	31.7	380
8	0.50	45.8	550	0.54	49.2	590
10	0.63	72.1	865	0.68	77.1	925
12	0.68	92.1	1105	0.73	97.9	1175
14	0.79	123.3	1480	0.85	131.7	1580
16	0.85	152.1	1825	0.92	162.9	1955
18	0.92	184.2	2210	0.99	196.7	2360
20	0.97	216.7	2600	1.05	232.1	2785
24	1.17	309.6	3715	1.26	346.2	4155
30	1.39	462.1	5545	1.50	511.3	6135
36	1.54	617.1	7405	1.79	727.9	8735
42	1.71	802.9	9635			
48	2.02	1077.1	12925			
54	2.21	1333.8	16005			
60	2.38	1594.6	19135			

Size, in.	Class 350 350 Lb. Pressure 808 Ft. Head		
	Thick-ness, in.	Wt. Based on 12-Ft. Length †	
		Avg. per Foot	Per Length
3	0.37	14.2	170
4	0.40	19.2	230
6	0.50	34.2	410
8	0.58	53.8	645
10	0.73	81.7	980
12	0.79	105.0	1260
14	0.92	148.3	1780
16	0.99	181.7	2180
18	1.07	220.8	2650
20	1.22	277.1	3325
24	1.36	370.0	4440
30	1.62	557.9	6695
36	1.93	794.2	9530

* Table reproduced from A.S.A. Spec. A21.2-1953 by permission of the American Water Works Association.
† Including bell-and-spigot bead. Calculated weight of pipe rounded off to nearest 5 lb.

TABLE 61. CAST-IRON BELL-AND-SPIGOT PIPE, CENTRIFUGALLY CAST, TYPE II

Nominal Inside Diameter, in.	100-lb. Class * or Max. Working Pressure		150-lb. Class † or Max. Working Pressure		200-lb. Class * or Max. Working Pressure		250-lb. Class † or Max. Working Pressure	
	Thickness, in.	Approximate Weight per Ft., lb.	Thickness, in.	Approximate Weight per Ft., lb.	Thickness, in.	Approximate Weight per Ft., lb.	Thickness, in.	Approximate Weight per Ft., lb.
3	0.32	12.4	0.33 *	12.5 †	0.34	13.0		
4	0.32	15.3	0.34	16.1	0.36	16.9	0.38	18.1
6	0.34	23.5	0.37	25.7	0.40	27.1	0.43	28.7
8	0.38	34.6	0.42	38.6	0.46	41.0	0.50	44.6
10	0.42	47.1	0.47	52.2	0.52	57.4	0.57	62.3
12	0.44	39.0	0.50	66.1	0.57	75.0	0.62	81.1
14	0.48	74.6	0.55	86.9	0.62	96.9	0.69	108.0
16	0.52	91.9	0.60	108.1	0.68	121.0	0.75	133.6
18	0.56	111.0	0.65	130.4	0.74	147.2	0.83	164.9
20	0.58	127.9	0.68	152.0	0.78	172.2	0.88	193.6
24	0.64	168.9	0.76	202.9	0.88	231.5	1.00	262.1

* *From American Cast Iron Pipe Co.*

† *From Federal Specifications WW-P-421.*

Water hammer of ordinary intensity allowed for in the above table. Weights based on 16-ft. length. 100-lb. Class—weights for Class B fittings; 150-lb., 200-lb., 250-lb. Classes—weights for Class D fittings.

TABLE 62. STANDARD THICKNESSES AND WEIGHTS OF CAST-IRON PIT CAST PIPE *

Size Inches	Class 50 — 50 Lb. Pressure — 115 Feet Head Thickness, inches	Avg. Per Foot	Wt. Based on 12 Ft. Lgh.† Per Length	Class 100 — 100 Lb. Pressure — 231 Feet Head Thickness, inches	Avg. Per Foot	Wt. Based on 12 Ft. Lgh.† Per Length	Class 150 — 150 Lb. Pressure — 346 Feet Head Thickness, inches	Avg. Per Foot	Wt. Based on 12 Ft. Lgh.† Per Length	Class 200 — 200 Lb. Pressure — 462 Feet Head Thickness, inches	Avg. Per Foot	Wt. Based on 12 Ft. Lgh.† Per Length	Class 250 — 250 Lb. Pressure — 577 Feet Head Thickness, inches	Avg. Per Foot	Wt. Based on 12 Ft. Lgh.† Per Length	Class 300 — 300 Lb. Pressure — 693 Feet Head Thickness, inches	Avg. Per Foot	Wt. Based on 12 Ft. Lgh.† Per Length	Class 350 — 350 Lb. Pressure — 808 Feet Head Thickness, inches	Avg. Per Foot	Wt. Based on 12 Ft. Lgh.† Per Length
3	.37	14.2	170	.37	14.2	170	.37	14.2	170	.37	14.2	170	.37	14.2	170	.37	14.2	170	.37	14.2	170
4	.40	19.2	230	.40	19.2	230	.40	19.2	230	.40	19.2	230	.40	19.2	230	.40	19.2	230	.40	19.2	230
6	.43	30.0	360	.43	30.0	360	.43	30.0	360	.43	30.0	360	.43	30.0	360	.46	31.7	380	.50	34.2	410
8	.46	42.9	515	.46	42.9	515	.46	42.9	515	.46	42.9	515	.50	45.8	550	.54	49.2	590	.58	53.8	645
10	.50	57.1	685	.50	57.1	685	.54	60.8	730	.58	64.6	775	.63	72.1	865	.68	77.1	925	.73	81.7	980
12	.54	73.3	880	.54	73.3	880	.58	77.9	935	.63	83.8	1,005	.68	92	1,105	.73	97.9	1,175	.79	105.0	1,260
14	.54	85.4	1,025	.58	91.3	1,095	.63	100.8	1,210	.68	107.9	1,295	.79	123.3	1,480	.85	131.7	1,580	.92	148.3	1,780
16	.58	105.4	1,265	.63	113.3	1,360	.68	125.0	1,500	.79	142.5	1,710	.85	152.1	1,825	.92	162.7	1,955	.99	181.7	2,180
18	.63	127.9	1,535	.68	136.7	1,640	.73	150.4	1,805	.85	172.1	2,065	.92	184.2	2,210	.99	196.7	2,360	1.07	220.8	2,650
20	.66	148.8	1,785	.71	158.8	1,905	.83	188.8	2,265	.90	202.5	2,430	.97	216.7	2,600	1.05	232.1	2,785	1.22	277.1	3,325
24	.74	198.8	2,385	.80	212.9	2,555	.93	252.5	3,030	1.00	269.2	3,230	1.08	288.3	3,460	1.26	346.2	4,155	1.36	370.0	4,440
30	.87	288.3	3,460	.94	311.3	3,735	1.10	367.1	4,405	1.19	402.9	4,835	1.39	462.1	5,545	1.50	511.3	6,135	1.62	557.9	6,695
36	.97	384.2	4,610	1.05	420.8	5,050	1.22	491.3	5,895	1.43	578.3	6,940	1.54	617.1	7,405	1.79	727.9	8,735	1.93	794.2	9,530
42	1.07	497.5	5,970	1.25	579.2	6,950	1.35	637.9	7,655	1.58	749.2	8,990	1.71	802.9	9,635						
48	1.18	625.8	7,510	1.37	726.3	8,715	1.48	799.6	9,595	1.73	940.0	11,280	2.02	1,077.1	12,925						
54	1.30	777.1	9,325	1.51	906.3	10,875	1.63	997.1	11,965	1.90	1,168.8	14,025	2.21	1,333.8	16,005						
60	1.39	922.5	11,070	1.62	1,077.1	12,925	1.89	1,270.9	15,250	2.20	1,488.3	17,860	2.38	1,594.6	19,135						

* From American Standard Assn., Spec. A21.2–1939.

† Including bell and spigot bead. Calculated weight of pipe rounded off to nearest 5 pounds.

Note. These weights are for pipe laid without blocks, on flat bottom trench, with tamped backfill, under 5 feet of cover.

TABLE 63. APPROXIMATE QUANTITIES OF MATERIALS USED PER JOINT FOR WATER SERVICE *

Nominal Diameter, In.	Pounds of Joint Compound 2½″ Joint Depth †	Pounds of Hemp per Joint	Pounds of Lead in Joint 2″ Deep	Pounds of Lead in Joint 2¼″ Deep	Pounds of Lead in Joint 2½″ Deep
3		0.18	6.00	6.50	7.00
4	2.00	0.21	7.50	8.00	8.75
6	3.00	0.31	10.25	11.25	12.25
8	4.00	0.44	13.25	14.50	15.75
10	5.00	0.53	16.00	17.50	19.00
12	6.00	0.61	19.00	20.50	22.50
14	7.00	0.81	22.00	24.00	26.00
16	8.25	0.94	30.00	33.00	35.75
18	9.25	1.00	33.80	36.90	40.00
20	10.50	1.25	37.00	40.50	44.00
24	13.00	1.50	44.00	48.00	52.50

* *Adapted from U. S. Pipe and Foundry Co.*

† Approximate only; will vary with kind of material used.

Note. Weight of lead is based on std. wt. = 0.41 lb. per cu. in. This weight may vary 15% depending on purity.

TABLE 64. ASBESTOS-CEMENT SEWER PIPE

Pipe Size (inside diameter), in.	Class 1			Class 2		
	Shell Thickness, in.	Weight per Lin. Ft., lb.	Ultimate Strength 3-edge Bearing, lb. per lin. ft.	Shell Thickness, in.	Weight per Lin. Ft., lb.	Ultimate Strength 3-edge Bearing, lb. per lin. ft.
6	0.42	8.5	2660			
8	0.48	13.0	2500			
10	0.50	16.7	2200	0.56	18.6	2800
12	0.54	21.0	2200	0.64	25.1	3000
14	0.58	26.4	2200	0.73	33.2	3400
16	0.62	32.2	2200	0.82	42.0	3700
18	0.65	37.5	2100	0.90	51.7	4000
20	0.69	44.2	2200	0.94	59.8	4000
24	0.75	57.0	2200	1.06	80.3	4200
30	0.96	90.0	2800	1.24	116.0	4600
36	1.15	128.0	3300	1.41	158.0	5000

Standard laying length, 13 ft.

Furnished only in straight lengths.

Cast-iron fittings recommended for branch connections.

Ultimate strengths determined by tests made in accordance with procedure of A.S.T.M.

All data furnished by Johns-Manville Corp.

TABLE 65. CORRUGATED METAL CULVERT PIPE

Inside Pipe Diameter, in.	Weight per Linear Foot, lb.				
	16 Gage	14 Gage	12 Gage	10 Gage	8 Gage
4					
6					
8	7.6	9.3			
10	9.3	11.4			
12	10.8	13.3	18.5		
15	13.3	16.4	22.7		
18	15.8	19.5	27.0		
21	18.3	22.5	31.2	39.7	
24	21.0	26.0	35.9	45.7	
30		31.7	43.9	55.9	
36		37.9	52.4	66.7	81.1
42		44.4	61.5	78.3	95.1
48		50.5	70.0	89.1	108.3
54		57.8	80.1	102.0	123.9
60			88.2	112.3	136.4
66			96.6	123.1	149.5
72			105.1	133.9	162.6
84				156.6	190.3

Furnished in any length in multiples of 2 ft.
Data furnished for Armco Pipe by Shelt Co., Elmira, N. Y.

TABLE 66. DIMENSIONS AND WEIGHTS OF WELDED AND SEAMLESS STEEL PIPE BLACK AND GALVANIZED

Nominal Pipe Size, in.	Outside Diameter, in.	Number of Threads per Inch	Schedule 30 Wall, in.	Schedule 30 Weight per Linear Foot, lb.	Schedule 40 Wall, in.	Schedule 40 Weight per Linear Foot, lb.	Schedule 60 Wall, in.	Schedule 60 Weight per Linear Foot, lb.	Schedule 80 Wall, in.	Schedule 80 Weight per Linear Foot, lb.	Double-Extra Strong Pipe Wall, in.	Double-Extra Strong Pipe Weight per Linear Foot, lb.
½	0.840	14			0.109	0.85			0.147	1.09	0.294	1.714 *
¾	1.050	14			0.113	1.13			0.154	1.47	0.308	2.441 *
1	1.315	11½			0.133	1.68			0.179	2.17	0.358	3.659 *
1¼	1.660	11½			0.140	2.27			0.191	3.00	0.382	5.214 *
1½	1.900	11½			0.145	2.72			0.200	3.63	0.400	6.408 *
2	2.375	11½			0.154	3.65			0.218	5.02	0.436	9.03
2½	2.875	8			0.203	5.79			0.276	7.66	0.552	13.70
3	3.500	8			0.216	7.58			0.300	10.25	0.600	18.58
3½	4.000	8			0.226	9.11			0.318	12.51	....	.....
4	4.500	8			0.237	10.79			0.337	14.98	0.674	27.54
5	5.563	8			0.258	14.62			0.375	20.78	0.750	38.55
6	6.625	8			0.280	18.97			0.432	28.57	0.864	53.16
8	8.625	8	0.277	24.70	0.322	28.55	0.406	35.64	0.500	43.39	0.875	72.42
10	10.750	8	0.307	34.24	0.365	40.48	0.500	54.74	0.593	64.33		
12	12.750	8	0.330	43.77	0.406	53.53	0.562	73.22	0.687	88.51		

* Seamless pressure tubing.

Data in this table from National Tube Co.

Thicknesses in **bold-face type** in schedules 30 and 40 correspond with those of standard weight pipe; those in 60 and 80 with extra strong pipe; all weights are plain ends.

TABLE 67. ASBESTOS-CEMENT WATER PIPE

Pipe Size,* in.	Class 100		Class 150		Class 200	
	Shell Thickness, in.	Weight per Linear Foot, lb.	Shell Thickness, in.	Weight per Linear Foot, lb.	Shell Thickness, in.	Weight per Linear Foot, lb.
4	0.38	5.8	0.45	6.9	0.60	9.2
6	0.46	10.0	0.55	12.0	0.75	16.5
8	0.52	15.0	0.65	18.6	0.88	25.3
10	0.59	21.2	0.85	29.9	1.10	38.6
12	0.68	28.6	0.98	40.6	1.24	51.2
14	0.78	36.4	1.13	53.6	1.44	68.5
16	0.88	47.8	1.25	67.4	1.65	89.2
18	0.97	58.6	1.39	83.5	1.87	112.8
20	1.07	71.2	1.53	102.3	2.09	139.8
24	1.25	98.0	1.82	143.0	2.48	197.0
30	1.54	144.0	2.29	222.0	3.12	306.0
36	1.83	210.0	2.80	324.0	3.74	440.0

* Pipe size is inside diameter except sizes 4, 6, and 8 in. in Class 100 and Class 150, which are 3.95, 5.85, and 7.85 in., respectively.

Class of pipe is same as allowable working pressure in pounds per square inch. Furnished in straight lengths only, standard length = 13 ft.

Data from Johns-Manville Corp.

MISCELLANEOUS

INSPECTOR'S TIME RECORD

TUTTLE, SEELYE, PLACE & RAYMOND

Week ending____194__. Name____*Smith*____		Breakdown of Hours Worked on Each Job									Holidays, Vacation, and Sick Leave Allowed
Date	Description of Work Performed									Total Hours	
12/1	*Adm.*	6								6	
2		8								8	
3		10								10	
4		10								10	
5											
6		5								5	
7		15								15	
Total Hours		54								54	

Total Credit Hours ___ 61 ___

Breakdown in Hours Worked	
Job	Credit Hr
Holidays, Vacation, Sick Leave	
Total	

Employee_____

Approved_____

REPORT ON AIRFIELD RUNWAYS

INSPECTOR'S DAILY REPORT

AIRFIELD RUNWAYS

TUTTLE, SEELYE, PLACE & RAYMOND
Architect—Engineer
Fort Dix, New Jersey

Directive No. _____ Contract No. _____ Report No. _____
Prime contractor _____ Date _____
Subcontractor _____
Weather _____ Temp. 8 a.m. _____ 1 p.m. _____ 5 p.m. _____

| Items | Kind | Quantity | | | | Location |
		Units	Lin. Ft.	Sq. Yd.	Cu. Yd.	
Roads Excavation						
Borrow						
Fine grading						
Base						
Top						
Seal coat						
Shoulders						
Ditching						
Culverts and drains						
Foundation material						
Water mains						
Valves						
Hydrants						
Specials						
Sanitary sewer mains						
Manholes						
Specials						
Storm sewer mains						
Manholes						
Inlets						
Head walls						

Labor	Equipment	Inspector's Checking List					
		Sewers		Water		Roads	
		Material		Material		Material	
		Line		Blocking		Subgrade	
		Grade		Line		Consolidation	
		Joints		Grade		Surface	
		Backfill		Joints		Culverts	
		Manholes		Backfill		Head walls	
				Valves		Storm sewer	
Worked from	Worked from			Hydrant			

Remarks: _____

_____

_____

_____

_____

_____

_____

_____

_____

_____

_____

_____

_____

_____

_____

_____

_____

_____

_____

Inspector

INSPECTOR'S DAILY REPORT * GENERAL CONSTRUCTION

Report No. _____ Sheet No. _____　　　　Place _____

Specification No. _____　　　　Date _____ 19__

Contract No. _____ For _____

Contractor _____ Superintendent _____

Weather { A.M. _____ } Temperatures
{ P.M. _____ }

Tides (for all work affected thereby)

{ High _____ Elevation at _____ M. }
{ Low _____ Elevation at _____ M. }

Time of starting work—
　A.M. _____ _____ °F.
　NOON _____ _____ °F.
Time of stopping work—
　P.M. _____ _____ °F.

No.	Trade	Hours	Rate	Amount Wages	Class	Quantity	Remarks

CONTRACTOR'S FORCE INCLUDING SUPERVISORS AND SUBCONTRACTOR'S FORCES — WORK

MATERIAL RECEIVED THIS DATE

Item	Delivered	Passed	Rejected	Item	Delivered	Passed	Rejected

PLANT EQUIPMENT ON JOB		GOVERNMENT UTILITIES FURNISHED					
Equipment	Hours Worked	No.	Appliance or Service	Hr.	Duty	Rate	Amt.

Delays _____

Accidents _____

Defective work to be corrected later (enter in red) _____

Special instructions received or given _____

Tests _____

Items started this date _____
Items completed this date _____
Contractor's plant
  Items delivered _____
  Items removed _____
  Items out of commission (state time and cause) _____

Remarks: _____
_____
_____
_____
_____
_____
_____
_____
_____

_____, Inspector.

Instructions to inspectors. Make reports full and complete, and to include all work performed on contractor's plant. When the contractor, his chief engineer, general superintendent, or other responsible member of his organization visits the job, make a note, giving names, and also any instructions given by them to the superintendent on the job relative to the prosecution of the work. Note all accidents, delays, fires, etc., and give *your* opinion as to causes, and how the progress of the work is affected thereby.

* *From Navy Department—Bureau Yards and Docks.*

GENERAL CONTRACTOR'S DAILY REPORT

Date _____ 10-4-45 _____

Weather _____ Clear _____

Job _____

Temperature _____

	Mechanics		Laborers		Foreman		
	No.	Time	No.	Time	No.	Time	
Superintendent	1						
Watchman							
Timekeepers	1						
Excavation							
Engineer	1	7					Hoisting materials and taking down rubbish, etc.
Forms							
Laborers			17	119	1	8	Handling materials for ovens—7th floor
							Chipping and cutting cols. and beams.
Concrete							Cleaning and taking down rubbish, etc. from 7th, 8th, and 9th floors.
Cem. finish							Loading truck with rubbish. Helping carpenters.
Rein. steel							
Masonry							
Carpentry	11	77					Shoring for center line for ovens.
							Making up benches for lathers.
							Building forms for cols.
							Running power-saw and filing saws.
							Framing haunches.
							Shoring forms over ovens.

Equipment *Truck; hauling rubbish away.*

Subcontractors *Kalman.   Watering Floors.*

Excavation	Steel sash
Struct. steel	Calking
Misc. & orn. iron	Lathing
Cut stone	Plastering
Plumbing	Marble and tile
Heating	Floor covering
Electric	Weatherstripping
Waterproofing	Metal equipment
Hollow metal	Painting
Kalamein	Glazing
Rfg. & sheet metal	

Remarks *Wreckers—Cutting arches on 9th floor.   Cutting wood floor.   7th floor.   Removing rubbish, etc., from 9th floor.*

Visitors   Signed _____ Sheet No. _____

                                 Supt.

CONSTRUCTION JOB POWER

In order to give the field engineer a general perspective of job power, the following is submitted.

Air compressors used in construction are of various types and sizes.

The most common type for the usual construction job is the portable type mounted on wheels for easy moving.

Compressor should be placed in a safe location to avoid injury but as close to operations as possible in order to avoid expensive labor and material in pipe lines, and to avoid decreased efficiency due to line losses, leaky joints, and actual breakage of line resulting from accident or carelessness.

Compressor capacity is rated on the actual cubic feet of air delivered at a designated pressure, usually 100 p.s.i.

The usual capacities for portable compressors are 105 cu. ft., 210 cu. ft., 315 cu. ft., 365 cu. ft., and 500 cu. ft. per min.

There are many air tools for use with compressors; some of the more common are listed below:

Drills, jackhammers, wagon drills, drifters—for drilling holes in rock for use with explosives.

Breakers or busters—for breaking and chipping rock or loosening hard compact earth.

Air riveters (guns)—for driving rivets in steel bridge and building construction.

Plug drills—for plug and feather work, used generally in quarries for dimension stone such as granite, sandstone, and marble.

Air augers—for drilling holes in wood, in use on wooden piers, cofferdams, roof trusses, etc.

Bolt runners—for tightening bolts.

Tampers—to consolidate backfill.

Hoists, single and multiple drum—for use with derricks, mine scrapers, car haulage in industrial plants, etc.

Pneumatic sheeting hammers—in trenches or cofferdams to drive wood sheeting, usually up to about 3 in. thick.

Air spades—for digging hard clay or other compact material.

Air vibrators—for concrete.

Pile hammers—for driving any type of pile.

Air saws, air clamps, etc.

The above tools use a varying amount of air, depending on size, mechanical condition of tool, etc.

For tools in general use on a construction job, such as a drill, breaker, tamper, and spades, a figure of 50 cu. ft. can be taken to estimate the compressor capacity required.

For example, a 210 cu. ft. compressor will operate four average size drills, breakers, spades, or tampers, assuming that these tools are in fair mechanical condition.

The above figure is for practical field conditions.

Two or more compressors may be coupled together to increase the available amount of air. If this is done, the compressors should discharge into an air receiver or reservoir. This will increase efficiency, decrease wear on compressors, and insure an even flow of power to tools.

On any job it is good practice to have one spare tool for every four tools in use to avoid costly delays caused by mechanical failure.

Tools are expensive and should be well cared for; carelessness is an item that should not be on any report sheet.

Some attempts have been made to operate percussion tools (breakers) by gasoline or electricity, but this type of tool is not in general use as yet in the construction field.

Careful consideration should be given to weight of tool selected for various operations. For instance, a man can use a heavier, more powerful drill or breaker if he is drilling a down hole, i.e., a hole either vertical or on a slant away from him. But a much lighter tool should be provided for drilling or chipping a horizontal hole (breast hole), to avoid excessive fatigue. There is, however, a third leg or jack on the market which can be clamped to the drill or breaker which will relieve the operator of much of the weight of the tool and which adds considerably to the efficiency of the tool.

An air tool in operation is always cold owing to the expansion of the air out of the exhaust valve; hence, care must be taken to use a good grade of air oil for lubrication. One of the best of many ways to oil an air tool is by a line oiler. This is an oil reservoir holding about a pint of oil and can be set to provide oil drop by drop into the air line which is carried to the tool.

For several years manufacturers have provided a drill rod threaded on one end to receive a jack bit. This eliminates hand sharpening of steel on the job as the jack bit can be used until dull or until the gage is worn down, then it is simply unscrewed from the rod and replaced.

The gage of a bit is its width. As the drill rotates, the bit is worn down by the rock and gradually the bit becomes narrower until finally, in construction parlance, "the gage is gone."

The gage of a bit is of great importance. Drill rods usually provide for a depth of hole up to 10 ft. to 12 ft. or more by 2-ft. stages.

EXAMPLE.

GAGE

No. 1 or starter drill rod 2 ft.—Bit 2 in.
No. 2 drill rod 4 ft.—Bit 1¾ in.
No. 3 drill rod 6 ft.—Bit 1½ in.

Note that, on No. 1 bit, the gage is 2 in.

As the bit is worn or loses its gage, it is evident that No. 2 bit will not follow; that is, it will not seat at the bottom of the hole already drilled by No. 1. As a result, bit 2 will become fast in the hole resulting in loss of steel and time. The above bit sizes are arbitrary, but note that the gage for any following bit is ¼ in. smaller always.

Bits may be resharpened by special tools but always to a smaller gage; for example a 2-in. bit becomes 1¾ in., etc.

Bits are various shapes: X bits, cross bits + or six point or rose bits ⊕. The cross bit and six point are the more common. Although each shape has its strong supporters among rock men, in general it can be said that, in hard dense rock, the cross bit is superior, while in loosely stratified rock, the six point is superior. The six point bit is especially desirable in drilling concrete for demolition.

The use of goggles to protect the eyes is a wise precaution for men operating drills or breakers, and in enclosed places a simple dust mask can be provided to keep the nose and throat as free from dust as possible.

Electric tools such as saws, pumps, wood augers, vibrators, bolt runners, and drill presses have a place in construction. For many tools, electricity is more advantageous in that the primary power feeder is a distant power house and, after the feeder lines are run, power is available at the turn of a switch.

Gasoline- and diesel-fuel-driven motors are widely used as the primary power unit on all sizes of tools from the small compresser, table saws, pumps, vibrators, chain saws, electric generators, etc., to the giant locomotive.

TABLE 68. AIRPORT CONSTRUCTION. SUMMARY OF TESTS REQUIRED

A.A.S.H.O. = American Association of State Highway Officials

A.S.T.M. = American Society for Testing Materials

Section No.	Specification Item	Tests Required for	Type Test	A.A.S.H.O. Designations	A.S.T.M. Designations	Category
I	Excavation and grading	Soils: Density *	Physical	T 99	D 698	Field and lab.
		Mechanical analysis *	Physical	T 88	D 422	Field and lab.
		Elasticity index *	Physical	T 91	D 424	Field and lab.
II	Pipe for sewers, drains and culverts, etc.	Bituminous coated corrugated pipe	Physical and chemical	M-36, T 33	....	Cert.
		Asphalt	Carbon disulfide			Lab.
			Stability			Lab.
			Imperviousness			Lab.
		Concrete pipe	Physical	T 33, M 41, M 86,	C 14, C 75, C 76	Field and lab.
		Cast-iron pipe	Physical	M 64	A 142	Field and lab.
		Clay pipe	Physical and chemical	M 65, T 33	C 4, C 200	Field and lab.
		Perforated metal pipe	Physical	M 36, T 33	....	Cert.
		Jointing material	Metal coupling bands	M 36	....	Cert.
III	Manholes, inlets, headwalls, and miscellaneous structures	Brick	Physical	M 91, T 32	C 32	Lab.
		Cement	Physical and chemical	M 85, M 134	C 150, C 175	Lab.
		Sand *	Physical	M 6	C 33	Field and lab.
		Frames, covers and gratings	Gray iron	M 105	A 48	Cert.
			Malleable iron	M 106	A 47	Cert.
			Steel grates and frames	M 94	....	Cert.
			Galvanized	M 111	A 123	Cert.
IV	Portland cement concrete	Test cylinders made and tested	Physical	T 22	C 31	Field and lab.
		Test beams made and tested	Physical	T 97	C 31	Field and lab.
		Slump*	Physical	T 119	C 143	Field
		Sand, strength	Physical	T 71	C 109	Lab.
		Sand, organic impurities	Chemical	T 21	C 40	Lab.
		Sand, gradation	Physical	T 27	C 136	Field and lab
		Sand, soundness	Physical	T 104	C 88	Lab.
		Coarse aggregates, gradation *	Physical	T 27	C 136	Field and lab.
		Coarse aggregates, % wear	Physical	T 96	C 131	Lab.
		Coarse aggregates, soundness	Physical	T 104	C 88	Lab.
		Air content*	Physical	T 121	C 231, C 173	Field

		Reinforcing steel / Portland cement	Physical and chemical	M 31, M 54, M 55 / M 85, M 134	A 15, A 305 / C 150, C 175	Lab. / Lab.
		Reinforcing steel	Physical and chemical	M 31, M 54, M 55	A 15, A 305	Lab.
		Portland cement		M 85, M 134	C 150, C 175	Lab.
V	Underground electric duct	Bituminous fiber duct	Fed. Spec. WC-581			
VI	Subbase course	Gradation *	Physical	T 88	D 422	Field and lab.
		Liquid limit *	Physical	T 89	D 423	Field and lab.
		Plasticity index *	Physical	T 91	D 424	Field and lab.
		Density *	Physical	T 99	D 698	Field and lab.
VII	Dry bound macadam	Gradation *	Physical	T 27	C 136	Field and lab.
		Abrasion	Physical	T 96	C 131	Lab.
		Soundness	Chemical	T 104	C 88	Lab.
VIII	Penetration macadam	Gradation *	Physical	T 27	C 136	Field and lab.
		Bituminous materials	Same as for Section IX			
		Surface tests *	Straight edge			Field
IX	Binder and wearing courses	Asphalt content *	Physical	T 58	D 762	Lab.
		Filler *	Physical	M 17	D 242	Lab.
		Aggregates, gradation *	Physical	M 79	D 1073	Field and lab.
		Aggregates, soundness	Chemica	T 104	C 88	Lab.
		Aggregates, swell	Physical	T 101		Lab.
		Aggregates, stripping *	Physical	Special tes	C 131	Field and lab.
		Aggregates, % wear	Physical	T 96		Lab.
		Sampling bituminous materials	Physical	T 2, T 40, T 41	D 75, D 140, D 979	Field
		Asphalt penetration *	Physical	T 49	D 5	Field and lab.
		Carbon tetrachloride	Physical	T 45	D 165	Lab.
		Ductility	Physical	T 51	D 113	Lab.
		Flash point	Physical	T 48	D 92	Lab.
		Loss on heating	Physical	T 47	D 6	Lab.
		Lime	Physical		C 141	Lab.
		Stability	Physical	Hubbard-Field or Marshall		Lab.
		Surface tests *	Straight and edge			Field
		Mechanical analysis of extracted aggregate	Physical	T 30		Lab.
		Joint filler		M 33, M 58, M 59, M 90	D 994, D 545, D 544	Lab.
X	Bituminous seal coat	Aggregates *	Same as for Section IX			
		Asphalt	Same as for Section IX			
		Per cent of water in asphalt	Physical	T 55	D 95, D 96	Lab.
		Furolviscosity	Physical	T 72	D 666	Lab.
		Partial distillates	Physical	T 78	D 402	Lab.
XI	Fences	Fabric	Physical and chemical		A 117, A 121	P-160, 161, 162, 163

* The more usual field tests are starred.

PART II

SURVEYING

TOPOGRAPHIC SURVEY

Traverse points should be selected with a view to economy of setups; e.g., so located that a maximum area can be seen by the instrument man. For accuracy the traverse should be run separately from the topography shots. For economy, where refined accuracy is not necessary, the traverse and the topography can be run simultaneously; i.e., the topography shots are taken as each traverse point is occupied.

Since stadia topography is normally plotted with a protractor, refinements greater than 15 minutes in the horizontal angle are not warranted. Considerable speed is attained when the horizontal angles are estimated to the nearest quarter degree and the vertical angles to the nearest minute.

SAMPLE NOTES

STADIA		TRAVERSE				
		All Horiz. ∡s	measured clockwise.			
Sta. Occ.	Sta. Obs.	Hor. ∡	Mag.B.	Rod Int.	Ver. ∡	Hor.Dist.
B	A	0°	N89°W	5.44	+2°40'	544'
	C	90°	N 1°E	6.05	-1°17'	606'
C	B	0°	S 0°W	6.07	+1°16'	608'
	D	201°30'	N22°E	6.95	-1°27'	696'
D	C	0°	S21°W	6.96	+1°27'	697'
	E	221°0'	N62°E	8.65	-1°35'	865'
E	D	0°	S61°W	8.64	+1°35'	864'
	F	102°25'	N16°W	10.21	-3°25'	1018'
F	E	0°	S15°E	10.21	+3°24'	1018'
	G	263°20'	N66°E	4.98	-2°16'	498'
G	F	0°	S67°W	4.99	+2°16'	499'

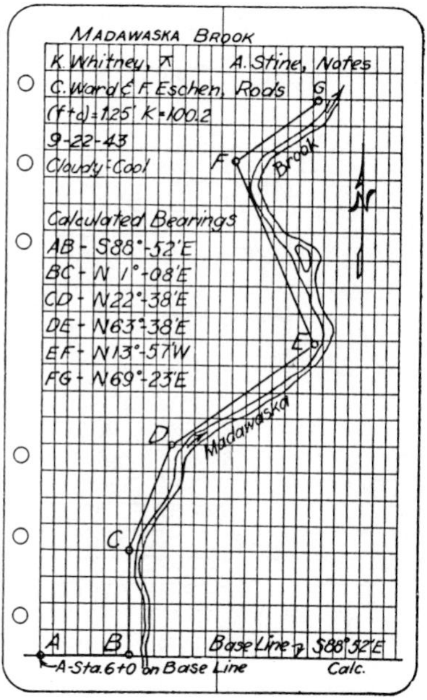

MADAWASKA BROOK

K Whitney, ✗   A. Stine, Notes

C. Ward & F. Eschen, Rods

(f+d)=1.25 K=100.2

9-22-43

Cloudy-Cool

Calculated Bearings

AB - S86°-52'E
BC - N 1°-08'E
CD - N22°-38'E
DE - N63°-38'E
EF - N13°-57'W
FG - N69°-23'E

Base Line - S88°52'E

A-Sta.6+0 on Base Line   Calc.

Fig. 1.

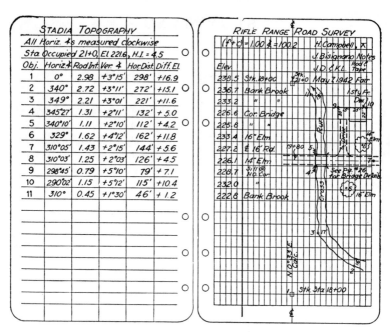

STADIA TOPOGRAPHY

All Horiz. ∢s measured clockwise

Sta. Occupied 21+0, El. 221.6, H.I. = 4.5

Obj.	Horiz ∢	Rod Int.	Ver. ∢	Hor. Dist.	Diff. El.
1	0°	2.98	+3°15'	298'	+16.9
2	340°	2.72	+3°11'	272'	+15.1
3	349°	2.21	+3°01'	221'	+11.6
4	345°27'	1.31	+2°11'	132'	+5.0
5	340°10'	1.11	+2°10'	112'	+4.2
6	329°	1.62	+4°12'	162'	+11.8
7	310°05'	1.43	+2°15'	144'	+5.6
8	310°03'	1.25	+2°03'	126'	+4.5
9	298°45'	0.79	+5°10'	79'	+7.1
10	290°02'	1.15	+5°12'	115'	+10.4
11	310°	0.45	+1°30'	46'	+1.2

RIFLE RANGE ROAD SURVEY

(f° + 0) = 1.00, ∮ = 100.2　　H. Campbell, X

J. Bisignano Notes

J. D. ∮ K. L. Rod ∮ Tape

Elev	
238.5	Stk. 18+00
236.7	Bank Brook
233.2	" "
226.6	Cor. Bridge
225.8	" "
233.4	16" Elm
227.2	℄ 16' Rd.
226.1	14" Elm
228.7	Sill @ N.O. Cor.
232.0	"
222.8	Bank Brook

May 7 1942 Fair

See pg. #26 for Bridge Details

Fig. 2.

CROSS SECTIONS

Sta.	B.S.	H.I.	F.S.	Elev.
B.M.6	6.07	212.27		206.20
25				
+50				
26				
+37				
27				
+50				
28				
+50				
29				
T.P.			3.51	208.76

PRELIM. - BRANDON ROAD

June 7 - Cloudy, Warm 75°-80°F

R. Shelton, X　　L. Vought, Rod, W. Bailey, Tape

Spike in stump - 24 +90 - 75' Rt.

Top of stake - Sta. 29 +45 - 50' Rt.

Fig. 3.

STADIA TABLES

Table 1. Stadia Reductions *

Differences in Elevation for 100 ft.   Inclined Distance

Min-utes	0°	1°	2°	3°	4°	5°	6°	7°	8°	9°	10°	11°	12°
0	0.00	1.74	3.49	5.23	6.96	8.68	10.40	12.10	13.78	15.45	17.10	18.73	20.34
2	0.06	1.80	3.55	5.28	7.02	8.74	10.45	12.15	13.84	15.51	17.16	18.78	20.39
4	0.12	1.86	3.60	5.34	7.07	8.80	10.51	12.21	13.89	15.56	17.21	18.84	20.44
6	0.17	1.92	3.66	5.40	7.13	8.85	10.57	12.26	13.95	15.62	17.26	18.89	20.50
8	0.23	1.98	3.72	5.46	7.19	8.91	10.62	12.32	14.01	15.67	17.32	18.95	20.55
10	0.29	2.04	3.78	5.52	7.25	8.97	10.68	12.38	14.06	15.73	17.37	19.00	20.60
12	0.35	2.09	3.84	5.57	7.30	9.03	10.74	12.43	14.12	15.78	17.43	19.05	20.66
14	0.41	2.15	3.90	5.63	7.36	9.08	10.79	12.49	14.17	15.84	17.48	19.11	20.71
16	0.47	2.21	3.95	5.69	7.42	9.14	10.85	12.55	14.23	15.89	17.54	19.16	20.76
18	0.52	2.27	4.01	5.75	7.48	9.20	10.91	12.60	14.28	15.95	17.59	19.21	20.81
20	0.58	2.33	4.07	5.80	7.53	9.25	10.96	12.66	14.34	16.00	17.65	19.27	20.87
22	0.64	2.38	4.13	5.86	7.59	9.31	11.02	12.72	14.40	16.06	17.70	19.32	20.92
24	0.70	2.44	4.18	5.92	7.65	9.37	11.08	12.77	14.45	16.11	17.76	19.38	20.97
26	0.76	2.50	4.24	5.98	7.71	9.43	11.13	12.83	14.51	16.17	17.81	19.43	21.03
28	0.81	2.56	4.30	6.04	7.76	9.48	11.19	12.88	14.56	16.22	17.86	19.48	21.08
30	0.87	2.62	4.36	6.09	7.82	9.54	11.25	12.94	14.62	16.28	17.92	19.54	21.13
32	0.93	2.67	4.42	6.15	7.88	9.60	11.30	13.00	14.67	16.33	17.97	19.59	21.18
34	0.99	2.73	4.48	6.21	7.94	9.65	11.36	13.05	14.73	16.39	18.03	19.64	21.24
36	1.05	2.79	4.53	6.27	7.99	9.71	11.42	13.11	14.79	16.44	18.08	19.70	21.29
38	1.11	2.85	4.59	6.33	8.05	9.77	11.47	13.17	14.84	16.50	18.14	19.75	21.34
40	1.16	2.91	4.65	6.38	8.11	9.83	11.53	13.22	14.90	16.55	18.19	19.80	21.39
42	1.22	2.97	4.71	6.44	8.17	9.88	11.59	13.28	14.95	16.61	18.24	19.86	21.45
44	1.28	3.02	4.76	6.50	8.22	9.94	11.64	13.33	15.01	16.66	18.30	19.91	21.50
46	1.34	3.08	4.82	6.56	8.28	10.00	11.70	13.39	15.06	16.72	18.35	19.96	21.55
48	1.40	3.14	4.88	6.61	8.34	10.05	11.76	13.45	15.12	16.77	18.41	20.02	21.60
50	1.45	3.20	4.94	6.67	8.40	10.11	11.81	13.50	15.17	16.83	18.46	20.07	21.66
52	1.51	3.26	4.99	6.73	8.45	10.17	11.87	13.56	15.23	16.88	18.51	20.12	21.71
54	1.57	3.31	5.05	6.79	8.51	10.22	11.93	13.61	15.28	16.94	18.57	20.18	21.76
56	1.63	3.37	5.11	6.84	8.57	10.28	11.98	13.67	15.34	16.99	18.62	20.23	21.81
58	1.69	3.43	5.17	6.90	8.63	10.34	12.04	13.73	15.40	17.05	18.68	20.28	21.87
60	1.74	3.49	5.23	6.96	8.68	10.40	12.10	13.78	15.45	17.10	18.73	20.34	21.92
$f + c$ .75	0.01	0.02	0.03	0.05	0.06	0.07	0.08	0.10	0.11	0.12	0.14	0.15	0.16
1.00	0.01	0.03	0.04	0.06	0.08	0.09	0.11	0.13	0.15	0.16	0.18	0.20	0.22
1.25	0.02	0.03	0.05	0.08	0.10	0.11	0.14	0.16	0.18	0.21	0.23	0.25	0.27

Corrections to Horizontal Distances

Min-utes	0°	1°	2°	3°	4°	5°	6°	7°	8°	9°	10°	11°	12°
0	......	0.03	0.12	0.27	0.49	0.76	1.09	1.49	1.94	2.45	3.02	3.64	4.32
10	......	0.04	0.14	0.31	0.53	0.81	1.15	1.56	2.02	2.54	3.12	3.75	4.44
20	......	0.05	0.17	0.34	0.57	0.86	1.22	1.63	2.10	2.63	3.22	3.86	4.56
30	0.01	0.07	0.19	0.37	0.62	0.92	1.28	1.70	2.18	2.72	3.32	3.97	4.68
40	0.01	0.08	0.22	0.41	0.66	0.98	1.35	1.78	2.27	2.82	3.42	4.09	4.81
50	0.02	0.10	0.24	0.45	0.71	1.03	1.42	1.86	2.36	2.92	3.53	4.21	4.93

Table 1.   Stadia Reductions (Continued) *

Differences in Elevation for 100 ft.   Inclined Distance

Min-utes	13°	14°	15°	16°	17°	18°	19°	20°	21°	22°	23°	24°	25°
0	21.92	23.47	25.00	26.50	27.96	29.39	30.78	32.14	33.46	34.73	35.97	37.16	38.30
2	21.97	23.52	25.05	26.55	28.01	29.44	30.83	32.18	33.50	34.77	36.01	37.20	38.34
4	22.02	23.58	25.10	26.59	28.06	29.48	30.87	32.23	33.54	34.82	36.05	37.23	38.38
6	22.08	23.63	25.15	26.64	28.10	29.53	30.92	32.27	33.59	34.86	36.09	37.37	38.41
8	22.13	23.68	25.20	26.69	28.15	29.58	30.97	32.32	33.63	34.90	36.13	37.31	38.45
10	22.18	23.73	25.25	26.74	28.20	29.62	31.01	32.36	33.67	34.94	36.17	37.35	38.49
12	22.23	23.78	25.30	26.79	28.25	29.67	31.06	32.41	33.72	34.98	36.21	37.39	38.53
14	22.28	23.83	25.35	26.84	28.30	29.72	31.10	32.45	33.76	35.02	36.25	37.43	38.56
16	22.34	23.88	25.40	26.89	28.34	29.76	31.15	32.49	33.80	35.07	36.29	37.47	38.60
18	22.39	23.93	25.45	26.94	28.39	29.81	31.19	32.54	33.84	35.11	36.33	37.51	38.64
20	22.44	23.99	25.50	26.99	28.44	29.86	31.24	32.58	33.89	35.15	36.37	37.54	38.67
22	22.49	24.04	25.55	27.04	28.49	29.90	31.28	32.63	33.93	35.19	36.41	37.58	38.71
24	22.54	24.09	25.60	27.09	28.54	29.95	31.33	32.67	33.97	35.23	36.45	37.62	38.75
26	22.60	24.14	25.65	27.13	28.58	30.00	31.38	32.72	34.01	35.27	36.49	37.66	38.78
28	22.65	24.19	25.70	27.18	28.63	30.04	31.42	32.76	34.06	35.31	36.53	37.70	38.82
30	22.70	24.24	25.75	27.23	28.68	30.09	31.47	32.80	34.10	35.36	36.57	37.74	38.86
32	22.75	24.29	25.80	27.28	28.73	30.14	31.51	32.85	34.14	35.40	36.61	37.77	38.89
34	22.80	24.34	25.85	27.33	28.77	30.19	31.56	32.89	34.18	35.44	36.65	37.81	38.93
36	22.85	24.39	25.90	27.38	28.82	30.23	31.60	32.93	34.23	35.48	36.69	37.85	38.97
38	22.91	24.44	25.95	27.43	28.87	30.28	31.65	32.98	34.27	35.52	36.73	37.89	39.00
40	22.96	24.49	26.00	27.48	28.92	30.32	31.69	33.02	34.31	35.56	36.77	37.93	39.04
42	23.01	24.55	26.05	27.52	28.96	30.37	31.74	33.07	34.35	35.60	36.80	37.96	39.08
44	23.06	24.60	26.10	27.57	29.01	30.41	31.78	33.11	34.40	35.64	36.84	38.00	39.11
46	23.11	24.65	26.15	27.62	29.06	30.46	31.83	33.15	34.44	35.68	36.88	38.04	39.15
48	23.16	24.70	26.20	27.67	29.11	30.51	31.87	33.20	34.48	35.72	36.92	38.08	39.18
50	23.22	24.75	26.25	27.72	29.15	30.55	31.92	33.24	34.52	35.76	36.96	38.11	39.22
52	23.27	24.80	26.30	27.77	29.20	30.60	31.96	33.28	34.57	35.80	37.00	38.15	39.26
54	23.32	24.85	26.35	27.81	29.25	30.65	32.01	33.33	34.61	35.85	37.04	38.19	39.29
56	23.37	24.90	26.40	27.86	29.30	30.69	32.05	33.37	34.65	35.89	37.08	38.23	39.33
58	23.42	24.95	26.45	27.91	29.34	30.74	32.09	33.41	34.69	35.93	37.12	38.26	39.36
60	23.47	25.00	26.50	27.96	29.39	30.78	32.14	33.46	34.73	35.97	37.16	38.30	39.40
f + c													
.75	0.17	0.19	0.20	0.21	0.23	0.24	0.25	0.26	0.27	0.29	0.30	0.31	0.32
1.00	0.23	0.25	0.27	0.28	0.30	0.32	0.33	0.35	0.37	0.38	0.40	0.41	0.43
1.25	0.29	0.31	0.34	0.36	0.38	0.40	0.42	0.44	0.46	0.48	0.50	0.52	0.54

Corrections to Horizontal Distances

Min-utes	13°	14°	15°	16°	17°	18°	19°	20°	21°	22°	23°	24°	25°
0	5.06	5.85	6.70	7.60	8.55	9.55	10.60	11.70	12.84	14.03	15.27	16.54	17.86
10	5.19	5.99	6.84	7.75	8.71	9.72	10.78	11.89	13.04	14.24	15.48	16.76	18.08
20	5.32	6.13	6.99	7.91	8.88	9.89	10.96	12.07	13.23	14.44	15.69	16.98	18.31
30	5.45	6.27	7.14	8.07	9.04	10.07	11.14	12.26	13.43	14.64	15.90	17.20	18.53
40	5.58	6.41	7.29	8.23	9.21	10.24	11.33	12.46	13.63	14.85	16.11	17.42	18.76
50	5.72	6.55	7.44	8.39	9.38	10.42	11.51	12.65	13.83	15.06	16.33	17.64	18.99

* *From Eshbach, Handbook of Engineering Fundamentals, John Wiley & Sons, 1936.*

STAKEOUT FOR STRUCTURES

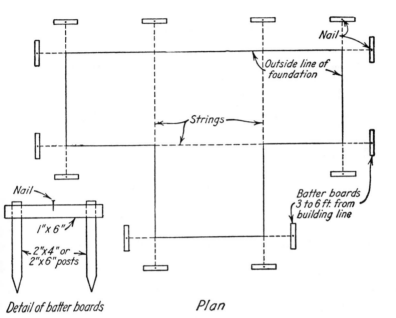

FIG. 4. Batter boards for structures.

Batter boards as illustrated are set on, or parallel to, the building or structure lines either before or after the rough excavation is completed. When set before excavating, the batter boards should be checked upon completion of the rough excavation. Points on the batter boards may be set on the outside foundation line or sometimes on the center line of columns. It is preferable to set the top of each batter board to some definite grade, such as the first-floor elevation or else some even foot above or below a working grade.

Before setting the batter boards a base line should be established and referenced in with ties. Targets may also be set on the base line projected. Angles turned from the base line should be established by the method of repetition (see p. 330) as an error of 1 minute in 300 ft. will throw the building line off 1 in.

From time to time during construction, the batter boards should be checked for disturbance or movement.

HIGHWAY CONSTRUCTION STAKEOUT

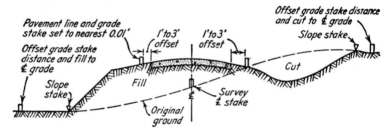

FIG. 5. Highway construction stakeout.

Before work begins, the construction centerline is staked out, usually on 50-ft. stations. Hubs are set at P.C.'s, P.T.'s, P.I.'s, and transit points. These hubs are tied in or offset, and the ties are recorded in the field book.

Offset grade stakes are set on 50-ft. stations far enough out to escape disturbance during operations where possible. Elevations of these stake tops are taken with a level, and the cut or fill to finish center-line grade is computed and marked on each stake. The distance to the toe or top of slope is marked on the offset grade stake or else the actual location of the toe or top of slope is marked with a slope stake. The station and the distance from the offset stake to center line are marked on the face of the offset stake. The superelevation plus or minus to edge of pavement and any pavement widening or curves are also marked on the offset stakes.

After rough grading is completed, blue tops or fine grade stakes are set every 50 ft. minimum. Blue tops are stakes set to fine grade and the top marked blue. Allowance for settlement or subsidence is sometimes made in setting these grades, or it may be made the contractor's responsibility, the engineer in the latter case setting the stakes to the grades shown on plan.

For concrete pavement, stakes are set usually every 50 ft. on tangents and straight grades and every 25 ft. on horizontal and vertical curves. These stakes are carefully aligned with a transit and tacks set on line. Either the tops are set to exact grade or the cut or fill is marked to finish grade.

Pavement stakes are set with a sufficient offset to allow room for the flanged bases of the forms, the offset usually being about 18 in. or 2 ft. from the edge of pavement. After the initial lane is placed, additional stakes may be set for other lanes or the forms may be set by leveling over with a line level.

For asphaltic pavements stakes are usually not set when the base has been constructed true to grade as the paving machines can be set for the required thickness. If the base is variable, steel pins for line and grade are usually set at 50- or 25-ft. intervals and offset enough to allow the machines to work. A 1-ft. offset is usually sufficient.

The amount of stakeout done for highway construction depends on the value and importance of the work, and judgment is required. For example, on cheap tertiary road construction only center-line stakes might be set at 100-ft. stations and a list of cuts and fill given to the foreman. The line and grade may then be transferred by the foreman, using a tape and hand level, to convenient trees, offset stakes, etc.

Through wooded country, stakes or marks are usually set at the clearing and grubbing limits. Trees to be saved are indicated by markings or signs.

In addition to line and grade stakes, right-of-way stakes may be necessary, also project markers and stakes set at intersection of right-of-way and adjoining property lines.

RAILROAD CONSTRUCTION STAKEOUT

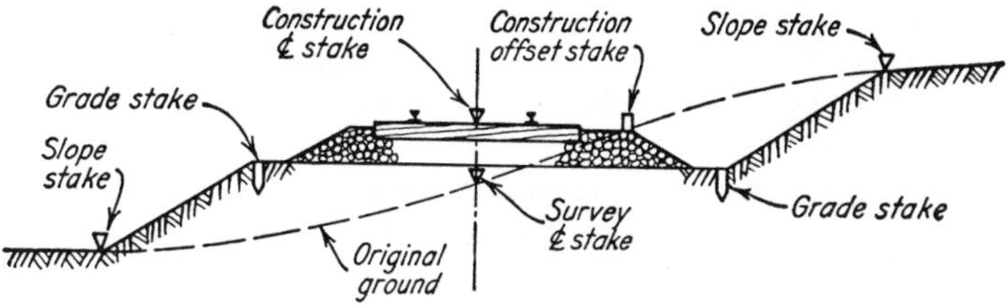

FIG. 6. Railroad construction stakeout.

Stakeout for the grading work is similar to highway stakeout.

After grading is finished, and the ballast, ties, and rails are being installed, stakes for exact alignment and grade of rails are set. These stakes are tacked for line and may be set on center line or offset about 2 ft. from one rail. The grade marked is usually finish grade to the near rail, superelevation being set for the other rail by using a track level.

AIRFIELD CONSTRUCTION STAKEOUT

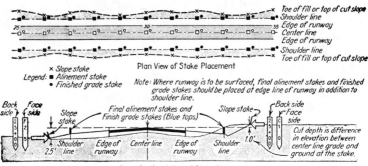

Fig. 7.   Airport stakeout.

The stakeout required differs from highway work in that the widths of runways and taxiways, together with their shoulders and graded areas, are so great that it is not practicable to set offset stakes to serve during construction.

The construction center line is staked out at 50-ft. stations and well referenced and tied in, and targets are set on the line extended.   During grading operations stakes are set continually day by day, at least one party usually being required at all times for each runway under construction.

For rough grading stakes at 50-ft. intervals both longitudinally and transversely are sufficient, but for fine grading stakes should be set at 25-ft. intervals.

Concrete pavement stakes are set exactly the same as for highways, but owing to the widths of runways and aprons it is not desirable to depend on a string level to transfer the grades for more than 2 or 3 lanes. Additional stake lines should be run in at intervals of 25 or 30 ft. transversely.

Stakeout for asphaltic pavements is the same as for highways.

Stakes for grading interior areas are usually set on 50- to 100-ft. grids and marked for cut and fill.

PIPELINE STAKEOUT *

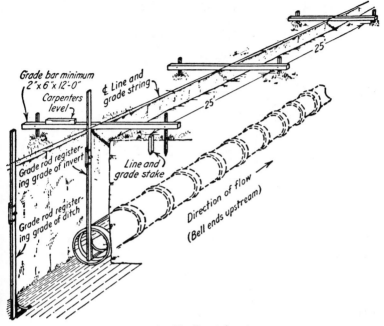

FIG. 8. Pipeline stakeout.

Before beginning the excavation, stakes should be set 25 ft. apart parallel to and offset from the center line of the drain on the side opposite to that on which earth will be thrown. Elevations of tops of stakes should be taken with a level and depth of cut marked on each. These stakes will serve as guides for the rough excavation.

Excavation should be begun at the outlet.

After the excavation is approximately to grade, batter boards should be placed across the trench opposite each stake with the top of each board at the same distance above the grade of the flow line. About 6.5 or 7 ft. above grade is good practice. The center line is then marked on the batter boards, and a string connecting these points will be directly above and parallel to the grade line. The center line at any point may then be obtained by dropping a plumb bob from the string, and the grade determined by measuring down from the string with a pole of proper length.

Laying of pipe should begin at the outlet and proceed upstream.

* *From Principles of Highway Construction Applied to Airports, Flight Strips and Other Landing Areas for Aircraft, Public Roads Administration.*

CIRCULAR CURVES

ARC DEFINITION

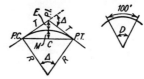

FORMULAS

$$R = \frac{5729.58}{D}$$

$$T = R \tan \frac{\Delta}{2}; \quad T = \frac{\tan 1° \text{ curve for } \Delta}{D}$$

$$L = \text{length} = \frac{100\Delta}{D}$$

$$M = R(1 - \cos \tfrac{1}{2}\Delta)$$

$$E = R\left(\frac{1}{\cos \tfrac{1}{2}\Delta} - 1\right); \quad E = \frac{\text{ext. } 1° \text{ curve for } \Delta}{D}$$

$$C = 2R \sin \frac{\Delta}{2}$$

DEFINITIONS

L = Length of circular curve.
P.I. = point of intersection.
P.C. = point of curvature.
P.T. = point of tangency.

EXAMPLE. *Given.* $\Delta = 54° 20'$; $D = 7° 40'$; P.I. = Sta. 125 + 39.88.
Required. R; T; L and Sta. of P.C. and P.T.
Solution.

$$R = \frac{5729.58}{7° 40'} = 747.34'.$$

$$T = 747.34 \ (\tan 27° 10') = 747.34(0.513195) = 383.53'.$$

Also, from p. 294 (funct. 1° curve) by interpolation, tan 1° curve for $\Delta 54° 20' = 2940.41$.

$$\therefore T = \frac{2940.41}{7° 40'} = 383.53'.$$

P.C. = Sta. 125 + 39.88 − 383.53 = Sta. 121 + 56.35.

$$L = \frac{100\Delta}{D} = \frac{100(54° 20')}{7° 40'} = 708.70'.$$

P.T. = Sta. 121 + 56.35 + 708.70 = Sta. 128 + 65.05.

288

DEFLECTIONS

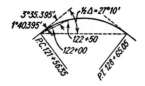

FORMULAS

Deflection angle $= \dfrac{D}{2}$ for 100'; $\dfrac{D}{4}$ for 50', etc.

For c feet (in minutes) $= 0.3\,cD$.

Deflection angle (in minutes) from P.C. to P.T. $= 0.3LD$.

Also, deflection angle (in degrees) from P.C. to P.T. $= \dfrac{\Delta}{2}$.

EXAMPLE. *Given.* $\Delta = 54° 20'$; $D = 7° 40'$; $L = 708.70$; P.C. $=$ **Sta.** **121** $+$ **56.35**; P.T. $=$ Sta. $128 + 65.05$.

Required. Deflection angle from P.C. to Sta. $122 + 00$; Sta. $122 + 50$ and P.T. Sta. $128 + 65.05$.

Solution.

Sta. $122 + 00 -$ P.C. Sta. $121 + 56.35 = 43.65'$.

∴ Deflection angle to Sta. $122 + 00 = 0.3 \times 43.65 \times 7° 40' = 100.395'$
$= 1° 40.395'$.

Deflection angle to Sta. $122 + 50 = 1° 40.395' + \dfrac{7° 40'}{4} = 1° 40.395'$
$+ 1° 55' = 3° 35.395'$.

Deflection angle to P.T. Sta. $128 + 65.05 = 0.3 \times 708.70 \times 7° 40'$
$= 27° 10'$.

Also, deflection angle to P.T. Sta. $128 + 65.05 = \dfrac{\Delta}{2} = \dfrac{54° 20'}{2}$
$= 27° 10'$.

EXTERNALS

Compound curve Avoid   Reverse curve Avoid   Broken-back curve Avoid

EXAMPLE. *Given.* $\Delta = 54° 20'$; $D = 7° 40'$; $R = 747.34'$.
Required. External "E".
Solution.

$$E = 747.34 \left(\frac{1}{.8896822} - 1 \right) = 92.67'.$$

Also, from p. 294 (funct. 1° curve) by interpolation, external 1° curve
for $\Delta 54° 20' = 710.48$.

$$\therefore E = \frac{710.48}{7° 40'} = 92.67'.$$

MINIMUM CURVATURE *

The curve should be at least 500 ft. long for $\Delta = 5$ degrees and increase 100 ft. in length for each decrease of 1 degree in the Δ.

Where topography permits, use simple 0° 20′ to 1° 00′ curves without superelevation or widening.

MAXIMUM CURVATURE *

| | DEGREE OF CURVE | |
ASSUMED DESIGN SPEED, M.P.H.	DESIRABLE MAXIMUM	ABSOLUTE MAXIMUM
30	20	25
40	11	14
50	7	9
60	5	6
70	3	4

TANGENT OFFSETS

The approximate offset from the tangent to the curve at any distance from the P.C. $= \dfrac{\text{distance}^2}{2R}$.

CHORD DEFINITION (R. R. CURVE)

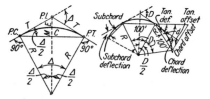

D (in degrees) subtends 100′ chord.

$$D = 100\,\Delta/L$$

$$D = \frac{\tan 1° \text{ curve}}{T} \quad \text{(approx.)}.$$

$$D = \frac{\text{ext. } 1° \text{ curve}}{E} \quad \text{(approx.)}.$$

Tan offset $= \dfrac{\text{chord}^2}{2R} = \text{chord} \cdot \sin \text{def.} = \left(\dfrac{\text{chord}^2}{100}\right)$ tan offset, Table 2.

Chord offset $= 2$ tan deflection for 100′ chord $= 100 \sin D°$.

* *From Geometric Design Standards by A.A.S.H.O.*

Tan def. = $\frac{1}{2}D\frac{chord}{100}$; for c feet = $0.3D \times c$ = def. for $1'$ in Table 2 $\times c$.

Chord def. = 2 tan def. = D for $100'$ chord.

FORMULAS

$$R = \frac{50}{\sin D/2} ; \quad R = T \cdot \cotan\frac{\Delta}{2} ; \quad R = \frac{E}{\text{exsec } \Delta/2} ; \quad T = R \cdot \tan\frac{\Delta}{2} ;$$

$$T = \frac{50 \tan \Delta/2}{\sin \Delta/2} ; \quad T = \frac{\tan 1° \text{ curve}}{D} + \text{corr.*} \quad L = 100\frac{\Delta}{D} ; \quad \Delta = \frac{DL}{100} ;$$

$$M = R\left(1 - \cos\frac{\Delta}{2}\right) ; \quad M = R \text{ vers}\frac{\Delta}{2} ; \quad E = T \cdot \tan\frac{\Delta}{4} ;$$

$$E = \frac{R}{\cos \Delta/2} - R ; \quad E = R \cdot \text{exsec}\frac{\Delta}{2}. \quad C = 2R \cdot \sin\frac{\Delta}{2} ;$$

$$E = \frac{\text{ext. } 1° \text{ curve}}{D} + \text{correction.*} \quad \sin\frac{D}{2} = \frac{50}{R} ; \quad \sin\frac{D}{2} = \frac{50 \tan \Delta/2}{T}$$

EXAMPLE. *Given.* $\Delta = 54° 20'$; $D = 7° 40'$, P.I. Sta. $125 + 39.88$.
Required. R, T, L, P.C., and P.T.
Solution.
$R = 50 \div \sin 3° 50' = 747.89$.
$T = 747.89 (\tan 27° 10') = 383.81$.
$L = 100\Delta \div D = 100 (54° 20') \div 7° 40' = 708.70$.
P.C. = P.I. Sta. $125 + 39.88 - 383.81 =$ Sta. $121 + 56.07$.
P.T. = Sta. $121 + 56.07 + 708.70 =$ Sta. $128 + 64.77$.

* See p. 295.

TABLE 2. RADII, DEFLECTIONS, OFFSETS, ORDINATES, CHORDS AND ARCS—100′ CHORDS *

D	Radius	Def. for 1 Ft.	Tan Off-set	Mid Ord.	For Subchords Add				Actual Arc per 100′ Sta.	Long Chords				*D*
					10′	20′	25′	50′		2 Sta.	3 Sta.	4 Sta.	5 Sta.	
30′	11,459.2	0.15	0.436	0.109					100.000	200.000	299.99	399.98	499.96	30′
1°	5,729.65	0.30	0.873	0.218					100.001	199.99	299.97	399.92	499.85	1°
30′	3,819.83	0.45	1.309	0.327					100.003	199.98	299.93	399.83	499.66	30′
2°	2,864.93	0.60	1.745	0.436					100.005	199.97	299.88	399.70	499.39	2°
30′	2,292.01	0.75	2.181	0.545					100.008	199.95	299.81	399.52	499.05	30′
3°	1,910.08	0.90	2.618	0.654					100.011	199.93	299.73	399.32	498.63	3°
30′	1,637.28	1.05	3.054	0.764					100.015	199.91	299.63	399.07	498.14	30′
4°	1,432.69	1.20	3.490	0.872				0.01	100.020	199.88	299.51	398.78	497.57	4°
30′	1,273.57	1.35	3.926	0.982				0.01	100.026	199.85	299.38	398.46	496.92	30′
5°	1,146.28	1.50	4.362	1.091			0.01	0.01	100.032	199.81	299.24	398.10	496.20	5°
30′	1,042.14	1.65	4.798	1.200			0.01	0.01	100.038	199.77	299.08	397.70	495.41	30′
6°	955.37	1.80	5.234	1.309		0.01	0.01	0.02	100.046	199.73	298.90	397.26	494.53	6°
30′	881.95	1.95	5.669	1.418		0.01	0.01	0.02	100.054	199.68	298.71	396.79	493.59	30′
7°	819.02	2.10	6.105	1.528		0.01	0.01	0.02	100.062	199.63	298.51	396.28	492.57	7°
8°	716.78	2.40	6.976	1.746	0.01	0.02	0.02	0.03	100.081	199.51	298.05	395.14	490.31	8°
10°	573.69	3.00	8.716	2.183	0.01	0.02	0.03	0.05	100.127	199.24	296.96	392.42	484.90	10°
12°	478.34	3.60	10.45	2.620	0.02	0.04	0.04	0.07	100.183	198.90	295.63	389.12	478.34	12°
14°	410.28	4.20	12.18	3.058	0.02	0.05	0.06	0.09	100.249	198.51	294.06	385.23	470.65	14°
16°	359.27	4.80	13.92	3.496	0.03	0.06	0.08	0.12	100.326	198.05	292.25	380.76	461.86	16°
18°	319.62	5.40	15.64	3.935	0.04	0.08	0.10	0.15	100.412	197.54	290.21	375.74	452.02	18°
20°	287.94	6.00	17.37	4.374	0.05	0.10	0.12	0.19	100.510	196.96	287.94	370.17	441.15	20°
22°	262.04	6.60	19.08	4.814	0.06	0.12	0.14	0.23	100.617	196.33	285.44	364.06	429.31	22°
24°	240.49	7.20	20.79	5.255	0.07	0.14	0.17	0.28	100.735	195.63	282.71	357.43	416.54	24°
30°	193.18	9.00	25.88	6.583	0.11	0.22	0.29	0.43	101.152	193.19	273.21	334.61	373.21	30°

* *Adapted from Railroad Curve Tables by Eugene Dietzgen Co.*

TABLE 3. MINUTES IN DECIMALS OF A DEGREE, SECONDS IN DECIMALS OF A MINUTE *

1	0.0167	11	0.1833	21	0.3500	31	0.5167	41	0.6833	51	0.8500
2	0.0333	12	0.2000	22	0.3667	32	0.5333	42	0.7000	52	0.8667
3	0.0500	13	0.2167	23	0.3833	33	0.5500	43	0.7167	53	0.8833
4	0.0667	14	0.2333	24	0.4000	34	0.5667	44	0.7333	54	0.9000
5	0.0833	15	0.2500	25	0.4167	35	0.5833	45	0.7500	55	0.9167
6	0.1000	16	0.2667	26	0.4333	36	0.6000	46	0.7667	56	0.9333
7	0.1167	17	0.2833	27	0.4500	37	0.6167	47	0.7833	57	0.9500
8	0.1333	18	0.3000	28	0.4667	38	0.6333	48	0.8000	58	0.9667
9	0.1500	19	0.3167	29	0.4833	39	0.6500	49	0.8167	59	0.9833
10	0.1667	20	0.3333	30	0.5000	40	0.6667	50	0.8333	60	1.0000

Proportional Part for $1'' = 0.000278$ of $1°$

USE OF TABLES 2 AND 3

Given	Required	Solution	
$D = 2° 30'$	Deflection for 35 ft.	$= 0.75 \times 35 = 26.25$	$= 26' 15''$
$D = 4°$	Tan offset for 125 ft.	$= 3.49(1.25/100)^2$	$= 5.45$ ft.
$D = 10°$	Mid ord. for 30 ft. chord	$= 0.0001 \times 30^2 \times 2.183$	$= 0.196$ ft.
$D = 14°$	Length of nominal 20 ft. sub chord	$= 20 + 0.05$	$= 20.05$ ft.
$D = 20°$	Actual length of arc for $L = 600$ ft. (6 Sta.)	$= 100.51 \times 6$	$= 603.06$ ft.
$D = 3°$	Long chord for 3 Sta.	= From Table 2	$= 299.73$ ft.
$\Delta = 27° 05' 11''$	Δ in decimals of $°$	From Table 3 $= 27 + 0.0833 + 11 \times 0.000278$	$= 27.086°$

* *Adapted from Railroad Curve Tables by Eugene Dietzgen Co.*

CIRCULAR CURVES

TABLE 4. FUNCTIONS OF 1° CURVE

See pp. 288, 289, 290 for use of table.

Central Angle	Tangent	External	Central Angle	Tangent	External	Central Angle	Tangent	External	Central Angle	Tangent	External
1°	50.00	0.22	31°	1588.95	216.25	61°	3374.98	920.1	91°	5830.46	2444.9
30'	75.00	0.49	30'	1615.91	223.51	30'	3408.74	937.3	30'	5881.58	2481.5
2°	100.01	0.87	32°	1642.93	230.90	62°	3442.68	954.8	92°	5933.15	2518.5
30'	125.02	1.36	30'	1670.02	238.43	30'	3476.79	972.4	30'	5985.20	2556.0
3°	150.03	1.96	33°	1697.18	246.08	63°	3511.09	990.2	93°	6037.72	2594.0
30'	175.05	2.67	30'	1724.41	253.87	30'	3545.57	1008.3	30'	6090.72	2632.6
4°	200.08	3.49	34°	1751.71	261.80	64°	3580.24	1026.6	94°	6144.22	2671.6
30'	225.12	4.42	30'	1779.08	269.86	30'	3615.09	1045.2	30'	6198.22	2711.2
5°	250.16	5.46	35°	1806.53	278.05	65°	3650.14	1063.9	95°	6252.74	2751.3
30'	275.21	6.61	30'	1834.05	286.39	30'	3685.39	1082.9	30'	6307.77	2792.0
6°	300.27	7.86	36°	1861.65	294.86	66°	3720.83	1102.2	96°	6363.34	2833.2
30'	325.35	9.23	30'	1889.33	303.47	30'	3756.48	1121.7	30'	6419.45	2875.0
7°	350.44	10.71	37°	1917.09	312.22	67°	3792.33	1141.4	97°	6476.11	2917.3
30'	375.54	12.29	30'	1944.93	321.11	30'	3828.38	1161.3	30'	6533.33	2960.3
8°	400.65	13.99	38°	1972.85	330.15	68°	3864.65	1181.6	98°	6591.13	3003.8
30'	425.78	15.80	30'	2000.86	339.32	30'	3901.13	1202.0	30'	6649.50	3047.9
9°	450.93	17.72	39°	2028.95	348.64	69°	3937.83	1222.7	99°	6708.47	3092.7
30'	476.09	19.75	30'	2057.13	358.11	30'	3974.75	1243.7	30'	6768.05	3138.1
10°	501.27	21.89	40°	2085.40	367.72	70°	4011.89	1265.0	100°	6828.25	3184.1
30'	526.47	24.14	30'	2113.75	377.47	30'	4049.27	1286.5	30'	6889.07	3230.8
11°	551.70	26.50	41°	2142.20	387.38	71°	4086.87	1308.2	101°	6950.53	3278.1
30'	576.94	28.97	30'	2170.74	397.43	30'	4124.71	1330.3	30'	7012.65	3326.1
12°	602.20	31.56	42°	2199.38	407.64	72°	4162.78	1352.6	102°	7075.44	3374.9
30'	627.49	34.26	30'	2228.11	417.99	30'	4201.10	1375.2	30'	7138.91	3424.3
13°	652.80	37.07	43°	2256.94	428.50	73°	4239.66	1398.0	103°	7203.07	3474.4
30'	678.14	39.99	30'	2285.87	439.16	30'	4278.48	1421.2	30'	7267.94	3525.2
14°	703.50	43.03	44°	2314.90	449.98	74°	4317.55	1444.6	104°	7333.53	3576.8
30'	728.89	46.18	30'	2344.03	460.95	30'	4356.87	1468.4	30'	7399.85	3629.2
15°	754.31	49.44	45°	2373.27	472.08	75°	4396.46	1492.4	105°	7466.93	3682.3
30'	779.76	52.82	30'	2402.61	483.37	30'	4436.31	1516.7	30'	7534.78	3736.2
16°	805.24	56.31	46°	2432.06	494.82	76°	4476.44	1541.4	106°	7603.41	3791.0
30'	830.75	59.91	30'	2461.62	506.42	30'	4516.83	1566.3	30'	7672.94	3846.5
17°	856.29	63.63	47°	2491.29	518.20	77°	4557.51	1591.6	107°	7743.08	3902.9
30'	881.87	67.47	30'	2521.07	530.13	30'	4598.47	1617.1	30'	7814.16	3960.1
18°	907.48	71.42	48°	2550.97	542.23	78°	4639.72	1643.0	108°	7886.09	4018.2
30'	933.12	75.49	30'	2580.99	554.50	30'	4681.26	1669.2	30'	7958.89	4077.2
19°	958.80	79.67	49°	2611.12	566.94	79°	4723.10	1695.8	109°	8032.57	4137.1
30'	984.52	83.97	30'	2641.37	579.54	30'	4765.24	1722.7	30'	8107.17	4197.9
20°	1010.28	88.39	50°	2671.75	592.32	80°	4807.69	1749.9	110°	8182.69	4259.7
30'	1036.08	92.92	30'	2702.24	605.27	30'	4850.45	1777.4	30'	8259.15	4322.4
21°	1061.91	97.58	51°	2732.87	618.39	81°	4893.52	1805.3	111°	8336.59	4386.1
30'	1087.79	102.35	30'	2763.62	631.69	30'	4936.92	1833.6	30'	8415.01	4450.9
22°	1113.72	107.24	52°	2794.50	645.17	82°	4980.65	1862.2	112°	8494.45	4516.6
30'	1139.68	112.25	30'	2825.52	658.83	30'	5024.71	1891.2	30'	8574.92	4583.4
23°	1165.70	117.38	53°	2856.66	672.66	83°	5069.10	1920.5	113°	8656.45	4651.3
30'	1191.75	122.63	30'	2887.95	686.68	30'	5113.84	1950.3	30'	8739.06	4720.3
24°	1217.86	128.00	54°	2919.37	700.89	84°	5158.93	1980.4	114°	8822.78	4790.4
30'	1244.01	133.50	30'	2950.93	715.28	30'	5204.38	2010.8	30'	8907.63	4861.7
25°	1270.22	139.11	55°	2982.63	729.85	85°	5250.19	2041.7	115°	8993.64	4934.1
30'	1296.47	144.85	30'	3014.48	744.62	30'	5296.37	2073.0	30'	9080.83	5007.8
26°	1322.78	150.71	56°	3046.47	759.58	86°	5342.92	2104.7	116°	9169.24	5082.7
30'	1349.14	156.70	30'	3078.61	774.73	30'	5389.85	2136.7	30'	9258.89	5158.8
27°	1375.55	162.81	57°	3110.91	790.08	87°	5437.17	2169.2	117°	9349.82	5236.2
30'	1402.02	169.04	30'	3143.35	805.62	30'	5484.88	2202.0	30'	9442.05	5315.0
28°	1428.54	175.41	58°	3175.96	821.37	88°	5532.99	2235.5	118°	9535.62	5395.1
30'	1455.13	181.89	30'	3208.72	837.31	30'	5581.51	2269.3	30'	9630.55	5476.5
29°	1481.77	188.51	59°	3241.64	853.46	89°	5630.44	2303.5	119°	9726.89	5559.4
30'	1508.47	195.25	30'	3274.72	869.82	30'	5679.79	2338.2	30'	9824.67	5643.8
30°	1535.24	202.12	60°	3307.97	886.38	90°	5729.58	2373.3	120°	9923.92	5729.7
30'	1562.06	209.12	30'	3341.39	903.15	30'	5779.80	2408.9	30'	10,024.68	5817.0

TABLE 5. CORRECTIONS FOR TANGENTS AND EXTERNALS

For railroad and highway curves laid out by the chord definition these corrections are to be added to the values found, using table on p. 294, in order to obtain the corrected tangents and external distances.

For Tangents Add *

Central Angle	Degree of Curve													
	5°	10°	15°	20°	25°	30°	35°	40°	45°	50°	55°	60°	65°	70°
10°	.03	.06	.09	.13	.16	.19	.22	.25	.28	.31	.34	.38	.42	.46
15°	.04	.10	.14	.19	.24	.29	.34	.39	.45	.51	.53	.58	.63	.68
20°	.06	.13	.19	.26	.32	.39	.45	.51	.58	.65	.72	.79	.84	.90
25°	.08	.16	.24	.33	.40	.49	.58	.67	.75	.83	.90	.99	1.06	1.14
30°	.10	.19	.29	.39	.49	.59	.69	.79	.89	.99	1.09	1.20	1.29	1.39
35°	.11	.22	.34	.47	.58	.69	.70	.81	.92	1.04	1.29	1.42	1.54	1.66
40°	.13	.26	.40	.53	.67	.80	.93	1.06	1.20	1.34	1.49	1.64	1.79	1.94
45°	.15	.30	.44	.60	.76	.91	1.06	1.21	1.37	1.52	1.70	1.87	2.04	2.21
50°	.17	.34	.51	.68	.85	1.02	1.19	1.36	1.54	1.72	1.91	2.10	2.29	2.48
55°	.19	.38	.57	.76	.95	1.14	1.32	1.52	1.72	1.92	2.14	2.35	2.56	2.77
60°	.21	.42	.63	.84	1.05	1.27	1.49	1.71	1.94	2.17	2.38	2.60	2.83	3.07
65°	.23	.46	.69	.93	1.16	1.40	1.64	1.88	2.13	2.38	2.63	2.88	3.13	3.39
70°	.25	.51	.76	1.02	1.28	1.54	1.80	2.06	2.33	2.60	2.88	3.16	3.44	3.72
75°	.27	.56	.83	1.12	1.40	1.69	1.98	2.27	2.57	2.87	3.16	3.47	3.78	4.09
80°	.30	.61	.91	1.22	1.53	1.84	2.15	2.46	2.78	3.10	3.44	3.78	4.12	4.46
85°	.33	.66	1.00	1.33	1.68	2.02	2.36	2.70	3.05	3.40	3.77	4.14	4.55	4.89
90°	.36	.72	1.09	1.45	1.83	2.20	2.57	2.94	3.32	3.70	4.10	4.50	4.91	5.32
95°	.39	.79	1.19	1.55	2.00	2.40	2.80	3.20	3.61	4.02	4.40	4.98	5.38	5.83
100°	.43	.86	1.30	1.74	2.18	2.62	3.06	3.50	3.95	4.40	4.88	5.37	5.85	6.34
110°	.51	1.03	1.56	2.08	2.61	3.14	3.67	4.21	4.76	5.31	5.86	6.43	7.01	7.60
120°	.62	1.25	1.93	2.52	3.16	3.81	4.45	5.11	5.77	6.44	7.12	7.80	8.50	9.22

For Externals Add *

Central Angle	Degree of Curve													
	5°	10°	15°	20°	25°	30°	35°	40°	45°	50°	55°	60°	65°	70°
10°	.001	.003	.004	.006	.007	.008	.009	.011	.012	.014	.015	.017	.018	.020
15°	.003	.007	.010	.014	.018	.023	.027	.029	.032	.035	.039	.043	.047	.051
20°	.006	.011	.017	.022	.028	.034	.038	.045	.051	.057	.063	.070	.076	.083
25°	.009	.018	.027	.036	.046	.056	.065	.074	.083	.093	.106	.120	.127	.135
30°	.013	.025	.038	.051	.065	.078	.090	.103	.116	.129	.149	.170	.179	.188
35°	.018	.035	.054	.072	.086	.109	.131	.153	.175	.197	.213	.230	.247	.264
40°	.023	.046	.070	.093	.117	.141	.172	.203	.234	.265	.277	.290	.315	.341
45°	.030	.060	.093	.119	.153	.184	.216	.254	.289	.325	.351	.378	.411	.445
50°	.037	.075	.116	.151	.189	.227	.266	.305	.345	.384	.425	.467	.508	.550
55°	.046	.093	.142	.188	.236	.283	.332	.381	.420	.479	.530	.582	.641	.700
60°	.056	.112	.168	.225	.283	.340	.398	.457	.516	.575	.636	.697	.774	.851
65°	.067	.135	.204	.273	.343	.412	.483	.554	.625	.697	.711	.845	.922	1.01
70°	.080	.159	.240	.321	.403	.485	.568	.652	.735	.819	.906	.994	1.08	1.17
75°	.095	.182	.286	.383	.480	.578	.678	.777	.877	.977	1.07	1.18	1.29	1.39
80°	.110	.220	.332	.445	.558	.671	.787	.903	1.02	1.13	1.25	1.38	1.50	1.62
85°	.128	.259	.391	.524	.657	.790	.926	1.06	1.20	1.34	1.47	1.62	1.76	1.91
90°	.149	.299	.450	.603	.756	.910	1.07	1.22	1.38	1.54	1.70	1.87	2.03	2.20
95°	.174	.350	.522	.706	.985	1.06	1.25	1.43	1.62	1.80	1.99	2.18	2.38	2.58
100°	.200	.401	.604	.809	1.01	1.22	1.43	1.64	1.85	2.06	2.28	2.50	2.73	2.96
110°	.268	.536	.806	1.08	1.35	1.63	1.91	2.20	2.48	2.76	3.05	3.35	3.66	3.96
120°	.360	.721	1.08	1.45	1.82	2.19	2.57	2.95	3.33	3.72	4.11	4.50	4.91	5.32

* Adapted from Dietzgen's Railroad Curve Tables by Eugene Dietzgen Co.

TABLE 6. DEFLECTIONS AND CHORD LENGTHS FOR CIRCULAR CURVES

For Laying Out Arc Definition Curves By Measured Chords

| Degree of Curve | Radius | Deflection for Arc Length | | | | Chord for Arc Length | | |
| | | Deflection = arc length (0.3° of curve) | | | | Chord = 2R sin def. | | |
		1'	25'	50'	100'	25'	50'	100'
0° 30'	11,459.16	0° 00.15'	0° 03.75'	0° 07.50'	0° 15.00'	25.00'	50.00'	100.00'
1°	5,729.58	0° 00.30'	0° 07.50'	0° 15.00'	0° 30.00'	25.00'	50.00'	100.00'
30'	3,819.72	0° 00.45'	0° 11.25'	0° 22.50'	0° 45.00'	25.00'	50.00'	100.00'
2°	2,864.79	0° 00.60'	0° 15.00'	0° 30.00'	1° 00.00'	25.00'	50.00'	100.00'
30'	2,291.83	0° 00.75'	0° 18.75'	0° 37.50'	1° 15.00'	25.00'	50.00'	99.99'
3°	1,909.86	0° 00.90'	0° 22.50'	0° 45.00'	1° 30.00'	25.00'	50.00'	99.99'
30'	1,637.02	0° 01.05'	0° 26.25'	0° 52.50'	1° 45.00'	25.00'	50.00'	99.98'
4°	1,432.40	0° 01.20'	0° 30.00'	1° 00.00'	2° 00.00'	25.00'	50.00'	99.98'
30'	1,273.24	0° 01.35'	0° 33.75'	1° 07.50'	2° 15.00'	25.00'	50.00'	99.97'
5°	1,145.92	0° 01.50'	0° 37.50'	1° 15.00'	2° 30.00'	25.00'	50.00'	99.97'
30'	1,041.74	0° 01.65'	0° 41.25'	1° 22.50'	2° 45.00'	25.00'	50.00'	99.96'
6°	954.93	0° 01.80'	0° 45.00'	1° 30.00'	3° 00.00'	25.00'	50.00'	99.95'
30'	881.47	0° 01.95'	0° 48.75'	1° 37.50'	3° 15.00'	25.00'	50.00'	99.95'
7°	818.51	0° 02.10'	0° 52.50'	1° 45.00'	3° 30.00'	25.00'	50.00'	99.94'
30'	763.94	0° 02.25'	0° 56.25'	1° 52.50'	3° 45.00'	25.00'	49.99'	99.93'
8°	716.20	0° 02.40'	1° 00.00'	2° 00.00'	4° 00.00'	25.00'	49.99'	99.92'
30'	674.07	0° 02.55'	1° 03.75'	2° 07.50'	4° 15.00'	25.00'	49.99'	99.91'
9°	636.62	0° 02.70'	1° 07.50'	2° 15.00'	4° 30.00'	25.00'	49.99'	99.90'
30'	603.11	0° 02.85'	1° 11.25'	2° 22.50'	4° 45.00'	25.00'	49.99'	99.89'
10°	572.96	0° 03.00'	1° 15.00'	2° 30.00'	5° 00.00'	25.00'	49.98'	99.87'
11°	520.87	0° 03.30'	1° 22.50'	2° 45.00'	5° 30.00'	25.00'	49.98'	99.85'
12°	477.46	0° 03.60'	1° 30.00'	3° 00.00'	6° 00.00'	25.00'	49.98'	99.82'
13°	440.74	0° 03.90'	1° 37.50'	3° 15.00'	6° 30.00'	25.00'	49.97'	99.79'
14°	409.26	0° 04.20'	1° 45.00'	3° 30.00'	7° 00.00'	25.00'	49.97'	99.75'
15°	381.97	0° 04.50'	1° 52.50'	3° 45.00'	7° 30.00'	25.00'	49.96'	99.72'
16°	358.10	0° 04.80'	2° 00.00'	4° 00.00'	8° 00.00'	25.00'	49.96'	99.68'
17°	337.03	0° 05.10'	2° 07.50'	4° 15.00'	8° 30.00'	25.00'	49.95'	99.63'
18°	318.31	0° 05.40'	2° 15.00'	4° 30.00'	9° 00.00'	24.99'	49.95'	99.59'
19°	301.56	0° 05.70'	2° 22.50'	4° 45.00'	9° 30.00'	24.99'	49.94'	99.54'
20°	286.48	0° 06.00'	2° 30.00'	5° 00.00'	10° 00.00'	24.99'	49.94'	99.49'
21°	272.84	0° 06.30'	2° 37.50'	5° 15.00'	10° 30.00'	24.99'	49.93'	99.44'
22°	260.44	0° 06.60'	2° 45.00'	5° 30.00'	11° 00.00'	24.99'	49.92'	99.39'
23°	249.11	0° 06.90'	2° 52.50'	5° 45.00'	11° 30.00'	24.99'	49.92'	99.33'
24°	238.73	0° 07.20'	3° 00.00'	6° 00.00'	12° 00.00'	24.99'	49.91'	99.27'
38° 12'	150	0° 11.45'	4° 46.48'	9° 32.96'	19° 05.92'	24.97'	49.77'	98.16'
28° 39'	200	0° 08.59'	3° 34.86'	7° 09.72'	14° 19.44'	24.98'	49.87'	98.96'
25° 28'	225	0° 07.64'	3° 10.99'	6° 21.97'	12° 43.94'	24.99'	49.90'	99.18'
22° 55'	250	0° 06.88'	2° 51.89'	5° 43.78'	11° 27.55'	24.99'	49.92'	99.34'
20° 50'	275	0° 06.25'	2° 36.26'	5° 12.52'	10° 25.04'	24.99'	49.93'	99.45'
19° 06'	300	0° 05.73'	2° 23.24'	4° 46.48'	9° 32.96'	24.99'	49.94'	99.54'
17° 38'	325	0° 05.29'	2° 12.22'	4° 24.44'	8° 48.88'	24.99'	49.95'	99.61'
16° 22'	350	0° 04.91'	2° 02.78'	4° 05.55'	8° 11.11'	25.00'	49.96'	99.66'
15° 17'	375	0° 04.58'	1° 54.59'	3° 49.18'	7° 38.37'	25.00'	49.96'	99.70'
14° 19'	400	0° 04.30'	1° 47.43'	3° 34.86'	7° 09.72'	25.00'	49.97'	99.74'
12° 44'	450	0° 03.82'	1° 35.49'	3° 10.99'	6° 21.97'	25.00'	49.97'	99.79'
11° 28'	500	0° 03.44'	1° 25.94'	2° 51.89'	5° 43.77'	25.00'	49.98'	99.83'
10° 25'	550	0° 03.13'	1° 18.13'	2° 36.26'	5° 12.52'	25.00'	49.98'	99.86'
9° 33'	600	0° 02.86'	1° 11.62'	2° 23.24'	4° 46.48'	25.00'	49.99'	99.89'
8° 50'	650	0° 02.64'	1° 06.11'	2° 12.22'	4° 24.44'	25.00'	49.99'	99.90'
8° 11'	700	0° 02.46'	1° 01.39'	2° 02.78'	4° 05.55'	25.00'	49.99'	99.92'
7° 38'	750	0° 02.29'	0° 57.30'	1° 54.59'	3° 49.18'	25.00'	50.00'	99.93'
7° 10'	800	0° 02.15'	0° 53.71'	1° 47.43'	3° 34.86'	25.00'	50.00'	99.93'
6° 44'	850	0° 02.02'	0° 50.56'	1° 41.11'	3° 22.22'	25.00'	50.00'	99.94'
6° 22'	900	0° 01.91'	0° 47.75'	1° 35.49'	3° 10.99'	25.00'	50.00'	99.95'
6° 02'	950	0° 01.81'	0° 45.23'	1° 30.47'	3° 00.93'	25.00'	50.00'	99.95'
5° 44'	1000	0° 01.72'	0° 42.97'	1° 25.94'	2° 51.89'	25.00'	50.00'	99.96'

Deflection for curves of even radii $= \dfrac{1718.873}{R}$ arc length.

TABLE 7. LENGTHS OF CIRCULAR ARCS FOR UNIT RADIUS *

By the use of this table, the length of any arc may be found if the length of the radius and the angle of the segment are known. EXAMPLE. *Required:* The length of arc of segment of 32° 15′ 27″ with radius of 24 ft. 3 in.

From table: Length of arc (radius 1) for 32° = 0.5585054

15′ = 0.0043633

27″ = 0.0001309

—————————

0.5629996

0.5629996 × 24.25 (length of radius) = 13.65 ft.

Degrees						Minutes		Seconds	
°		°		°		′		″	
1	.017 4533	61	1.064 6508	121	2.111 8484	1	.000 2909	1	.000 0048
2	.034 9066	62	1.082 1041	122	2.129 3017	2	.000 5818	2	.000 0097
3	.052 3599	63	1.099 5574	123	2.146 7550	3	.000 8727	3	.000 0145
4	.069 8132	64	1.117 0107	124	2.164 2083	4	.001 1636	4	.000 0194
5	.087 2665	65	1.134 4640	125	2.181 6616	5	.001 4544	5	.000 0242
6	.104 7198	66	1.151 9173	126	2.199 1149	6	.001 7453	6	.000 0291
7	.122 1730	67	1.169 3706	127	2.216 5682	7	.002 0362	7	.000 0339
8	.139 6263	68	1.186 8239	128	2.234 0214	8	.002 3271	8	.000 0388
9	.157 0796	69	1.204 2772	129	2.251 4747	9	.002 6180	9	.000 0436
10	.174 5329	70	1.221 7305	130	2.268 9280	10	.002 9089	10	.000 0485
11	.191 9862	71	1.239 1838	131	2.286 3813	11	.003 1998	11	.000 0533
12	.209 4395	72	1.256 6371	132	2.303 8346	12	.003 4907	12	.000 0582
13	.226 8928	73	1.274 0904	133	2.321 2879	13	.003 7815	13	.000 0630
14	.244 3461	74	1.291 5436	134	2.338 7412	14	.004 0724	14	.000 0679
15	.261 7994	75	1.308 9969	135	2.356 1945	15	.004 3633	15	.000 0727
16	.279 2527	76	1.326 4502	136	2.373 6478	16	.004 6542	16	.000 0776
17	.296 7060	77	1.343 9035	137	2.391 1011	17	.004 9451	17	.000 0824
18	.314 1593	78	1.361 3568	138	2.408 5544	18	.005 2360	18	.000 0873
19	.331 6126	79	1.378 8101	139	2.426 0077	19	.005 5269	19	.000 0921
20	.349 0659	80	1.396 2634	140	2.443 4610	20	.005 8178	20	.000 0970
21	.366 5191	81	1.413 7167	141	2.460 9142	21	.006 1087	21	.000 1018
22	.383 9724	82	1.431 1700	142	2.478 3675	22	.006 3995	22	.000 1067
23	.401 4257	83	1.448 6233	143	2.495 8208	23	.006 6904	23	.000 1115
24	.418 8790	84	1.466 0766	144	2.513 2741	24	.006 9813	24	.000 1164
25	.436 3323	85	1.483 5299	145	2.530 7274	25	.007 2722	25	.000 1212
26	.453 7856	86	1.500 9832	146	2.548 1807	26	.007 5631	26	.000 1261
27	.471 2389	87	1.518 4364	147	2.565 6340	27	.007 8540	27	.000 1309
28	.488 6922	88	1.535 8897	148	2.583 0873	28	.008 1449	28	.000 1357
29	.506 1455	89	1.553 3430	149	2.600 5406	29	.008 4358	29	.000 1406
30	.523 5988	90	1.570 7963	150	2.617 9939	30	.008 7266	30	.000 1454
31	.541 0521	91	1.588 2496	151	2.635 4472	31	.009 0175	31	.000 1503
32	.558 5054	92	1.605 7029	152	2.652 9005	32	.009 3084	32	.000 1551
33	.575 9587	93	1.623 1562	153	2.670 3538	33	.009 5993	33	.000 1600
34	.593 4119	94	1.640 6095	154	2.687 8070	34	.009 8902	34	.000 1648
35	.610 8652	95	1.658 0628	155	2.705 2603	35	.010 1811	35	.000 1697
36	.628 3185	96	1.675 5161	156	2.722 7136	36	.010 4720	36	.000 1745
37	.645 7718	97	1.692 9694	157	2.740 1669	37	.010 7629	37	.000 1794
38	.663 2251	98	1.710 4227	158	2.757 6202	38	.011 0538	38	.000 1842
39	.680 6784	99	1.727 8760	159	2.775 0735	39	.011 3446	39	.000 1891
40	.698 1317	100	1.745 3293	160	2.792 5268	40	.011 6355	40	.000 1939
41	.715 5850	101	1.762 7825	161	2.809 9801	41	.011 9264	41	.000 1988
42	.733 0383	102	1.780 2358	162	2.827 4334	42	.012 2173	42	.000 2036
43	.750 4916	103	1.797 6891	163	2.844 8867	43	.012 5082	43	.000 2085
44	.767 9449	104	1.815 1424	164	2.862 3400	44	.012 7991	44	.000 2133
45	.785 3982	105	1.832 5957	165	2.879 7933	45	.013 0900	45	.000 2182
46	.802 8515	106	1.850 0490	166	2.897 2466	46	.013 3809	46	.000 2230
47	.820 3047	107	1.867 5023	167	2.914 6999	47	.013 6717	47	.000 2279
48	.837 7580	108	1.884 9556	168	2.932 1531	48	.013 9626	48	.000 2327
49	.855 2113	109	1.902 4089	169	2.949 6064	49	.014 2535	49	.000 2376
50	.872 6646	110	1.919 8622	170	2.967 0597	50	.014 5444	50	.000 2424
51	.890 1179	111	1.937 3155	171	2.984 5130	51	.014 8353	51	.000 2473
52	.907 5712	112	1.954 7688	172	3.001 9663	52	.015 1262	52	.000 2521
53	.925 0245	113	1.972 2221	173	3.019 4196	53	.015 4171	53	.000 2570
54	.942 4778	114	1.989 6753	174	3.036 8729	54	.015 7080	54	.000 2618
55	.959 9311	115	2.007 1286	175	3.054 3262	55	.015 9989	55	.000 2666
56	.977 3844	116	2.024 5819	176	3.071 7795	56	.016 2897	56	.000 2715
57	.994 8377	117	2.042 0352	177	3.089 2328	57	.016 5806	57	.000 2763
58	1.012 2910	118	2.059 4885	178	3.106 6861	58	.016 8715	58	.000 2812
59	1.029 7443	119	2.076 9418	179	3.124 1394	59	.017 1624	59	.000 2860
60	1.047 1976	120	2.094 3951	180	3.141 5927	60	.017 4533	60	.000 2909

* *From War Department, Surveying Tables.*

TABLE 8. METRIC CURVES

Deflection Angle 20-m. Chord	Radius in Meters	Log of Radius	Mid. Ordinate	Tangent Offset	Degree of Equivalent U. S. Curve	Deflection Angle 20-m. Chord
0° 10′	3437.75	3.536274	.015	0.058	0° 30′	0° 10′
20	1718.89	3.235246	.029	0.116	1 01	20
30	1145.93	3.059158	.044	0.175	1 31	30
40	859.46	2.934224	.058	0.233	2 02	40
50	687.57	2.837319	.073	0.291	2 32	50
1 00	572.99	2.758145	.087	0.349	3 03	1 00
10	491.14	2.691206	.102	0.407	3 33	10
20	429.76	2.633223	.116	0.465	4 04	20
30	382.02	2.582081	.131	0.524	4 34	30
40	343.82	2.536335	.145	0.582	5 05	40
50	312.58	2.494955	.160	0.640	5 35	50
2 00	286.54	2.457181	.175	0.698	6 06	2 00
10	264.51	2.422434	.189	0.756	6 36	10
20	245.62	2.390266	.204	0.814	7 07	20
30	229.26	2.360320	.218	0.872	7 37	30
40	214.94	2.332311	.233	0.931	8 08	40
50	202.30	2.306002	.247	0.989	8 38	50
3 00	191.07	2.281200	.262	1.047	9 09	3 00
10	181.03	2.257741	.276	1.105	9 40	10
20	171.98	2.235489	.291	1.163	10 10	20
30	163.80	2.214325	.306	1.221	10 41	30
40	156.37	2.194148	.320	1.279	11 11	40
50	149.58	2.174870	.335	1.337	11 42	50
4 00	143.36	2.156416	.349	1.395	12 12	4 00
10	137.63	2.138717	.364	1.453	12 43	10
20	132.35	2.121715	.378	1.511	13 13	20
30	127.45	2.105357	.393	1.569	13 44	30
40	122.91	2.089596	.407	1.627	14 15	40
50	118.68	2.074391	.422	1.685	14 45	50
5 00	114.737	2.059704	.437	1.743	15 16	5 00
10	111.045	2.045501	.451	1.801	15 47	10
20	107.585	2.031751	.466	1.859	16 17	20
30	104.334	2.018427	.480	1.917	16 48	30
40	101.275	2.005503	.495	1.975	17 19	40
50	98.391	1.992956	.509	2.033	17 49	50
6 00	95.668	1.980765	.524	2.091	18 20	6 00
10	93.092	1.968911	.539	2.148	18 51	10
20	90.652	1.957375	.553	2.206	19 21	20
30	88.337	1.946141	.568	2.264	19 52	30
40	86.138	1.935194	.582	2.322	20 •23	40
50	84.047	1.924520	.597	2.380	20 54	50
7 00	82.055	1.914105	.612	2.437	21 24	7 00
10	80.156	1.903938	.626	2.495	21 55	10
20	78.344	1.894008	.641	2.553	22 26	20
30	76.613	1.884302	.655	2.611	22 57	30
40	74.957	1.874813	.670	2.668	23 28	40
50	73.372	1.865530	.685	2.726	23 59	50
8 00	71.853	1.856445	.699	2.783	24 29	8 00
10	70.396	1.847549	.714	2.841	25 00	10
20	68.998	1.838836	.729	2.899	25 31	20
30	67.655	1.830298	.743	2.956	26 02	30
40	66.363	1.821928	.758	3.014	26 33	40
50	65.121	1.813720	.772	3.071	27 04	50
9 00	63.925	1.805668	.787	3.129	27 35	9 00
10	62.772	1.797766	.802	3.186	28 06	10
20	61.6e1	1.790008	.816	3.244	28 37	20
30	60.589	1.782391	.831	3.301	29 08	30
40	59.554	1.774908	.846	3.358	29 39	40
50	58.554	1.767556	.860	3.416	30 10	50
10° 00′	57.588	1.760330	.875	3.473	30 41	10° 00′

SHORT-RADIUS CURVES

Note. The degree of curve is not usually used for the curves involved in street intersections, curbs, road intersections, runway and taxiway fillets, and turnarounds, traffic circles, rotaries, cloverleafs, etc. These curves are defined by the radius R, and central angle, Δ or θ.

NOTATION

T = tangent length P.C. or P.T. to P.l.
L = arc length P.C. to P.T.
l = arc length for any subchord
C = long chord P.C. to P.T.
c = any subchord.
d = deflection to any point.
Δ = central angle in degrees.
θ = central angle in radians.

One radian $= \dfrac{360°}{2\pi} = \dfrac{180°}{\pi}$

$ = 57.2958°$
$ = 57°\ 17'\ 44.8''$

$\pi = 3.14159.$

M = mid. ordinate; m for subchords.
E = external; e for subchords.

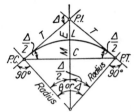

Short-radius Curve

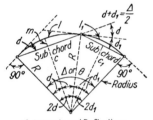

Subchords and Deflections

$$R = \frac{L}{\theta} = \frac{L \cdot 180/\Delta}{\pi} = \frac{L}{\Delta}\,57.2958 = T\cdot\cot\frac{\Delta}{2} = \frac{C}{2\sin\Delta/2}.$$

$$\frac{4M^2 + C^2}{8M} = \frac{M^2 + (C/2)^2}{2M}.$$

$$L = R\theta = \frac{\Delta R\pi}{180} = 0.017453\Delta R = \text{circum.}\cdot\frac{\Delta}{360}.$$

$$T = R\cdot\tan\frac{\Delta}{2} = E\cdot\cot\frac{\Delta}{4} = \frac{C}{2\cos\Delta/2}.$$

$$C = 2R\cdot\sin\frac{\Delta}{2} = 2T\cdot\cos\frac{\Delta}{2} = 2\sqrt{M(2R - M)}$$

$$M = R\cdot\text{vers}\frac{\Delta}{2} = E\cdot\cos\frac{\Delta}{2} = R\left(1 - \cos\frac{\Delta}{2}\right).$$

$$E = R\cdot\text{exsec}\frac{\Delta}{2} = T\cdot\tan\frac{\Delta}{4} = \frac{R}{\cos\Delta/2} - R.$$

$$\Delta = \frac{180L}{\pi R} = 57.2958\,\frac{L}{R} = \theta\cdot57.2958.$$

$$\theta = \frac{L}{R} = \frac{\Delta\pi}{180} = \Delta \cdot 0.017453.$$

$$\sin\frac{\Delta}{2} = \frac{C}{2R} \; ; \quad \cos\frac{\Delta}{2} = \frac{R - M}{R} = \frac{C}{2T} \; ; \quad \tan\frac{\Delta}{2} = \frac{T}{R}$$

Subcord $= 2R \cdot \sin d = 2(R - M) \cdot \tan d.$

d(in minutes) $= 1718.873 \dfrac{l}{R} \cdot$ \qquad Radius $= \dfrac{C}{2\sin d} \cdot$

Length $= \dfrac{\pi R d}{90} = 0.034906 R d (d$ in degrees$).$

Mid. ordinate $= R(1 - \cos d) = 2R \cdot \sin^2 \dfrac{d}{2}$

Tan $d = \dfrac{\frac{1}{2}C}{R - m} \; ; \quad \sin d = \dfrac{\frac{1}{2}C}{R}$

Excess of l over $c = l - c = l - 2R \cdot \sin d.$

Sum of deflection angles, $d_1 + d_2 + \cdots d_n = \dfrac{\Delta}{2}$

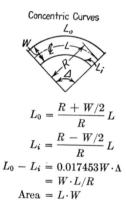

Concentric Curves

$$L_0 = \frac{R + W/2}{R} L$$

$$L_i = \frac{R - W/2}{R} L$$

$$L_0 - L_i = 0.017453 W \cdot \Lambda$$
$$= W \cdot L/R$$
$$\text{Area} = L \cdot W$$

EXAMPLE. *Given.* $R = 50'; \Delta = 110°(\theta = 1.9195); l = 50'.$
Required. $L, l_1, d, d_1, c,$ and $c_1.$
Solution.

$L = 50 \times 1.9195 = 95.98'; l_1 = 95.98 - 50 = 45.98'.$
$d = 1718.873 \times 50/50 = 28° 39'.$
$d_1 = 1718.873 \times \dfrac{45.98}{50} = 26° 21'.$

$c = 2R \sin 28° 39' = 47.946'.$
$c_1 = 2R \sin 26° 21' = 44.385'.$

TABLE 9. DEFLECTIONS (d) AND MIDDLE ORDINATES (m) FOR SUBCHORDS *

	Radius → Chord	10'	12'	15'	18'	20'	25'	30'	35'	40'	45'	50'	60'	70'	80'	90'	100'	120'	150'
Deflection	5'	14°29'	12°01'	9°36'	7°59'	7°11'	5°44'	4°47'	4°06'	3°35'	3°11'	2°52'	2°23'	2°03'	1°47'	1°35'	1°26'	1°12'	0°57'
	10'	30°00'	24°37'	19°28'	16°08'	14°29'	11°32'	9°36'	8°13'	7°11'	6°23'	5°44'	4°47'	4°06'	3°35'	3°11'	2°52'	2°23'	1°55'
	20'		56°26'	41°49'	33°45'	30°00'	23°35'	19°28'	16°36'	14°29'	12°50'	11°32'	9°36'	8°13'	7°11'	6°23'	5°44'	4°47'	3°49'
	25'			56°27'	43°59'	38°41'	30°00'	24°37'	20°55'	18°13'	16°08'	14°29'	12°01'	10°17'	8°59'	7°59'	7°11'	5°59'	4°47'
	50'							56°27'	45°35'	38°41'	33°45'	30°00'	24°37'	20°55'	18°13'	16°08'	14°29'	12°01'	9°36'
M.	10'	1.34	1.09	0.86	0.71	0.64	0.51	0.42	0.36	0.31	0.28	0.25	0.21	0.18	0.16	0.14	0.13	0.10	0.08
	20'		5.37	3.82	3.03	2.68	2.09	1.72	1.46	1.27	1.12	1.01	0.84	0.72	0.63	0.56	0.50	0.42	0.33

* Adapted from Lefax Society, Inc., Philadelphia, Pa.

Circle

$$\text{Area} = \pi R^2 = \frac{\pi D^2}{4}$$

$$\text{Circumference} = 2\pi R = \pi D.$$

$$R = \frac{\text{Cir.}}{2\pi} = \frac{D}{2} = \sqrt{\frac{\text{Area}}{\pi}}$$

$$D = 2R = \text{cir.}/\pi$$

Sector of Circle

$$\text{Area} = 0.008727 R^2 \Delta$$
$$= \frac{l}{2} \cdot R = \pi R^2 \frac{\Delta}{360}$$
$$= R^2 \cdot \frac{\theta}{2}$$

when
$$\Delta = 90°: A = 0.3927C^2; \ 0.7854R^2$$

Segment of Circle

$$A_1 = R^2 \left(\tan \frac{\Delta}{2} - \frac{\Delta \pi}{360} \right) = R \left(T - \frac{l}{2} \right).$$

$$A_2 = \frac{lR - c(R - M)}{2} = \left(\pi R^2 \frac{\Delta}{360} \right) - \left[\left(R \sin \frac{\Delta}{2} \right) \right.$$
$$\left. \left(R \cos \frac{\Delta}{2} \right) \right].$$

$$A_2 = \left(\pi R^2 \frac{\Delta}{360} \right) - \frac{1}{2}c(R - M)$$

$$A_2 = \frac{2}{3} Mc \begin{cases} \text{Correct for parabolic segment, approximate} \\ \quad \text{for circular segment.} \end{cases}$$

$$A_2 = \frac{1}{2}R^2(\theta - \sin \Delta) = \frac{2}{3}Mc + \frac{M^3}{2c}$$

$$A_3 = \frac{1}{2}R^2 \sin \Delta = \frac{1}{2}c(R - M) = \left(R \sin \frac{\Delta}{2} \right)\left(R \cos \frac{\Delta}{2} \right).$$

When $\Delta = 90°: A_1 = 0.2146R^2$
$$= 1.2594E^2$$

Fig. 9.   Formulas for areas.

TRANSITION CURVES *

Formulas

$$T_s = (R_c + p) \tan \frac{\Delta}{2} + k.$$

$$E_s = (R_c + p) \operatorname{exsec} \frac{\Delta}{2} + p = \frac{R_c + p}{\cos \frac{\Delta}{2}} - R_c.$$

$$P = y_c - R_c(1 - \cos \theta_s) = \frac{y_c}{4} \text{ (approx.).}$$

$$k = x_c - R_c \sin \theta_s = \frac{L_s}{2} \text{ (approx.).}$$

$$\theta_s = \frac{L_s D_c}{200} \; ; \; \theta = \left(\frac{L}{L_s}\right)^2 \theta_s.$$

$$\theta = \frac{L^2 D_c}{200 \, L_s}.$$

$$L_c = \frac{100 \, \Delta_c}{D_c} \; ; \; \text{L.C.} = \frac{X_c}{\cos \phi_c}.$$

$$\Delta_c = \Delta - \frac{L_s D_c}{100}.$$

$$D = \frac{L}{L_s} D_c.$$

$$D_c = \frac{200 \, \theta_s}{L_s}.$$

Offsets to x and y

$$y = \frac{L^3}{L_s} y_c = L(y \text{ for } L_s = 1).$$
$$y_c = L_s(y \text{ for } L_s = 1).$$
$$x = L(x \text{ for } L_s = 1);$$
$$x_c = L_s(x \text{ for } L_s = 1).$$

Offsets to ¼ Points

$$y \text{ at } \tfrac{1}{4} \text{ point} = y_c/4^3$$
$$y \text{ at } \tfrac{1}{2} \text{ point} = y_c/2^3 = P/2$$
$$\text{(approx.)}$$
$$y \text{ at } \tfrac{3}{4} \text{ point} = y_c/(\tfrac{4}{3})^3$$

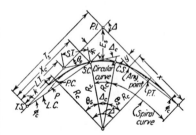

Note. At the P.C. the spiral approximately bisects P.

Total Length of Curve

$$T_s \text{ to S.T.} = 2L_s + 100 \frac{\Delta_c}{D_c}$$

$$\phi_c = \theta/3 - c; \; \phi = (L/L_s)^2 \phi_c.$$

Fig. 10. Circular curves with spiral transitions.

* *Adapted from Transition Curves for Highways by Joseph Barnett, P.R.A.*

303

Notes for Fig. 10. With L_s given or selected from Table 11 below, p, k, x, y, L.T., S.T., and L.C. may be computed for any spiral by multiplying functions for $L_s = 1$ in Table 12, p. 310, by L_s or L in feet. Interpolate for values of θ or θ_s between even degrees. For circular curve layout see pp. 288, 289, 290.

Circular curve may be omitted and curve made transitional throughout in which case S.C. and C.S. coincide at S.C.S., $\theta = \Delta/2$, $\Delta_c = 0$, and T_s and E_s are computed from Table 13, p. 311.

NOTATION

R_c = radius of the circular curve.

P = offset distance from tangent to the P.C. of the circular curve produced.

k = distance from T.S. to P.C. along tangent.

T_s = tangent distance.

E_s = external distance.

x_c, y_c = coordinates from T.S. to S.C. and S.T. to C.S.

θ = spiral angle at any point on spiral.

θ_s = spiral angle at S.C. or C.S.

L = length of spiral, T.S. to any point on spiral.

L_s = length of spiral, T.S. to S.C. or S.T. to C.S.

D_c = degree of circular curve (arc definition).

D = degree of curve at any point on spiral.

x, y = coordinates from T.S. or S.T. to any point on spiral.

ϕ_c = deflection from tangent at T.S. to S.C.

ϕ = deflection from tangent at T.S., S.T. or any point on spiral to any other point on spiral.

L.T., S.T. = long tangent, short tangent.

L.C. = long chord of spiral transition.

Δ = intersection and central angle of entire curve.

Δ_c = intersection and central angle of circular curve.

L_c = length of circular curve, S.C. to C.S.

Note. The degree of curvature varies directly as the length, from zero curvature at T.S. to the maximum of Dc at the S.C. The spiral departs from the circular curve at the same rate as from the tangent.

SPIRAL LAYOUT (See pp. 307, 308, 309 also.)

Method I: Deflections to even stations by formula $\phi = \theta/3 = 1/3\theta_s(L/L_s)^2$. Correct ϕ for c when $\theta > 20°$.

TABLE 10.　C IN FORMULA, $\phi = \theta/3 - C$

(For curves with θ over 20°)

θ in degrees	20	25	30	35	40	45	50
c in minutes	0.4	0.8	1.4	2.2	3.4	4.8	5.6

Method II: Offsets from tangent. Establish by measuring x distances from T.S. and y distances from tangent. Compute θ for each point and then compute x and y coordinates from Table 12, p. 310, or use ¼ point formulas above.

Method III: Deflection angle from T.S. or S.T. to any point on spiral with coordinates x and y is the angle whose tangent = y/x.

Method IV: Deflection angles from T.S. to points of 10 equal divisions (10 chord spiral) are: $0.01\phi_c$; $0.04\phi_c$; $0.16\phi_c$; $0.25\phi_c$; $0.36\phi_c$; $0.49\phi_c$; $0.64\phi_c$; $0.81\phi_c$ and ϕ_c.

TABLE 11.　MINIMUM TRANSITION LENGTHS

D_c	30 M.P.H. L_s	40 M.P.H. L_s	50 M.P.H. L_s	60 M.P.H. L_s	70 M.P.H. L_s	D_c
1° 30′	150′	150′	150′	150′	150′	1° 30′
2°	150′	150′	150′	150′	200′	2°
2° 30′	150′	150′	150′	150′	250′	2° 30′
3°	150′	150′	150′	150′	300′	3°
3° 30′	150′	150′	150′	200′	350′	3° 30′
4°	150′	150′	150′	250′	400′	4°
5°	150′	150′	150′	300′		
6°	150′	150′	200′	350′		
7°	150′	150′	250′		Based on	
8°–9°	150′	150′	300′		$L_s = \dfrac{1.6V^3}{R_c}$	
10°–12°	150′	200′				
13°–14°	150′	250′				
15°–23°	150′			Where: $V = 0.75$ design speed in M.P.H.		
24°	200′			Min. $L_s = 150$ ft.		

INSERTION OF SPIRALS INTO EXISTING ALIGNMENT OF CIRCULAR CURVES

L_S = Length of spiral select from table 11, page 305.

θ_S = Spiral angle = $\frac{L_S D_C}{200}$, where D_C = Degree of curvature (arc definition).

P = Offset of curve at P.C. to permit spiral introduction from table, page 310 knowing θ_S.

CASE I – Radius of original circular curve reduced by value of "P" to provide space to insert spiral transition.

1st trial : Assume $D_C = D$, find trial "P" as above.

2nd trial : Compute $D_C = \frac{5727.58}{R_C}$, find correct "P."

CASE II – Radius of original curve retained and curve center "O" shifted inward.

Note : Degree of curve retained.

$R_C = R$
$D_C = D$

CASE III – Original circular curve location retained and tangents shifted outward to insert spiral.

Trial and error adjustment may be necessary to reconcile P, θ_S and D_C.

Sharper curves "D_C" at ends provide offset "P" to insert spiral.

CASE IV – Original alignment retained as closely as possible by compounding circular curve at both ends.

θ_a = Equivalent spiral angle. Use in table on page 310 to find "P_a".

Common R bisects spiral

$D_C = D_{C1} - D_{C2}$

$\Delta_1 = \frac{L_a}{200} D_{C1}$

$\Delta_2 = \frac{L_a}{200} D_{C2}$

$\theta_a = \Delta_2 - \Delta_1$

CASE V – To insert a spiral in a compound curve.

Given : R_1, R_2, T, P_1 & P_2

Required : Angle Δ

Solution : $\tan \Delta_1 = \frac{T}{R_1 + R_2}$

$A - B = \frac{R_1 + R_2}{\cos \Delta_1}$

$\cos \Delta_2 = \frac{R_1 + P_1 + P_2 + R_2}{A \cdot B}$

$\Delta = \Delta_1 - \Delta_2$

CASE VI – To insert spirals between simple reverse curves separated by a tangent.

PROPERTIES AND EXAMPLES *

Properties of Spiral

1. Offsets, y, vary as the cube of L, or length of spiral. $\therefore$ y at any point $= (L/L_s)^3 y_c$. See Fig. 11.

2. Spiral angle θ varies as L^2. $\therefore$ θ at any point on spiral $= (L/L_s)^2 \theta_s$.

3. Deflection angle ϕ varies as L^2. $\therefore$ $\phi = (L/L_s)^2 \phi_c$. $\phi_c = \frac{1}{3}\theta_s - c$, c being a constant; see Table 10, p. 305. (May be neglected for ordinary problems.)

4. D, or degree of curve of spiral at any point, varies directly as L. $\therefore$ $D = L/L_s D_c$.

5. Spiral bisects P very nearly and k approximately $= \frac{1}{2}L_s$. $\therefore$ Offset from circular curve or tangent to midpoint of spiral is $\frac{1}{2}P$ very nearly.

6. Spiral departs from the circular curve between S.C. and P.C. at the same rate as from the tangent. $\therefore$ Radial offsets from circular curve between S.C. and P.C. to the spiral are the same as perpendicular offsets from the tangent between T.S. and P.C.

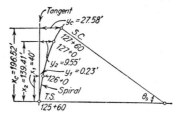

Given. Spiral $L_s = 200'$; $\theta_s = 24°$; T.S. at Sta. 125 + 60.
Required. Offsets to even stations.
Solution. Compute θ and read x and y for $L_s = 1$ from table on p. 319.

Sta.	L	θ	x, $L_s = 1$	y, $L_s = 1$	x	y
126 + 0	40	0° 58′	0.99997	0.00559	40.0	0.22
127 + 0	140	11° 46′	0.99578	0.06821	139.41	9.55
127 + 60	200	24° 0′	0.98260	0.13789	196.52	27.58

Fig. 11. Offsets to even stations.

* *Reference Transition Curves for Highways by Joseph Barnett, P.R.A.*

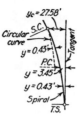

Given. Spiral, $L_s = 200'$; $\theta_s = 24°$.

Required. Offsets to $\frac{1}{4}$ points.

Solution. From Fig. 11, $y_c = 27.58'$. By formula, y at any point $= (L/L_s)^3 y_c$

At $\frac{1}{4}$ points, $y = 27.58 \times \frac{1}{64} = 0.43'$.

At $\frac{1}{2}$ points, $y = 27.58 \times \frac{1}{8} = 3.45'$.

FIG. 12. Offsets to $\frac{1}{4}$ points.

Given. Spiral with $L_s = 200'$ and $\theta_s = 24°$.

Required. Deflection angles ϕ, to Sta. $126 + 0$; ϕ_2 to Sta. $127 + 0$; ϕ_c to Sta. $127 + 60$.

Solution. By formulas, $\phi_c = \theta_s/3 - c$ and $\phi = \frac{1}{3}(L/L_s)^2 \theta_s - c$.

Sta. $127 + 60$: $\phi_c = 24/3 - 0.8 = 7.9866° = 7° 69.2'$

Sta. $126 + 00$: $\phi_1 = (L/L_s)^2\phi_c = (40/200)^2 \times 7.9866$
$= 0.3195° = 0° 19'$

Sta. $127 + 00$: $\phi_2 = (L/L_s)^2\phi_c = (140/200)^2 \times 7.9866$
$= 3.9134° = 3° 55'$

Layout. With transit at T.S., foresight along tangent with vernier at $0°$ Turn ϕ_1 and measure 40 ft. to Sta. $126 + 0$. Turn ϕ_2 and measure 100 ft. from Sta. $126 + 0$ to Sta. $127 + 0$. Turn ϕ_c and measure $60'$ from Sta. $127 + 0$ to S.C.

FIG. 13. Deflections to even stations.

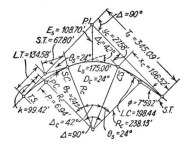

Given. $\Delta = 90°$; $D_c = 24°$; $L_s = 200'$. Formulas from p. 303. Functions of spiral for $L_s = 1$ from p. 310.

For layout of circular curve, see pp. 288, 289, 290.

Layout of Control Points *

Establish T.S. by measuring k from P.O.T. normal to P.C. or by T_s from P.I. Establish S.C. by L.T., θ_s, and S.T. or by x_c and y_c from T.S., or by ϕ_c and L.C. from T.S.

Note. Figures 11–13 give all dimensions usually necessary to plot or locate the spiral. The following example is a curve fully worked out.

Required	Formula	Solution
θ_s	$L_s D_c \div 200$	$\theta_s = 24°$
Δ_c	$\Delta - (L_s D_c \div 100)$	$\Delta_c = 42°$
L_c	$100 \Delta_c \div D_c$	$L_c = 175.00$
ϕ_c	$\tfrac{1}{3}\theta_s - c$	$\phi_c = 7° 59.2'$
y_c	$(y$ for $L_s = 1) \cdot L_s$	$y_c = 27.58'$
x_c	$(x$ for $L_s = 1) \cdot L_s$	$x_c = 196.52'$
P	$y_c - R_c(1 - \cos \theta_s)$	$P = 6.94'$
k	$x_c - R_c \sin \theta_s$	$k = 99.42'$
E_s	$(R_c + P)$ exsec $\Delta/2 + P$	$E_s = 108.70'$
T_s	$(R_c + P) \tan \Delta/2 + k$	$T_s = 345.09'$
L.T.	(L.T. for $L_s = 1)L_s$; $\theta = 24°$	L.T. $= 134.58'$
S.T.	(S.T. for $L_s = 1)L_s$; $\theta = 24°$	S.T. $= 67.80'$
L.C.	(L.C. for $L_s = 1)L_s$; $\theta = 24°$	L.C. $= 198.44$

Fig. 14. Computations for spiral transitions to circular curves.

* *Adapted from O'Rourke, General Engineering Handbook, McGraw-Hill.*

TABLE 12. FUNCTIONS OF TRANSITION FOR $L_s = 1$ *

Enter table with value of θ or θ_s, and multiply function by L or L_s. See pp. 304–309 for use of table.

θ	p	k	x	y	L.T.	S.T.	L.C.	θ
0°	.00000	.50000	1.00000	.00000	.36667	.33333	1.00000	0°
1°	.00146	.49999	.99997	.00582	.66668	.33334	.99999	1°
2°	.00291	.49998	.99988	.01163	.66671	.33337	.99995	2°
3°	.00435	.49995	.99973	.01745	.66676	.33342	.99988	3°
4°	.00581	.49992	.99951	.02326	.66684	.33349	.99978	4°
5°	.00727	.49987	.99924	.02907	.66693	.33358	.99966	5°
6°	.00872	.49982	.99890	.03488	.66705	.33368	.99951	6°
7°	.01018	.49975	.99851	.04068	.66719	.33381	.99934	7°
8°	.01163	.49967	.99805	.04648	.66735	.33395	.99913	8°
9°	.01308	.49959	.99754	.05227	.66753	.33412	.99890	9°
10°	.01453	.49949	.99696	.05805	.66773	.33430	.99865	10°
11°	.01598	.49939	.99632	.06383	.66796	.33451	.99836	11°
12°	.01743	.49927	.99562	.06959	.66821	.33473	.99805	12°
13°	.01887	.49914	.99486	.07535	.66847	.33498	.99771	13°
14°	.02032	.49901	.99405	.08110	.66877	.33524	.99735	14°
15°	.02176	.49886	.99317	.08684	.66908	.33553	.99696	15°
16°	.02320	.49870	.99223	.09257	.66941	.33583	.99654	16°
17°	.02465	.49854	.99123	.09828	.66977	.33615	.99609	17°
18°	.02608	.49836	.99018	.10398	.67015	.33650	.99562	18°
19°	.02752	.49817	.98906	.10967	.67055	.33687	.99512	19°
20°	.02896	.49798	.98788	.11535	.67097	.33725	.99460	20°
21°	.03040	.49777	.98665	.12101	.67142	.33766	.99404	21°
22°	.03183	.49755	.98536	.12665	.67189	.33809	.99346	22°
23°	.03326	.49733	.98401	.13228	.67238	.33854	.99286	23°
24°	.03469	.49709	.98260	.13789	.67290	.33901	.99222	24°
25°	.03611	.49684	.98113	.14348	.67344	.33950	.99157	25°
26°	.03753	.49658	.97960	.14905	.67400	.34001	.99088	26°
27°	.03896	.49632	.97802	.15461	.67459	.34055	.99017	27°
28°	.04037	.49605	.97638	.16014	.67520	.34111	.98943	28°
29°	.04179	.49576	.97469	.16565	.67584	.34169	.98866	29°
30°	.04321	.49546	.97293	.17114	.67650	.34229	.98787	30°
31°	.04462	.49516	.97112	.17661	.67719	.34292	.98705	31°
32°	.04602	.49484	.96926	.18206	.67790	.34356	.98621	32°
33°	.04743	.49452	.96733	.18748	.67863	.34424	.98534	33°
34°	.04883	.49419	.96536	.19288	.67939	.34493	.98444	34°
35°	.05023	.49385	.96332	.19826	.68018	.34565	.98351	35°
36°	.05163	.49349	.96124	.20361	.68100	.34640	.98257	36°
37°	.05301	.49313	.95910	.20893	.68184	.34717	.98159	37°
38°	.05441	.49276	.95690	.21423	.68271	.34796	.98059	38°
39°	.05579	.49238	.95466	.21949	.68360	.34878	.97956	39°
40°	.05718	.49199	.95235	.22473	.68452	.34962	.97851	40°
41°	.05855	.49159	.95000	.22994	.68547	.35049	.97743	41°
42°	.05993	.49118	.94759	.23513	.68645	.35139	.97632	42°
43°	.06130	.49075	.94513	.24028	.68746	.35232	.97519	43°
44°	.06267	.49032	.94262	.24540	.68850	.35327	.97404	44°
45°	.06403	.48990	.94005	.25049	.68957	.35424	.97285	45°
46°	.06538	.48945	.93744	.25555	.69066	.35525	.97165	46°
47°	.06674	.48900	.93477	.26057	.69179	.35629	.97041	47°
48°	.06809	.48852	.93206	.26556	.69295	.35735	.96916	48°
49°	.06944	.48805	.92930	.27052	.69414	.35844	.96787	49°
50°	.07078	.48757	.92649	.27544	.69536	.35957	.96656	50°

* *Adapted from Transition Curves for Highways by Joseph Barnett, P.R.A.*

TABLE 13. FUNCTIONS OF CURVES TRANSITIONAL THROUGHOUT TANGENTS AND EXTERNALS FOR $L_s = 1$*

$\Delta°$	T_s	E_s	$\Delta°$	T_s	E_s	$\Delta°$	T_s	E_s
6°	1.00064	0.01747	38°	1.02682	0.11599	70°	1.10214	0.24203
7	1.00087	0.02040	39	1.02832	0.11936	71	1.10561	0.24681
8	1.00114	0.02332	40	1.02987	0.12275	72	1.10917	0.25167
9	1.00144	0.02625	41	1.03146	0.12617	73	1.11281	0.25660
10	1.00178	0.02918	42	1.03310	0.12962	74	1.11654	0.26161
11	1.00216	0.03213	43	1.03479	0.13309	75	1.12036	0.26669
12	1.00257	0.03507	44	1.03653	0.13660	76	1.12427	0.27186
13	1.00302	0.03802	45	1.03831	0.14012	77	1.12828	0.27710
14	1.00350	0.04098	46	1.04015	0.14370	78	1.13240	0.28244
15	1.00402	0.04396	47	1.04204	0.14730	79	1.13661	0.28786
16	1.00458	0.04693	48	1.04399	0.15094	80	1.14092	0.29337
17	1.00518	0.04992	49	1.04598	0.15460	81	1.14535	0.29898
18	1.00581	0.05292	50	1.04804	0.15831	82	1.14988	0.30468
19	1.00648	0.05593	51	1.05014	0.16206	83	1.15453	0.31048
20	1.00719	0.05895	52	1.05230	0.16584	84	1.15930	0.31639
21	1.00794	0.06198	53	1.05452	0.16966	85	1.16418	0.32241
22	1.00873	0.06502	54	1.05680	0.17352	86	1.16919	0.32854
23	1.00955	0.06808	55	1.05913	0.17742	87	1.17433	0.33478
24	1.01042	0.07115	56	1.06153	0.18137	88	1.17960	0.34115
25	1.01132	0.07424	57	1.06399	0.18536	89	1.18500	0.34763
26	1.01226	0.07734	58	1.06651	0.18940	90	1.19054	0.35425
27	1.01324	0.08045	59	1.06909	0.19348	91	1.19623	0.36099
28	1.01427	0.08358	60	1.07174	0.19762	92	1.20207	0.36788
29	1.01533	0.08674	61	1.07446	0.20181	93	1.20806	0.37490
30	1.01644	0.08990	62	1.07724	0.20604	94	1.21421	0.38207
31	1.01758	0.09309	63	1.08010	0.21034	95	1.22052	0.38940
32	1.01877	0.09630	64	1.08302	0.21468	96	1.22700	0.39688
33	1.02000	0.09952	65	1.08602	0.21908	97	1.23366	0.40453
34	1.02128	0.10277	66	1.08909	0.22355	98	1.24050	0.41234
35	1.02260	0.10604	67	1.09223	0.22807	99	1.24753	0.42034
36	1.02396	0.10933	68	1.09546	0.23266	100	1.25475	0.42852
37	1.02537	0.11265	69	1.09876	0.23731			

* *Adapted from Transition Curves for Highways by Joseph Barnett, P.R.A.*

Case VII. Given Δ and an external or tangent distance; to determine a curve transitional throughout.

Enter Table 13 at known Δ and read T_s and E_s values. Then $L_s = E_s/E_s$ value and $T_s = L_s \cdot$ tangent value, or $L_s = T_s/T_s$ value and $E_s = L_s \cdot$ external value.

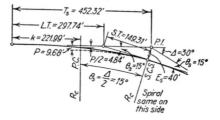

EXAMPLE. *Given.* $\Delta = 30°$ and $E_s = 40'$.

Required. L_s, T_s, θ_s, L.T., S.T., D_c, P, and k.

Solution. $L_s = 40 \div 0.08990 = 444.9$, say $445'$. $T_s = 1.01644 \times 445 = 452.32'$. $\theta_s = \dfrac{\Delta}{2} = 15°$. $D_c = \dfrac{200\theta_s}{L_s} = 6.47'$. L.T. $= 0.66908 \times 445 = 297.74'$. S.T. $= 0.33553 \times 445 = 149.31'$. $p = 0.02176 \times 445 = 9.68'$. $k = 498.86 \times 445 = 221.99'$.

FIG. 15. Spiral layout by offsets or deflections (same as for spiral transitions to a circular curve).

VERTICAL CURVES (Parabolic)

FORMULAS

A = algebraic difference of grades = $+ g_1\% - (-g_2\%)$.

$e = AL/8$.

$d = l^2A/2L$; $d = 4e\,(l/L)^2$.

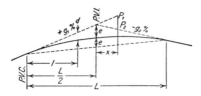

VERTICAL SUMMIT CURVE

Length of vertical summit curves should provide required sight distance. See Vol. I, p. 3-61.

Note. All horizontal distances shown on this page—L, l, l_1, l_2, x, x_1, x_2— are expressed in 100 ft. stations.

Where L, length of vertical curve, is not determined by sight distance criteria, the minimum value for comfort is

$$L = \frac{AV^2}{10,000} *$$

EXAMPLE. *Given.* $g_1\% = +3.00\%$; $g_2\% = -2.00\%$; $L = 3.00$; $l = 0.50$.
Required. A, e, and d.
Solution.

$$A = 3.00 - (-2.00) = 5.00$$

$$e = \frac{5.00 \times 3.00}{8} = 1.875'$$

$$d = \frac{0.50^2 \times 5.00}{2 \times 3.00} = 0.208'$$

Also,

$$d = 4(1.875)\left(\frac{0.50}{3.00}\right)^2 = 0.208'.$$

To find Sta. of P.V.I. when elevations of P_1 and P_2 are known.

FORMULA

$$x = \frac{\text{elev. } P_1 - \text{elev. } P_2}{A}$$

EXAMPLE. *Given.* Elev. $P_1 = 154.50$; elev. $P_2 = 150.00$; $A = 5.00$.
Required. x = distance in 100' stations from known point to P.V.I.

* *From O'Rourke, General Engineering Handbook, McGraw-Hill.*

Solution.

$$x = \frac{154.50 - 150.00}{5.00} = 0.90(100' \text{ stations})$$

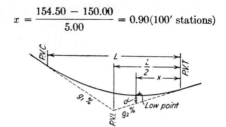

To find low point on sag curve.

VERTICAL SAG CURVE

Length of vertical sag curve should provide headlight illumination for a safe stopping distance. See Vol. I, p. 3-63.

FORMULAS

$x = g(\text{lesser gradient}) \, L/A$.
$d(\text{at low point}) = x^2 A/2L$.

EXAMPLE. *Given.* $g_1\% = -3.00\%$; $g_2 = +2.00\%$; $L = 3.00$; $A = 5.00$.
Required. x and d.
Solution.

$$x = 2.00 \times \frac{3.00}{5.00} = 1.20'$$

$$d = \frac{1.20^2 \times 5.00}{2 \times 3.00} = 1.20'$$

Note. High point on summit curve can be found by same method.

FIG. 16.   Symmetrical vertical curves.

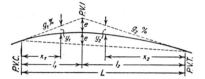

FORMULAS

$$e = \frac{l_1 l_2}{2(l_1 + l_2)} (g_1 - g_2); \; y_1 = e\left(\frac{x_1}{l_1}\right)^2; \; y_2 = e\left(\frac{x_2}{l_2}\right)^2$$

EXAMPLE. *Given.* $g_1 = 3.00\%$; $g_2 = 2.00\%$; $L = 4.00$; $l_1 = 1.50$; $l_2 = 2.50$; $x_1 = 0.50$; $x_2 = 1.00$.

Required. e, y_1, and y_2.

Solution.

$$e = \frac{1.50 \times 2.50}{2(1.50 + 2.50)} (3.00 + 2.00) = 2.35'$$

$$y_1 = 2.35 \left(\frac{0.50}{1.50}\right)^2 = 0.26'$$

$$y_2 = 2.35 \left(\frac{1.00}{2.50}\right)^2 = 0.38'$$

FIG. 17. Unsymmetrical vertical curves used to fit unusual conditions.

PARABOLIC CROWN ORDINATES

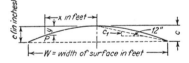

FORMULAS

SYMMETRICAL CROWN

Used for roads and for streets where gutters are same elevation.

$$c = c_1 \left(\frac{W}{2}\right); \; y = 4c\left(\frac{x}{W}\right)^2.$$

EXAMPLE. *Given.* $c_1 = \frac{1}{8}''$; $W = 22'$; and $x = 6'$.

Required. c; y (at any point P).

Solution.

$$c = \frac{1}{8} \times \frac{22}{2} = 1.375'' = 1\frac{3}{8}''$$

$$y = 4 \times 1.375 \left(\frac{6}{22}\right)^2 = 0.409'' = \frac{13}{32}''$$

Ordinates—Any Parabolic Curve

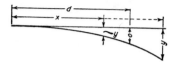

Unsymmetrical Crown

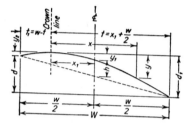

Used for city streets where conditions necessitate different gutter elevations. If slope per foot is over ½ in., a stepped curb or retaining wall should be used on uphill side of street.

Also used for off-center crowns on three-lane roads to provide symmetrical crown for future four lanes.

Also used for transition onto superelevated curves.

Offsets from tangent to curve vary directly as the squares of the tangent distances.

Formula

$$d^2:x^2 = o:y. \quad \therefore \quad y = \frac{ox^2}{d^2}$$

Example. *Given.* $d = 10'$; $o = 6''$; and $x = 5'$.
Required. y.
Solution.

$$y = \frac{6 \times 5^2}{10^2} = 1.50'' = 1\tfrac{1}{2}''$$

Alternative Method

Divide the distance from center line or high point to edge of pavement into 10 equal spaces. Multiply figures in chart by total crown to get ordinates from crown elevation to pavement surface for points shown.

EXAMPLE. *Given.* Total crown = 6″.
Required. Ordinates at fifth and eighth points.
Solution.
Ordinate at fifth point = 0.25 × 6 = 1.50″ = 1½″.
Ordinate at eighth point = 0.64 × 6 = 3.84″ = 3¹³⁄₁₆″.

FORMULAS

$$x_1 = \frac{dw}{8h} \; ; y_1 = \frac{d^2}{16h} \; ; d_1 = \frac{d}{2} + h + y_1; y = \frac{d_1 x^2}{t^2}$$

$$y_2 = d_1 - d; t = x_1 + \frac{w}{2} \; ; t_1 = W - t$$

EXAMPLE. *Given.* $h = 0.5'$; $w = 40'$; $d = 0.5'$; $x = 10'$.
Required. x_1; y_1; d_1; y_2; y; t and t_1.
Solution.

$$x_1 = \frac{0.5 \times 40}{8 \times 0.5} = 5.0'$$

$$y_1 = \frac{0.5^2}{16 \times 0.5} = 0.0312' = 0.375'' = \tfrac{3}{8}''$$

$$d_1 = \frac{0.5}{2} + 0.5 + 0.0312 = 0.7812' = 9.375'' = 9\tfrac{3}{8}''$$

$$y_2 = 0.7812 - 0.5 = 0.2812' = 3.375'' = 3\tfrac{3}{8}''$$

$$t = 5.0 + \frac{40}{2} = 25.0'$$

$$y = \frac{0.7812 \times 10^2}{25^2} = 0.125'' = 1\tfrac{1}{2}''$$

$$t_1 = 40 - 25 = 15'$$

TABLE 14. PARABOLIC CROWN ORDINATES

ORDINATES FROM GRADE TANGENT TO SURFACE FOR EACH FOOT OF WIDTH
SURFACE WIDTH OF STREET OR ROAD

20′

#	1/2″:1′ Crown		#	3/8″:1′ Crown		#	1/4″:1′ Crown	
1	1/16″	.004′	1	1/16″	.003′	1		.002′
2	3/16″	.017′	2	1/8″	.012′	2	1/16″	.008′
3	3/8″	.037′	3	1/4″	.028′	3	3/16″	.019′
4	13/16″	.067′	4	3/8″	.050′	4	1/4″	.033′
5	1 1/4″	.104′	5	5/8″	.078′	5	5/8″	.052′
6	1 13/16″	.150′	6	1 3/8″	.112′	6	7/8″	.075′
7	2 7/16″	.204′	7	1 13/16″	.153′	7	1 1/4″	.102′
8	3 3/16″	.267′	8	2 3/8″	.200′	8	1 9/16″	.133′
9	4 1/16″	.337′	9	3 1/16″	.253′	9	2 1/4″	.190′
10	5″	.417′	10	3 3/4″	.312′	10	2 1/2″	.208′

22′

#	3/8″:1′ Crown		#	1/4″:1′ Crown		#	1/8″:1′ Crown	
1	1/16″	.003′	1		.002′	1		.001′
2	1/8″	.011′	2	1/8″	.008′	2	1/16″	.004′
3	5/16″	.026′	3	3/16″	.017′	3	1/8″	.009′
4	9/16″	.045′	4	3/8″	.030′	4	3/16″	.015′
5	7/8″	.071′	5	9/16″	.047′	5	5/16″	.024′
6	1 1/4″	.102′	6	13/16″	.068′	6	7/16″	.034′
7	1 11/16″	.139′	7	1 1/16″	.093′	7	9/16″	.046′
8	2 3/16″	.182′	8	1 7/16″	.121′	8	3/4″	.061′
9	2 3/4″	.230′	9	1 13/16″	.153′	9	15/16″	.077′
10	3 7/16″	.284′	10	2 1/4″	.189′	10	1 1/8″	.095′
11	4 1/8″	.344′	11	2 3/4″	.229′	11	1 3/8″	.115′

34′

#	1/4″:1′ Crown		#	1/8″:1′ Crown	
1		.003′	1		.001′
2	1/16″	.005′	2	1/16″	.002′
3	1/8″	.011′	3	1/8″	.006′
4	3/8″	.020′	4	3/16″	.010′
5	3/8″	.031′	5	1/4″	.015′
6	9/16″	.044′	6	3/8″	.022′
7	3/4″	.060′	7	7/8″	.030′
8	1 5/16″	.078′	8	3/4″	.039′
9	1 3/16″	.099′	9	3/16″	.050′
10	1 1/2″	.123′	10	1 1/4″	.061′
11	1 3/4″	.148′	11	1 1/4″	.074′
12	2 1/2″	.176′	12	1 7/16″	.088′
13	2 1/2″	.207′	13	1 11/16″	.104′
14	2 7/8″	.240′	14	1 7/8″	.120′
15	3 5/16″	.276′	15	1 5/8″	.138′
16	3 3/4″	.314′	16	1 7/8″	.157′
17	4 1/4″	.354′	17	2 1/8″	.177′

40′

#	1/4″:1′ Crown		#	1/8″:1′ Crown	
1	1/16″	.000′	1		.001′
2	3/16″	.004′	2	1/16″	.002′
3	1/8″	.009′	3	1/8″	.005′
4	3/16″	.017′	4	3/16″	.008′
5	5/16″	.026′	5	1/4″	.013′
6	7/16″	.038′	6	3/8″	.019′
7	5/8″	.051′	7	3/8″	.026′
8	13/16″	.067′	8	1/2″	.033′
9	1 1/16″	.084′	9	5/8″	.042′
10	1 1/4″	.104′	10	3/4″	.052′
11	1 1/2″	.126′	11	7/8″	.063′
12	1 13/16″	.150′	12	1″	.075′
13	2 1/16″	.176′	13	1 1/8″	.088′
14	2 13/16″	.204′	14	1 1/4″	.102′
15	2 13/16″	.234′	15	1 3/8″	.117′
16	3 3/16″	.267′	16	1 5/8″	.133′
17	3 5/8″	.301′	17	1 13/16″	.151′
18	4 1/16″	.338′	18	2″	.169′
19	4 1/2″	.376′	19	2 1/4″	.188′
20	5″	.417′			

44′

#	3/8″:1′ Crown	
1	0″	.002′
2	0″	.002′
3	3/16″	.004′
4	1/8″	.008′
5	3/16″	.012′
6	3/16″	.017′
7	3/8″	.023′
8	7/8″	.030′
9	9/16″	.038′
10	11/16″	.047′
11	13/16″	.057′
12	15/16″	.068′
13	1 3/8″	.080′
14	1 5/16″	.093′
15	1 7/16″	.107′
16	1 5/16″	.121′
17	1 11/16″	.137′
18	1 13/16″	.153′
19	2 1/16″	.171′
20	2 1/4″	.189′
21	2 1/2″	.209′
22	2 3/4″	.229′

60′

#	1/8″:1′ Crown	
1	0″	.000′
2	0″	.001′
3	1/16″	.003′
4	1/16″	.006′
5	1/8″	.009′
6	1/8″	.012′
7	3/16″	.017′
8	1/4″	.022′
9	1/4″	.028′
10	5/16″	.035′
11	1/2″	.042′
12	5/8″	.050′
13	5/16″	.059′
14	13/16″	.068′
15	15/16″	.078′
16	11/16″	.089′
17	1 3/16″	.100′
18	1 3/8″	.112′
19	1 1/16″	.125′
20	1 1/2″	.139′
21	1 11/16″	.153′
22	1 3/16″	.168′
23	2 3/16″	.184′
24	2 3/8″	.200′
25	2 5/8″	.217′
26	2 13/16″	.235′
27	3 1/16″	.253′
28	3 1/4″	.272′
29	3 1/2″	.292′
30	3 3/4″	.312′

22′-0″ Surface width
7′-11″-0″ — 7′-0″
Crown = 3/8″:1′
C or high point
1 11/16″ = .139′
Grade tangent
Surface
1 3/8″ = .344′

RAILROAD TURNOUTS AND CROSSOVERS

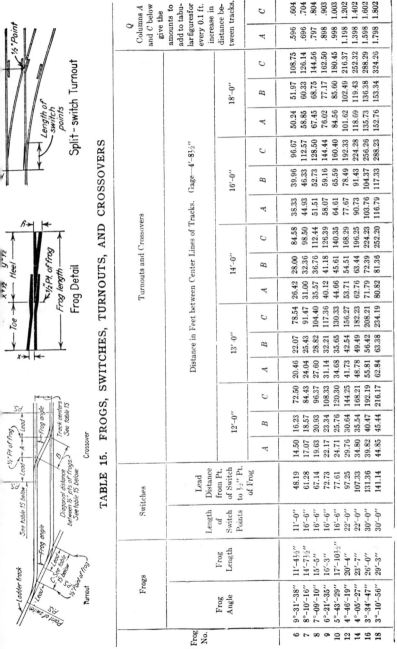

Split - switch Turnout

Frog Detail

Crossover

Turnout

TABLE 15. FROGS, SWITCHES, TURNOUTS, AND CROSSOVERS

Frogs			Switches			Turnouts and Crossovers															Q
						Distance in Feet between Center Lines of Tracks. Gage—4′-8½″															Columns A and C below give the amounts to add to tabular figures for every 0.1 ft. increase in distance between tracks.
Frog No.	Frog Angle	Frog Length	Length of Switch Points	Lead Distance from Pt. of Switch to ½″ Pt. of Frog		12′-0″			13′-0″			14′-0″			16′-0″			18′-0″				
						A	B	C	A	B	C	A	B	C	A	B	C	A	B	C	A	C
6	9°-31′-38″	11′-4½″	11′-0″	48.19		14.50	16.23	72.50	20.46	22.07	78.54	26.49	28.00	84.58	38.33	39.96	96.67	50.24	51.97	108.75	.596	.604
7	8°-10′-16″	14′-7½″	16′-6″	61.28		17.07	18.57	84.43	24.04	25.43	91.47	31.00	32.36	98.50	44.93	46.33	112.57	58.85	60.33	126.14	.696	.704
8	7°-09′-10″	15′-5″	16′-6″	67.14		19.63	20.93	96.37	27.60	28.82	104.40	35.57	36.76	112.44	51.51	52.73	128.50	67.45	68.75	144.56	.797	.804
9	6°-21′-35″	16′-3″	16′-6″	72.73		22.17	23.34	108.33	31.14	32.21	117.36	40.12	41.18	126.39	58.07	59.16	144.44	76.02	77.17	162.50	.898	.903
10	5°-43′-29″	17′-10½″	16′-6″	77.61		24.71	25.76	120.30	34.68	35.65	130.33	44.66	45.61	140.35	64.61	65.59	160.40	84.56	85.60	180.45	.998	1.003
12	4°-46′-19″	20′-4″	22′-0″	97.25		29.76	30.64	144.25	41.73	42.54	156.27	53.71	54.51	168.29	77.67	78.49	192.33	101.62	102.49	216.37	1.198	1.202
14	4°-05′-27″	23′-7″	22′-0″	107.33		34.80	35.54	168.21	48.78	49.49	182.23	62.76	63.44	196.25	90.73	91.43	224.28	118.69	119.43	252.32	1.398	1.402
16	3°-34′-47″	26′-0″	30′-0″	131.36		39.82	40.47	192.19	55.81	56.42	208.21	71.79	72.39	224.23	103.76	104.37	256.26	135.73	136.38	288.29	1.598	1.602
18	3°-10′-56″	29′-3″	30′-0″	141.14		44.85	45.44	216.17	62.84	63.38	234.19	80.82	81.36	252.20	116.79	117.33	288.23	152.76	153.34	324.26	1.798	1.802

319

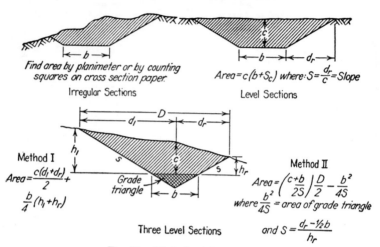

Fig. 18. Methods of finding areas.

1. **By average end areas:** * Volume in cubic yards $= \dfrac{A_0 + A_1}{2} \cdot \dfrac{l}{27}$, where $l =$ distance in feet between section A_0 and A_1. Compute end areas as indicated in Fig. 18. Use Tables 16 and 17; also see example on p. 321.

2. **By prismoidal formula:** Volume in cubic yards $= \dfrac{A_0 + 4M + A_1}{6} \cdot \dfrac{l}{27}$, where $l =$ distance in feet between sections A_0 and A_1, $M =$ area at section midway between section A_0 and A_1.

3. **Using prismoidal corrections:** Subtract volume in Table 18, p. 326, from volume found using average end areas method.

4. To find volume of excavation on curves use average end area method with l between sections as indicated below. Fill volumes can be computed similarly.

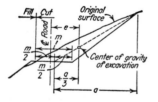

$l =$ distance between centers of gravity of adjacent sections.

Locate c.g. as shown on left; plot e on plan, and scale l along curve as indicated at right.

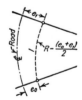

Fig. 19. Methods of finding volumes.

* Used by most state highway departments and Public Roads Administration. Recommended for roads and airports.

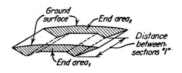

EXAMPLE 1. *Given.* End area₁ = 97 sq. ft.; end area₂ = 120 sq. ft.; $l = 50'$.

Required. Cubic yards between sections.

Solution. D.A. = 97 + 120 = 217 sq. ft. Enter D.A. column, and to right of 217 find C.Y. = 201 in C.Y. column.

Use Table 17 for D.A. of from 500 to 1000 cu. yd.

EXAMPLE 2. *Given.* D.A. = 2751 sq. ft.; $l = 50'$.

Required. Cubic yards between stations.

Solution. D.A. of 2000 = 1852 cu. yd. Find at bottom of Table 16; D.A. of 751 sq. ft. = 695 cu. yd. Therefore cubic yards for D.A. of 2751 sq. ft. = 1852 + 695 = 2547 cu. yd.

EXAMPLE 3. When l is less than 50'.

Given. D.A. = 217 sq. ft.; $l = 37'$.

Required. C.Y. between sections.

Solution. Enter column "Distance between Sections" and to right of · 37 find "Constant" .6852. Then .6852 × 217 = 149 C.Y.

DOUBLE END AREA VOLUMES

TABLE 16.   CUBIC YARDS FOR SUM OF END AREAS FOR DISTANCE BETWEEN STATIONS OF 50 FT. *

D.A. = sum of end areas in square feet.

D.A.	C.Y.	D.A.	C.Y.	D.A.	C.Y.	D.A.	C.Y.	D.A.	C.Y.	D.A.	C.Y.	D.A.	C.Y.	D.A.	C.Y.	D.A.	C.Y.	D.A.	C.Y.	Distance between Sections	Constant
0	0	50	46	100	93	150	139	200	185	250	231	300	278	350	324	400	370	450	417	0'	.0000
1	1	51	47	101	94	151	140	201	186	251	232	301	279	351	325	401	371	451	418	1'	.0185
2	2	52	48	102	94	152	141	202	187	252	233	302	280	352	326	402	372	452	419	2'	.0370
3	3	53	49	103	95	153	142	203	188	253	234	303	281	353	327	403	373	453	419	3'	.0556
4	4	54	50	104	96	154	143	204	189	254	235	304	281	354	328	404	374	454	420	4'	.0741
5	5	55	51	105	97	155	144	205	190	255	236	305	282	355	329	405	375	455	421	5'	.0926
6	6	56	52	106	98	156	144	206	191	256	237	306	283	356	330	406	376	456	422	6'	.1111
7	6	57	53	107	99	157	145	207	192	257	238	307	284	357	331	407	377	457	423	7'	.1296
8	7	58	54	108	100	158	146	208	193	258	239	308	285	358	331	408	378	458	424	8'	.1482
9	8	59	55	109	101	159	147	209	194	259	240	309	286	359	332	409	379	459	425	9'	.1667
10	9	60	56	110	102	160	148	210	194	260	241	310	287	360	333	410	380	460	426	10'	.1852
11	10	61	56	111	103	161	149	211	195	261	242	311	288	361	334	411	381	461	427	11'	.2037
12	11	62	57	112	104	162	150	212	196	262	243	312	289	362	335	412	381	462	428	12'	.2222
13	12	63	58	113	105	163	151	213	197	263	244	313	290	363	336	413	382	463	429	13'	.2407
14	13	64	59	114	106	164	152	214	198	264	244	314	291	364	337	414	383	464	430	14'	.2593
15	14	65	60	115	106	165	153	215	199	265	245	315	292	365	338	415	384	465	431	15'	.2778
16	15	66	61	116	107	166	154	216	200	266	246	316	293	366	339	416	385	466	431	16'	.2963
17	16	67	62	117	108	167	155	217	201	267	247	317	294	367	340	417	386	467	432	17'	.3148
18	17	68	63	118	109	168	156	218	202	268	248	318	294	368	341	418	387	468	433	18'	.3333
19	18	69	64	119	110	169	156	219	203	269	249	319	295	369	342	419	388	469	434	19'	.3519
20	19	70	65	120	111	170	157	220	204	270	250	320	296	370	343	420	389	470	435	20'	.3704
21	19	71	66	121	112	171	158	221	205	271	251	321	297	371	344	421	390	471	436	21'	.3889
22	20	72	67	122	113	172	159	222	206	272	252	322	298	372	344	422	391	472	437	22'	.4074

23	73	68	123	114	173	160	223	206	273	253	323	299	373	345	423	392	473	438	23'	.4259
24	74	69	124	115	174	161	224	207	274	254	324	300	374	346	424	393	474	439	24'	.4445
25	75	69	125	116	175	162	225	208	275	255	325	301	375	347	425	394	475	440	25'	.4630
26	76	70	126	117	176	163	226	209	276	256	326	302	376	348	426	395	476	441	26'	.4815
27	77	71	127	118	177	164	227	210	277	256	327	303	377	349	427	396	477	442	27'	.5000
28	78	72	128	119	178	165	228	211	278	257	328	304	378	350	428	397	478	443	28'	.5185
29	79	73	129	119	179	166	229	212	279	258	329	305	379	351	429	398	479	444	29'	.5370
30	80	74	130	120	180	167	230	213	280	259	330	306	380	352	430	399	480	444	30'	.5556
31	81	75	131	121	181	168	231	214	281	260	331	306	381	353	431	400	481	445	31'	.5741
32	82	76	132	122	182	169	232	215	282	261	332	307	382	354	432	401	482	446	32'	.5926
33	83	77	133	123	183	169	233	216	283	262	333	308	383	355	433	402	483	447	33'	.6111
34	84	78	134	124	184	170	234	217	284	263	334	309	384	356	434	403	484	448	34'	.6296
35	85	79	135	125	185	171	235	218	285	264	335	310	385	356	435	404	485	449	35'	.6482
36	86	80	136	126	186	172	236	219	286	265	336	311	386	357	436	405	486	450	36'	.6667
37	87	81	137	127	187	173	237	219	287	266	337	312	387	358	437	406	487	451	37'	.6852
38	88	81	138	128	188	174	238	220	288	267	338	313	388	359	438	406	488	452	38'	.7037
39	89	82	139	129	189	175	239	221	289	268	339	314	389	360	439	407	489	453	39'	.7222
40	90	83	140	130	190	176	240	222	290	269	340	315	390	361	440	408	490	454	40'	.7408
41	91	84	141	131	191	177	241	223	291	269	341	316	391	362	441	409	491	455	41'	.7593
42	92	85	142	131	192	178	242	224	292	270	342	317	392	363	442	410	492	456	42'	.7778
43	93	86	143	132	193	179	243	225	293	271	343	318	393	364	443	411	493	456	43'	.7963
44	94	87	144	133	194	180	244	226	294	272	344	319	394	365	444	412	494	457	44'	.8148
45	95	88	145	134	195	181	245	227	295	273	345	320	395	366	445	413	495	458	45'	.8333
46	96	89	146	135	196	181	246	228	296	274	346	321	396	367	446	414	496	459	46'	.8519
47	97	90	147	136	197	182	247	229	297	275	347	322	397	368	447	415	497	460	47'	.8704
48	98	91	148	137	198	183	248	230	298	276	348	323	398	369	448	416	498	461	48'	.8889
49	99	92	149	138	199	184	249	231	299	277	349		399		449		499	462	49'	.9074
																	500	463	50'	.9259

1000 = 926   2000 = 1852   3000 = 2778   4000 = 3704   5000 = 4630

*Based on average end area formula. Not as accurate as prismoidal formula, but as accurate as usual field measurements warrant. Specified for payment quantities by most state highway departments.

TABLE 17. CUBIC YARDS FOR SUM OF END AREAS FOR DISTANCE BETWEEN STATIONS OF 50 FT.*

D.A. = sum of end areas in square feet (double end area).

D.A.	C.Y.	D.A.	C.Y.	D.A.	C.Y.	D.A.	C.Y.	D.A.	C.Y.	D.A.	C.Y.	D.A.	C.Y.	D.A.	C.Y.	D.A.	C.Y.	D.A.	C.Y.	Distance between Sections	Constant
500	463	550	509	600	556	650	602	700	648	750	694	800	741	850	787	900	833	950	880	1	0.0185
501	464	551	510	601	556	651	603	701	649	751	695	801	742	851	788	901	834	951	881	2	0.0370
502	465	552	511	602	557	652	604	702	650	752	696	802	743	852	789	902	835	952	881	3	0.0556
503	466	553	512	603	558	653	605	703	651	753	697	803	744	853	790	903	836	953	882	4	0.0741
504	467	554	513	604	559	654	606	704	652	754	698	804	744	854	791	904	837	954	883	5	0.0926
505	468	555	514	605	560	655	606	705	653	755	699	805	745	855	792	905	838	955	884	6	0.1111
506	469	556	515	606	561	656	607	706	654	756	700	806	746	856	793	906	839	956	885	7	0.1296
507	469	557	516	607	562	657	608	707	655	757	701	807	747	857	794	907	840	957	886	8	0.1482
508	470	558	517	608	563	658	609	708	656	758	702	808	748	858	794	908	841	958	887	9	0.1667
509	471	559	518	609	564	659	610	709	656	759	703	809	749	859	795	909	842	959	888	10	0.1852
510	472	560	519	610	565	660	611	710	657	760	704	810	750	860	796	910	843	960	889	11	0.2037
511	473	561	519	611	566	661	612	711	658	761	705	811	751	861	797	911	844	961	890	12	0.2222
512	474	562	520	612	567	662	613	712	659	762	706	812	752	862	798	912	844	962	891	13	0.2407
513	475	563	521	613	568	663	614	713	660	763	706	813	753	863	799	913	845	963	892	14	0.2593
514	476	564	522	614	569	664	615	714	661	764	707	814	754	864	800	914	846	964	893	15	0.2778
515	477	565	523	615	569	665	616	715	662	765	708	815	755	865	801	915	847	965	894	16	0.2963
516	478	566	524	616	570	666	617	716	663	766	709	816	756	866	802	916	848	966	894	17	0.3148
517	479	567	525	617	571	667	618	717	664	767	710	817	756	867	803	917	849	967	895	18	0.3333
518	480	568	526	618	572	668	619	718	665	768	711	818	757	868	804	918	850	968	896	19	0.3519
519	481	569	527	619	573	669	619	719	666	769	712	819	758	869	805	919	851	969	897	20	0.3704
520	481	570	528	620	574	670	620	720	667	770	713	820	759	870	806	920	852	970	898	21	0.3889
521	482	571	529	621	575	671	621	721	668	771	714	821	760	871	806	921	853	971	899	22	0.4074
522	483	572	530	622	576	672	622	722	669	772	715	822	761	872	807	922	854	972	900	23	0.4259
523	484	573	531	623	577	673	623	723	669	773	716	823	762	873	808	923	855	973	901	24	0.4445

524	485	574	531	624	578	674	624	724	670	774	717	824	763	874	809	924	856	974	902	25	0.4630
525	486	575	532	625	579	675	625	725	671	775	718	825	764	875	810	925	856	975	903	26	0.4815
526	487	576	533	626	580	676	626	726	672	776	719	826	765	876	811	926	857	976	904	27	0.5000
527	488	577	534	627	581	677	627	727	673	777	719	827	766	877	812	927	858	977	905	28	0.5185
528	489	578	535	628	581	678	628	728	674	778	720	828	767	878	813	928	859	978	906	29	0.5370
529	490	579	536	629	582	679	629	729	675	779	721	829	768	879	814	929	860	979	906	30	0.5556
530	491	580	537	630	583	680	630	730	676	780	722	830	769	880	815	930	861	980	907	31	0.5741
531	492	581	538	631	584	681	631	731	677	781	723	831	769	881	816	931	862	981	908	32	0.5926
532	493	582	539	632	585	682	631	732	678	782	724	832	770	882	817	932	863	982	909	33	0.6111
533	494	583	540	633	586	683	632	733	679	783	725	833	771	883	818	933	864	983	910	34	0.6296
534	494	584	541	634	587	684	633	734	680	784	726	834	772	884	819	934	865	984	911	35	0.6482
535	495	585	542	635	588	685	634	735	681	785	727	835	773	885	819	935	866	985	912	36	0.6667
536	496	586	543	636	589	686	635	736	681	786	728	836	774	886	820	936	867	986	913	37	0.6852
537	497	587	544	637	590	687	636	737	682	787	729	837	775	887	821	937	868	987	914	38	0.7037
538	498	588	544	638	591	688	637	738	683	788	730	838	776	888	822	938	869	988	915	39	0.7222
539	499	589	545	639	592	689	638	739	684	789	731	839	777	889	823	939	869	989	916	40	0.7408
540	500	590	546	640	593	690	639	740	685	790	731	840	778	890	824	940	870	990	917	41	0.7593
541	501	591	547	641	593	691	640	741	686	791	732	841	779	891	825	941	871	991	918	42	0.7778
542	502	592	548	642	594	692	641	742	687	792	733	842	780	892	826	942	872	992	919	43	0.7963
543	503	593	549	643	595	693	642	743	688	793	734	843	781	893	827	943	873	993	919	44	0.8148
544	504	594	550	644	596	694	643	744	689	794	735	844	781	894	828	944	874	994	920	45	0.8333
545	505	595	551	645	597	695	644	745	690	795	736	845	782	895	829	945	875	995	921	46	0.8519
546	506	596	552	646	598	696	645	746	691	796	737	846	783	896	830	946	876	996	922	47	0.8704
547	506	597	553	647	599	697	646	747	692	797	738	847	784	897	831	947	877	997	923	48	0.8889
548	507	598	554	648	600	698	646	748	693	798	739	848	785	898	831	948	878	998	924	49	0.9074
549	508	599	555	649	601	699	647	749	694	799	740	849	786	899	832	949	879	999	925	50	0.9259

1000 = 926	2000 = 1852	3000 = 2778	4000 = 3704	5000 = 4630	6000 = 5556	7000 = 6481	8000 = 7407

* Based on average end area formula. Not as accurate as prismoidal, but as accurate as usual field measurements warrant. Specified for payment quantities by most state highway departments.

For examples illustrating use of table see p. 321.

TABLE 18. PRISMOIDAL CORRECTIONS FOR L = 100′ STATIONS

$c_1 - c_2 =$	1	2	3	4	5	6	7	8	9
$D_1 - D_2$									
0.1	0.03	0.06	0.09	0.12	0.15	0.19	0.22	0.25	0.28
0.2	0.06	0.12	0.19	0.25	0.31	0.37	0.43	0.49	0.56
0.3	0.09	0.19	0.28	0.37	0.46	0.56	0.65	0.74	0.83
0.4	0.12	0.25	0.37	0.49	0.62	0.74	0.86	0.99	1.11
0.5	0.15	0.31	0.46	0.62	0.77	0.93	1.08	1.23	1.39
0.6	0.19	0.37	0.56	0.74	0.93	1.11	1.30	1.48	1.67
0.7	0.22	0.43	0.65	0.86	1.08	1.30	1.51	1.73	1.94
0.8	0.25	0.49	0.74	0.99	1.23	1.48	1.73	1.98	2.22
0.9	0.28	0.56	0.83	1.11	1.39	1.67	1.94	2.22	2.50
1.0	0.31	0.62	0.93	1.23	1.54	1.85	2.16	2.47	2.78
1.1	0.34	0.68	1.02	1.36	1.70	2.04	2.38	2.72	3.06
1.2	0.37	0.74	1.11	1.48	1.85	2.22	2.59	2.96	3.33
1.3	0.40	0.80	1.20	1.60	2.01	2.41	2.81	3.21	3.61
1.4	0.43	0.86	1.30	1.73	2.16	2.59	3.02	3.46	3.89
1.5	0.46	0.93	1.39	1.85	2.31	2.78	3.24	3.70	4.17
1.6	0.49	0.99	1.48	1.98	2.47	2.96	3.46	3.95	4.44
1.7	0.52	1.05	1.57	2.10	2.62	3.15	3.67	4.20	4.72
1.8	0.56	1.11	1.67	2.22	2.78	3.33	3.89	4.44	5.00
1.9	0.59	1.17	1.76	2.35	2.93	3.52	4.10	4.69	5.28
2.0	0.62	1.23	1.85	2.47	3.09	3.70	4.32	4.94	5.56
2.1	0.65	1.30	1.94	2.59	3.24	3.89	4.54	5.19	5.83
2.2	0.68	1.36	2.04	2.72	3.40	4.07	4.75	5.43	6.11
2.3	0.71	1.42	2.13	2.84	3.55	4.26	4.97	5.68	6.39
2.4	0.74	1.48	2.22	2.96	3.70	4.44	5.19	5.93	6.67
2.5	0.77	1.54	2.31	3.09	3.86	4.63	5.40	6.17	6.94
2.6	0.80	1.60	2.41	3.21	4.01	4.81	5.62	6.42	7.22
2.7	0.83	1.67	2.50	3.33	4.17	5.00	5.83	6.67	7.50
2.8	0.86	1.73	2.59	3.46	4.32	5.19	6.05	6.91	7.78
2.9	0.90	1.79	2.69	3.58	4.48	5.37	6.27	7.16	8.06
3.0	0.93	1.85	2.78	3.70	4.63	5.56	6.48	7.41	8.33
3.1	0.96	1.91	2.87	3.83	4.78	5.74	6.70	7.65	8.61
3.2	0.99	1.98	2.96	3.95	4.94	5.93	6.91	7.90	8.89
3.3	1.02	2.04	3.06	4.07	5.09	6.11	7.13	8.15	9.17
3.4	1.05	2.10	3.15	4.20	5.25	6.30	7.35	8.40	9.44
3.5	1.08	2.16	3.24	4.32	5.40	6.48	7.56	8.64	9.72
3.6	1.11	2.22	3.33	4.44	5.56	6.67	7.78	8.89	10.00
3.7	1.14	2.28	3.43	4.57	5.71	6.85	7.99	9.14	10.28
3.8	1.17	2.35	3.52	4.69	5.86	7.04	8.21	9.38	10.56
3.9	1.20	2.41	3.61	4.81	6.02	7.22	8.43	9.63	10.83
4.0	1.23	2.47	3.70	4.94	6.17	7.41	8.64	9.88	11.11
4.1	1.27	2.53	3.80	5.06	6.33	7.59	8.86	10.12	11.39
4.2	1.30	2.59	3.89	5.19	6.48	7.78	9.07	10.37	11.67
4.3	1.33	2.65	3.98	5.31	6.64	7.96	9.29	10.62	11.94
4.4	1.36	2.72	4.07	5.43	6.79	8.15	9.51	10.86	12.22
4.5	1.39	2.78	4.17	5.56	6.94	8.33	9.72	11.11	12.50
4.6	1.42	2.84	4.26	5.68	7.10	8.52	9.94	11.36	12.78
4.7	1.45	2.90	4.35	5.80	7.25	8.70	10.15	11.60	13.06
4.8	1.48	2.96	4.44	5.93	7.41	8.89	10.37	11.85	13.33
4.9	1.51	3.02	4.54	6.05	7.56	9.07	10.50	12.10	13.61
5.0	1.54	3.09	4.63	6.17	7.72	9.26	10.80	12.35	13.89
$c_1 - c_2 =$	1	2	3	4	5	6	7	8	9

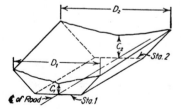

EXAMPLE. *Given.* $c_1 = 4′$, $D_1 = 130′$, $c_2 = 8′$, $D_2 = 138′$.

Required. Prismoidal correction value.

Solution. $c_1 - c_2 = 4$; $D_1 - D_2 = 8$. Enter table at 8.0; read correction = 9.88 cu. yd. $(c_2 - c_1)(D_2 - D_1) = (8 - 4)(138 - 130) = +$. Subtract correction from volume by average end area method. See p. 320.

TABLE 18.   PRISMOIDAL CORRECTIONS FOR L = 100' STATIONS,[*]
Continued

$c_1 - c_2 =$	1	2	3	4	5	6	7	8	9
$D_1 - D_2$									
5.1	1.57	3.15	4.72	6.30	7.87	9.44	11.02	12.59	14.17
5.2	1.60	3.21	4.81	6.42	8.02	9.63	11.23	12.84	14.44
5.3	1.64	3.27	4.91	6.54	8.18	9.81	11.45	13.09	14.72
5.4	1.67	3.33	5.00	6.67	8.33	10.00	11.67	13.33	15.00
5.5	1.70	3.40	5.09	6.79	8.49	10.19	11.88	13.58	15.28
5.6	1.73	3.46	5.19	6.91	8.64	10.37	12.10	13.83	15.56
5.7	1.76	3.52	5.28	7.04	8.80	10.56	12.31	14.07	15.83
5.8	1.79	3.58	5.37	7.16	8.95	10.74	12.53	14.32	16.11
5.9	1.82	3.64	5.46	7.28	9.10	10.93	12.75	14.57	16.39
6.0	1.85	3.70	5.56	7.41	9.26	11.11	12.96	14.81	16.67
6.1	1.88	3.77	5.65	7.53	9.41	11.30	13.18	15.06	16.94
6.2	1.91	3.83	5.74	7.65	9.57	11.48	13.40	15.31	17.22
6.3	1.94	3.89	5.83	7.78	9.72	11.67	13.61	15.56	17.50
6.4	1.98	3.95	5.93	7.90	9.88	11.85	13.83	15.80	17.78
6.5	2.01	4.01	6.02	8.02	10.03	12.04	14.04	16.05	18.06
6.6	2.04	4.07	6.11	8.15	10.19	12.22	14.26	16.30	18.33
6.7	2.07	4.14	6.20	8.27	10.34	12.41	14.48	16.54	18.61
6.8	2.10	4.20	6.30	8.40	10.49	12.59	14.69	16.79	18.89
6.9	2.13	4.26	6.39	8.52	10.65	12.78	14.91	17.04	19.17
7.0	2.16	4.32	6.48	8.64	10.80	12.96	15.12	17.28	19.44
7.1	2.19	4.38	6.57	8.77	10.96	13.15	15.34	17.53	19.72
7.2	2.22	4.44	6.67	8.89	11.11	13.33	15.56	17.78	20.00
7.3	2.25	4.51	6.76	9.01	11.27	13.52	15.77	18.02	20.28
7.4	2.28	4.57	6.85	9.14	11.42	13.70	15.99	18.27	20.56
7.5	2.31	4.63	6.94	9.26	11.57	13.89	16.20	18.52	20.83
7.6	2.35	4.69	7.04	9.38	11.73	14.07	16.42	18.77	21.11
7.7	2.38	4.75	7.13	9.51	11.88	14.26	16.64	19.01	21.39
7.8	2.41	4.81	7.22	9.63	12.04	14.44	16.85	19.26	21.67
7.9	2.44	4.88	7.31	9.75	12.19	14.63	17.07	19.51	21.94
8.0	2.47	4.94	7.41	9.88	12.35	14.81	17.28	19.75	22.22
8.1	2.50	5.00	7.50	10.00	12.50	15.00	17.50	20.00	22.50
8.2	2.53	5.06	7.59	10.12	12.65	15.19	17.72	20.25	22.78
8.3	2.56	5.12	7.69	10.25	12.81	15.37	17.93	20.49	23.06
8.4	2.59	5.19	7.78	10.37	12.96	15.56	18.15	20.74	23.33
8.5	2.62	5.25	7.87	10.49	13.12	15.74	18.36	20.99	23.61
8.6	2.65	5.31	7.96	10.62	13.27	15.93	18.58	21.23	23.89
8.7	2.69	5.37	8.06	10.74	13.43	16.11	18.80	21.48	24.17
8.8	2.72	5.43	8.15	10.86	13.58	16.30	19.01	21.73	24.44
8.9	2.75	5.49	8.24	10.99	13.73	16.48	19.23	21.97	24.72
9.0	2.78	5.56	8.33	11.11	13.89	16.67	19.44	22.22	25.00
9.1	2.81	5.62	8.43	11.23	14.04	16.85	19.66	22.47	25.28
9.2	2.84	5.68	8.52	11.36	14.20	17.04	19.88	22.72	25.56
9.3	2.87	5.74	8.61	11.48	14.35	17.22	20.09	22.96	25.83
9.4	2.90	5.80	8.70	11.60	14.51	17.41	20.31	23.21	26.11
9.5	2.93	5.86	8.80	11.73	14.66	17.59	20.52	23.46	26.39
9.6	2.96	5.93	8.89	11.85	14.81	17.78	20.74	23.70	26.67
9.7	2.99	5.99	8.98	11.98	14.97	17.96	20.96	23.95	26.94
9.8	3.02	6.05	9.07	12.10	15.12	18.15	21.17	24.20	27.22
9.9	3.06	6.11	9.17	12.22	15.28	18.33	21.39	24.44	27.50
10.0	3.09	6.17	9.26	12.35	15.43	18.52	21.60	24.69	27.78
$c_1 - c_2 =$	1	2	3	4	5	6	7	8	9

c_1, c_2, D_1, and D_2 are shown for a three-level section.   Volume by average end area ± prismoidal correction = volume by prismoidal formula.

When $(c_2 - c_1)(D_2 - D_1)$ is +, subtract correction.

When $(c_2 - c_1)(D_2 - D_1)$ is −, add correction.

Irregular sections are generally treated the same as three-level sections.

[*] *From American Civil Engineers Handbook by Merriman and Wiggin.*

LEVELING

SAMPLE NOTES

April 26, 1944 - Cloudy, Cool, 50-60°F.

BENCH LEVELS

STA.	B.S.+	H.I.	F.S.	Elev.	Estab. Elev.
B.M.#6	4.57	110.13			105.56
T.P.#1	3.18	107.18	6.13	104.00	
T.P.#2	2.56	104.07	5.67	101.51	
T.P.#3	5.06	105.05	4.08	99.99	
B.M.#7	4.17	103.02	6.20	98.85	
T.P.#4	8.11	105.94	5.19	97.83	
T.P.#5	7.16	107.03	6.07	99.87	
B.M.#8			5.55	101.48	101.47
	34.81		38.89	105.56	
			34.81	4.08	
			4.08	check	

E. Kroyer, Level   J. Lenart, Rod

MYRTLE STREET

K&E Wye Level 2167
Concrete Monument Sta 4+16 - 56' Rt.
Nail Head in Tel. Pole Sta 6+50 Lt.
Top of stake   " 8+25 -50' Lt.
Top Nut Hydrant   " 10+15 Rt.
□ Cut on Top N.E. Corner of Retaining wall
    Sta 12+25 -55.7' Rt.
Point on Curb   Sta 14+06 -30' Rt.
Cor. Conc. Step   " 16+19 -60' Rt.
U.S.G.S. Mon.   " 18+35 -45' Lt.

FIG. 20.

June 17, 1943 - Warm, Humid Overcast 70°-75°F.

PROFILE LEVELS

STA.	B.S.+	H.I.	F.S.	Elev.	Estab. Elev.
B.M.2	4.15	99.82			95.67
0+00			6.2	93.6	
0+50			5.9	93.9	
0+70			4.7	95.1	
1+00			5.5	94.3	
1+25			7.0	92.8	
1+50			6.0	93.8	
2+00			5.3	94.5	
2+20			5.5	94.3	
2+28			9.7	90.1	
2+34			9.9	89.9	
2+44			5.0	94.8	
2+50			4.5	95.3	
T.P. 1	5.05	100.88	3.99	95.83	
3+00			5.6	95.3	
3+50			6.6	94.3	
4+00			6.9	94.0	
4+50			7.2	93.7	
T.P. 2	7.42	101.79	6.51	94.37	
B.M.3			8.33	93.46	93.47
	16.62		18.83	95.67	
			16.62	2.21	
			2.21	check	

Berger Level 3197   CRANDALL
& RUNWAY 3-A-   FIELD
Whitten, Level; Wiesendanger, Rod

Concrete Mon. Sta 0+10 - 265' Rt. of &

Rock Out crop

Top of Bank
Bottom of Ditch
    "   "   "
Top of Bank
Top of & Stake - Sta. 2+50

Top of & Stake - Sta. 4+50
Concrete Mon. Sta 5+25 - 275' Rt. of &

FIG. 21.

328

TRANSIT PROBLEMS

1. Determination of Distance to Inaccessible Point

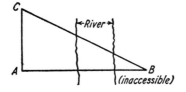

Required. *AB.*

Procedure. Set transit at *A*, sight on *B*. Turn 90° and set *C* at a point at least equal to ½*AB*. Measure length *AC*. Set up at *C* and measure angle *ACB*. *AB = AC ×* tangent *ACB*.

2. Angles by Repetition *

Required. A more accurate determination of an angle than possible by a single measurement.

Procedure. (1) Set the transit very carefully over the point. (2) Set the *A* vernier at zero, read the *B* vernier, and record the readings. (3) With the telescope in its normal position, measure one of the angles in a clockwise direction, and record both vernier readings to the smallest reading of the vernier. (4) Leaving the upper motion clamped, again set on the first point and again measure the angle in a clockwise direction (thus doubling the angle). (5) Continue until six repetitions have been secured. Record both vernier readings and the total angle turned. (6) In a like manner, setting the *B* vernier at zero, measure the explement of the angle in a counterclockwise direction with the bubble down, but read the horizontal circle as though the angle itself had been measured clockwise. (7) Go through the same process for all other angles about the point. (8) Compute the value of each of the angles for each direction turned, and compare with the single measurement. (9) Find the mean of each of these sets of single angles. For a transit reading to single minutes the total error should not exceed $10''\sqrt{n}$, in which n is the number of observations. (10) Adjust the angles so that their sum shall equal 360° by distributing the error equally among the mean values.

Hints and Precautions. (1) Level the transit very carefully before each repetition, but do not disturb the leveling screws while a measurement is being made. (2) The mean of each set of single angles should furnish a value free from instrumental errors. The station adjustment is an attempt to distribute the accidental errors so that the condition that there are 360° about a point shall be fulfilled. (3) Do not become

* *Adapted from Davis, Manual of Surveying, McGraw-Hill.*

confused when calculating the total angle turned. Observe how the horizontal limb is graduated, and do not omit 360°. (4) The instrument should be handled very carefully. When turning on the lower motion the hands should be in contact with the lower plate (not the alidade), and when making an exact setting on a point the last movement of the tangent screw should be clockwise or against the opposing spring. (5) After each repetition the instrument should be turned on its lower motion in a direction opposite to that of the measurement. (6) The single measurement is taken as a check on the number of repetitions. It should agree closely with the mean value.

Practical Applications. This method is used in triangulation work to measure any angle accurately. The number of sets of readings and the number of repetitions in each set observed depend upon the desired accuracy.

3. Laying off Angles by Repetition *

Required. To lay off a given horizontal angle more accurately than by a single setting of the vernier.

Procedure. (1) Set the transit carefully over the point and lay off the angle. (2) Set a stake on the line of sight, preferably at least 500 ft. from the instrument, and carefully set a tack. (3) By repetition measure the angle laid off, as in the previous problem, making six repetitions in each direction. (4) Find the angular discrepancy between the angle laid off and the required angle. Move the tack perpendicular to the line of sight, a distance equal to the sine of the angular discrepancy times the measured distance between the stakes. (5) Set the tack accordingly.

Practical Applications. This method is of use in laying out large buildings, valuable city lots or right of ways, important highway work such as viaducts and bridges, and airport runway center lines. With a transit vernier reading to 1 minute, an error of 30 sec. in a single reading might easily occur; in 300 ft. this would amount to approximately ½ in.

4. Area by Double Meridian Distance *

Required. Area of a closed traverse.
Rules.
Latitude = distance times cosine bearing angle.
Departure = distance times sine bearing angle.

Latitudes and departures are positive or negative according as they are north and east or south and west.

In any closed traverse the algebraic sum of the latitudes (or departures) must equal zero.

Compass rule for balancing. The correction to be applied to the lati-

* *Adapted from Davis, Manual of Surveying, McGraw-Hill.*

tude (or departure) of any course is to the total error in latitude (or departure) as the length of the course is to the perimeter of the field.

Transit rule for balancing. The correction to be applied to the latitude (or departure) of any course is to the total error in latitude (or departure) as the latitude (or departure) of that course is to the arithmetical sum of all the latitudes (or departures).

Rules for double meridian distances. (1) The D.M.D. of the first course equals the departure of that course.

(2) The D.M.D. of any other course equals the D.M.D. of the preceding course plus the departure of the preceding course plus the departure of the course itself.

(3) The D.M.D. of the last course is numerically equal to the departure of that course, but with opposite sign.

Procedure. (1) Transcribe necessary data from the field book into a form similar to that shown below. Check the copy.

(2) Calculate the latitude and departure of each course, using logarithms as shown in sample computations or more quickly and accurately with natural functions and a calculating machine if one is available. Check results with the slide rule.

(3) Determine the total error in latitude and in departure, and compute the error of closure.

(4) Determine the latitude and departure corrections by one of the preceding rules for balancing.

(5) Apply these corrections, and check by taking the algebraic sum of the corrected latitudes and the algebraic sum of the corrected departures. Each of these sums should equal zero.

(6) From the corrected departures compute the D.M.D.'s, applying the preceding rules and starting from the most westerly point in the survey. If the last D.M.D. is not numerically equal to the last corrected departure, it will indicate that a mistake in addition has been made.

(7) Compute double areas by the preceding rule paying special attention to signs. Check computations.

(8) Sum up the double areas, divide by 2, and transform into acres.

Hints and Precautions. (1) Use tables of logarithms or natural functions with number of places consistent with the precision of the field measurements. If the bearings have been determined with the surveyor's compass, four places will be sufficient; if angles have been taken to the nearest minute (in error less than 30 seconds) with the transit, five-place tables should be used.

(2) Checks should be applied after each of the steps in the computations. An absolute check on the work can, of course, be had only by recomputation, by methods that will give as many significant figures in the final result as the original computations gave. However, the slide rule will furnish an approximate check, which is very desirable.

(3) If, after having calculated the latitudes and departures and after having checked them against large errors, the error of closure is found to be larger than that allowable, the computer may frequently locate the mistake, whether it be in computations or field work, through the relation of total error in latitudes and total error in departures. Thus, if the mistake is in the length of one line and there are no other large errors, the ratio of the total error in departures to the total error in latitudes will approximately express the tangent of the bearing angle of that line, or if a mistake has been made in the latitude of a line the departures may nearly close. The computer should, therefore, conduct a critical examination of results and should then recompute those values that seem most likely to contain the mistake. If the mistake is not brought to light when all latitudes and departures have been rechecked, then, and only then, may he be warranted in concluding that the mistake occurred in the field.

(4) The compass rule or transit rule will be used for balancing latitudes and departures according as the error is assumed to be as much in angles as in distances or as the error is assumed to be mostly due to erroneous lengths.

(5) When the error of closure is small, the latitudes and departures may usually be balanced by inspection without computing the corrections by either of the preceding rules. When the computer knows the conditions surrounding the field work, he may often distribute the error according to his own judgment rather than by any fixed rule.

(6) Often neither calculated nor magnetic bearings of lines are shown in the transit notes. If deflection or interior angles were taken, it will be convenient to assume one of the lines in the traverse as the meridian and calculate the bearings of other lines accordingly. If magnetic bearings are recorded in the field notes, they should not be confused with calculated bearings and used as the basis of computations, for their precision will not warrant such use.

(7) Corrections for erroneous length of chain or tape should not be overlooked. Constant errors of this sort will have no effect on the error of closure.

(8) By starting with the most westerly point in the survey all the D.M.D.'s become positive; it is not necessary for the solution of the problem that this point be chosen, but it is customary.

Practical Applications. The double meridian distance method of calculating the area within a closed traverse is universally followed in preference to subdividing into triangles. It is generally agreed that it takes less time, is more systematic, and offers more easy checks; through the use of latitudes and departures, the error of closure is readily determined.

Some surveyors favor the method of double parallel distances, which is the same in principle as the preceding method, the only difference being that in double parallel distances (D.P.D.'s) the bases of trapezoids are

Line	Cal. Bear.	Dist. 66' Ch.	Latitudes		Departures		Corrected		D.M.Ds.	Double Areas	
			N	S	E	W	Lats.	Deps.		+	−
A–B	S 80°29½'W	34.464		5.694		33.991	− 5.693	−33.990	61.812		351.89
B–C	S 33°04' W	25.493		21.364		13.911	−21.361	−13.911	13.911		297.15
C–D	S 33°46¾'E	33.934		28.205	18.867		−28.201	+18.867	18.867		532.06
D–E	N 87°58¼'E	28.625	1.013		28.607		+ 1.013	+28.608	66.342	67.21	
E–A	N 0°27' E	54.235	54.234		0.426		+54.242	+ 0.426	95.376	5173.51	
		176.751	55.247	55.263	47.900	47.902	ΣL=0	ΣD=0		5240.72	1181.10
				55.247		47.900				1181.10	
				.016		.002			2	4059.62	

E. of C. = $\frac{.016}{176.751}$ = $\frac{1}{11,000}$ E = $\sqrt{.016^2 + .002^2}$ = 0.016 Chains.

2029.81 Sq. Ch.
or 202.981 Ac.

Line	A–B	B–C	C–D	D–E	E–A
Lat.	5.694	21.364	28.205	1.013	54.234
Log. Lat.	0.75542	1.32968	1.45032	0.00584	1.73427
Log. Cos.	9.21805	9.92326	9.91969	8.54899	9.99999
Log. Dist.	1.53737	1.40642	1.53063	1.45674	1.73428
Log. Sin.	9.99399	9.73689	9.74509	9.99973	7.89535
Log. Dep.	1.53136	1.14331	2.27572	1.45647	9.62964
Dep.	33.991	13.911	18.867	28.607	0.426
Log. Cor. Lat.	0.75534	1.32962	1.45026	0.00584	1.73434
Log. D.M.D.	1.79107	1.14336	1.27570	1.82179	1.97944
Leg. D.A.	2.54641	2.47298	2.72596	1.82763	3.71378
Double Area	351.89	297.15	532.06	67.21	5173.51

Note:
Survey Balanced
by Transit Rule.

Fig. 22.

along a line perpendicular to the meridian, whereas in double meridian distances they lie on the meridian itself. Thus, the rules for finding D.M.D.'s may be changed to rules for D.P.D.'s by substituting the word "latitude" for "departure"; and the rule for finding double areas will then be as follows: The double area of any trapezoid equals the product of its D.P.D. and its corrected departure.

5. Omitted Side *

Required. Length and bearing of one side of a traverse, this side not accessible in field. (It is assumed that errors in measured sides are negligible; all errors are thrown into computed side.)

Procedure. (1) Calculate the latitudes and departures of the known lines as in the previous problem, and find their totals. (2) On the preceding assumption, and since the algebraic sum of latitudes and of departures for any closed traverse is zero, it follows that the latitude and departure of the unknown line are numerically equal to the sums of corresponding quantities for the known lines, but with opposite sign. Therefore, determine the bearing and length of the unknown line by the equations:

$$\text{Tan bearing angle} = \frac{\text{departure of line}}{\text{latitude of line}}$$

and

* *Adapted from Davis, Manual of Surveying, McGraw-Hill.*

Length of line = $\sqrt{\text{latitude}^2 + \text{departure}^2}$

$$= \frac{\text{latitude of line}}{\cos \text{ bearing angle}} = \frac{\text{departure of line}}{\sin \text{ bearing angle}}$$

Precaution. Plot known sides, and graphically check omitted side and bearing.

6. Prolongation of a Line by Double Sighting with Transit * (Double Centering)

Required. To produce a straight line with precision.

Procedure. (1) Set the instrument carefully over the forward point on the line with the telescope normal and backsight on line. Use the lower horizontal motion, the upper motion being clamped. (2) Plunge the telescope, and set a stake on the line in advance. Mark a point on the stake exactly on line. (3) Take a second backsight on line in the same manner as before, with the telescope inverted. Plunge the telescope again, and mark a second point on the advance stake. (4) If this point is not coincident with the first point set, a point midway between them is on the line. (5) Set the transit over this point, and advance by the same process, backsighting upon the next point in the rear. Continue in this way for the desired distance.

Hints and Precautions. (1) Be sure that one backsight from each station is taken with the *telescope inverted* and one with the *telescope direct*. (2) Tacks should be set in all stakes, and after being set should be checked. A finely divided scale should be used for bisecting the distances. (3) Whenever an opportunity arises, take backsights as far back on a line as possible to check the line.

Practical Applications. The method of double sighting is used when it is desired to set a point in advance accurately on line. The process of double sighting eliminates instrumental errors. It is used in prolonging lines of a considerable length or setting points accurately ahead on line. Frequently a line prolonged by simply plunging the telescope with a transit supposed to be in perfect adjustment has later been found to be not a straight line but a curve of large radius. The same method should be used when setting transit points ahead on a curve.

7. Establishing a Line by Balancing-in with Transit (Bucking-in)

Required. To establish an intermediate transit point on a line when the two ends of the line are not intervisible.

Procedure. (1) Set up the transit where the intermediate point is required, and as near as can be estimated, on the line. (2) Backsight with telescope normal on the point marking one end of the line, and plunge

* *Adapted from Davis, Manual of Surveying, McGraw-Hill.*

the telescope. (3) Move the transit a proportionate amount of the distance by which the line of sight fails to strike the point at the opposite end of the line. (4) Repeat the procedure until the line of sight is coincident with the line. (5) Establish the point by lowering the plumb bob of the transit. (6) Repeat the process with the telescope inverted as in double centering. If the instrument is not in adjustment a second point will be found; the correct point is set midway between the two.

Hints and Precautions. The final movement of the transit can usually be made with the shifting head. Until near the correct point, it is unnecessary to level the transit carefully. Additional points on the line can be set by direct sighting.

8. Layout of Circular Curve

Required. To establish the P.C. and P.T. of a simple curve and set points at intervals along the curve.

Procedure. (1) Lay off both tangents from the P.I., thus locating the P.C. and P.T. (2) Set up the transit over the P.C.; set vernier at zero and foresight on P.I. Unclamp the upper motion and sight at the P.T. if visible; the deflection angle of the long chord should equal ½ the external angle Δ. (3) From the previously computed list of deflections, lay out the points on the curve using the proper deflection angle and subchord or full chord as required.

Hints and Precautions. (1) If the back tangent has been stationed the P.C. may be set from the nearest station. (2) When the survey is to be carried ahead the transit may be set up over the P.T. and the curve laid out from it, thus saving a set-up. (3) When setting a transit point or an accurate point on the curve (P.O.C.), the backsight should be checked and the deflection turned with the telescope plunged in both the inverted and direct positions, the point being set as in double centering for a straight line.

Set-up on Curve. When all the stations of a curve are not visible from either the P.C. or P.T., a transit point must be set at some point on the curve (P.O.C.) and the transit moved up to it. (1) Locate the P.O.C. (2) Set up over the P.O.C. backsight on the P.C. with a zero reading on the vernier. (3) Plunge the telescope, and turn the telescope inward until the vernier reading (deflection) for the P.O.C. is reached. The line of sight will then be tangent to the curve. (4) Lay off the deflections for the points to be set as computed in the original list.

Note. Any other station than the P.C. may be sighted provided the proper deflection is used. The following rules apply:

Rule I. When the transit is set on any point on a curve, an auxiliary tangent to the curve at that point may be found by sighting at any station on the curve with the deflection of the station sighted laid off on the proper side of zero and turning the upper motion until the vernier reading (deflection) for the point occupied is reached.

Rule II. When the transit is set on any point on a curve (including the P.C. or P.T.), any other point on the curve may be set by sighting at any point on the curve with the deflection for the point sighted laid off on the proper side of the vernier and turning the upper motion in the proper direction until the vernier reading (deflection) for the point to be set is reached.

SAMPLE NOTES

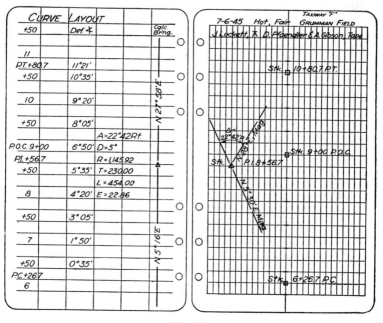

F<small>IG.</small> 23.

ALLOWABLE ERRORS

Leveling *

Rough leveling for rapid reconnaissance or preliminary work; sights made up to 1000 ft.; rod readings to tenths; no attention paid to balancing backsights and foresights.

Suggested maximum error in feet $= \pm 0.4\sqrt{\text{distance in miles}}$.

Ordinary leveling as required for most engineering works; maximum sights 500 ft.; rod readings to hundredths; backsights and foresights roughly balance for both length of shots and uphill and downhill work; turning points on reasonably solid objects.

Suggested maximum error in feet $= \pm 0.1\sqrt{\text{distance in miles}}$.

    * *Adapted from Urquhart, Civil Engineering Handbook, McGraw-Hill.*

Accurate leveling, for principal bench marks; maximum sights 300 ft.; rod readings to thousandths; backsights and foresights paced and balanced; rod waved; bubble centered for each sight; turning points on very solid objects; level set very firmly.

Maximum error in feet $= \pm 0.05\sqrt{\text{distance in miles}}$.

This error is the same as allowed for third-order leveling, Corps of Engineers, U. S. Army.

Distances

By stadia, 1:750 maximum allowable error.

By tape, 1:5000 maximum allowable error for ordinary work.

Transit and Tape Traverses

Linear error of closure $= \sqrt{(\text{sum of latitudes})^2 + (\text{sum of departures})^2}$.

The precision of transit traverses is affected by both linear and angular errors of measurement. Many factors affect the precision, and it can be expressed only in very general terms. The following specifications give approximately the *maximum* linear and angular errors to be expected when the methods stated are followed. If the surveys are executed by well-trained men, with instruments in good adjustment, and under average field conditions, in general the error of closure should not exceed *half* the specified amount. The specifications apply to traverses of considerable length. It is assumed that a standardized tape is used.

Class 1. Precision sufficient for many preliminary surveys, for horizontal control of surveys plotted to intermediate scale, and for land surveys where the value of the land is low.

Transit angles read to the nearest minute. Sights taken on a range pole plumbed by eye. Distances measured with a 100-ft. steel tape. Pins or stakes set within 0.1 ft. of end of tape. Slopes under 3% disregarded. On slopes over 3%, distances either measured on the slope and corrections roughly applied, or measured with the tape held level and with an estimated standard pull.

Angular error of closure not to exceed $1'\,30''\sqrt{n}$, in which n is the number of observations. Total linear error of closure not to exceed 1/1000.

Class 2. Precision sufficient for most land surveys and for location of highways, railroads, etc. By far the greater number of transit traverses fall in this class.

Transit angles read carefully to the nearest minute. Sights taken on a range pole carefully plumbed. Pins or stakes set within 0.05 ft. of end of tape. Temperature corrections applied to the linear measurements if the temperature of air differs more than 15° F. from standard. Slopes under 2% disregarded. On slopes over 2%, distances either measured on the slope and corrections roughly applied, or measured with the tape held level and with a carefully estimated standard pull.

Angular error of closure not to exceed $1'\sqrt{n}$. Total linear error of closure not to exceed 1/3000.

Class 3. Precision sufficient for much of the work of city surveying, for surveys of important boundaries, and for the control of extensive topographic surveys.

Transit angles read twice with the instrument plunged between observations. Sights taken on a plumb line or on a range pole carefully plumbed. Pins set within 0.05 ft. of end of tape. Temperature of air determined within 10° F., and corrections applied to the linear measurements. Slopes determined within 2%, and corrections applied. Tape held level, the pull kept within 5 lb. of standard, and corrections for sag applied.

Angular error of closure not to exceed $30'' \sqrt{n}$. Total linear error of closure not to exceed 1/5000.

Class 4. Precision sufficient for accurate city surveying and for other especially important surveys.

Transit angles read twice with the instrument plunged between readings, each reading being taken as the mean of both A and B vernier readings. Verniers reading to 30″. Instrument in excellent adjustment. Sights taken with special care. Pins set within 0.02 ft. of end of tape. Temperature of tape determined within 5° F., and corrections applied. Slopes determined within 1%, and corrections applied. Tape held level, the pull kept within 3 lb. of standard, and corrections for sag applied.

Angular error of closure not to exceed $15'' \sqrt{n}$. Total linear error of closure not to exceed 1/10,000.*

DETERMINATION OF TRUE NORTH

OBSERVATION ON POLARIS

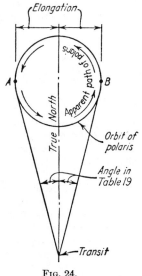

Procedure. Set up transit over a point. Observe Polaris at A or B, when the elongation remains constant—a 20-minute period during which Polaris appears to move vertically and actually varies not more than 0.1 minute from the elongation. Depress telescope and set a point ahead. Turn off the angle in Table 19 to give the true north.

Fig. 24.

* L. C. Urquhart, *Civil Engineering Handbook, McGraw-Hill.*

TABLE 19. AZIMUTHS OF POLARIS AT ELONGATION, FOR THE BEGINNING OF YEARS 1945–1955 *

(Computed by the U. S. Naval Observatory)

Latitude	1945	1946	1947	1948	1949	1950	1951	1952	1953	1954	1955
°	° ′	° ′	° ′	° ′	° ′	° ′	° ′	° ′	° ′	° ′	° ′
25	1 6.1	1 5.7	1 5.3	1 5.0	1 4.6	1 4.2	1 3.8	1 3.4	1 3.2	1 2.9	1 2.5
26	6.6	6.3	5.9	5.5	5.1	4.8	4.3	4.0	3.8	3.4	3.0
27	7.2	6.8	6.5	6.1	5.7	5.3	4.9	4.5	4.3	4.0	3.6
28	7.8	7.4	7.1	6.7	6.3	5.9	5.5	5.1	4.9	4.6	4.2
29	8.5	8.1	7.7	7.3	6.9	6.5	6.1	5.7	5.5	5.2	4.8
30	9.1	8.8	8.4	8.0	7.6	7.2	6.8	6.4	6.2	5.8	5.4
31	9.9	9.5	9.1	8.7	8.3	7.9	7.5	7.1	6.9	6.5	6.1
32	10.6	10.2	9.8	9.4	9.0	8.6	8.2	7.8	7.6	7.2	6.8
33	11.4	11.0	10.6	10.2	9.8	9.4	9.0	8.6	8.3	8.0	7.6
34	12.2	11.8	11.4	11.0	10.6	10.2	9.8	9.4	9.1	8.8	8.4
35	13.1	12.7	12.3	11.9	11.5	11.1	10.6	10.2	10.0	9.6	9.2
36	14.0	13.6	13.2	12.8	12.4	11.9	11.5	11.1	10.9	10.5	10.0
37	15.0	14.6	14.1	13.7	13.3	12.9	12.4	12.0	11.8	11.4	11.0
38	16.0	15.6	15.1	14.7	14.3	13.9	13.4	13.0	12.8	12.3	11.9
39	17.0	16.6	16.2	15.8	15.3	14.9	14.4	14.0	13.8	13.3	12.9
40	18.2	17.7	17.3	16.9	16.4	16.0	15.5	15.1	14.8	14.4	14.0
41	19.3	18.9	18.5	18.0	17.6	17.1	16.6	16.2	16.0	15.5	15.1
42	20.6	20.1	19.7	19.2	18.8	18.3	17.8	17.4	17.1	16.7	16.3
43	21.9	21.4	21.0	20.5	20.0	19.6	19.1	18.6	18.4	17.9	17.5
44	23.2	22.8	22.3	21.9	21.4	20.9	20.4	19.9	19.7	19.2	18.8
45	24.7	24.2	23.7	23.3	22.8	22.3	21.8	21.3	21.1	20.6	20.1
46	26.2	25.7	25.2	24.8	24.3	23.8	23.3	22.8	22.5	22.1	21.6
47	27.8	27.3	26.8	26.3	25.8	25.3	24.8	24.3	24.1	23.6	23.1
48	29.5	29.0	28.5	28.0	27.5	27.0	26.4	25.9	25.7	25.2	24.7
49	31.3	30.8	30.3	29.8	29.2	28.7	28.2	27.6	27.4	26.9	26.4
50	1 33.2	1 32.6	1 32.1	1 31.6	1 31.1	1 30.6	1 30.0	1 29.5	1 29.2	1 28.7	1 28.2

Corrections for middle of months

For middle of—	Correction
	′
Jan.	−0.5
Feb.	−0.4
Mar.	−0.3
Apr.	−0.1
May	+0.1
June	+0.2
July	+0.2
Aug.	+0.1
Sept.	−0.1
Oct.	−0.3
Nov.	−0.6
Dec.	−0.8

These data may be secured annually from the current *Nautical Ephemeris* or similar source.

* Data for 1945–1950 from *War Department, Surveying Tables.* Data for 1951–1955 extracted from the *American Ephemeris and Nautical Almanac* published by the U. S. Naval Observatory.

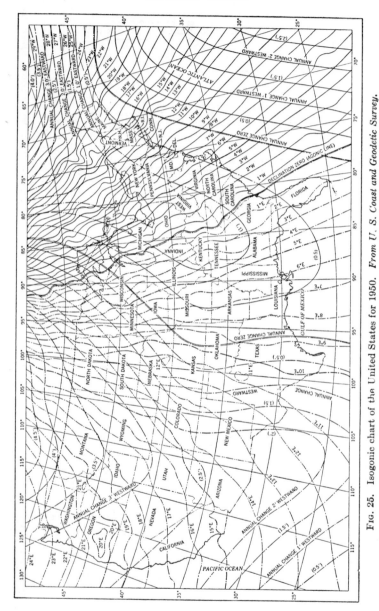

Fig. 25. Isogonic chart of the United States for 1950. *From U. S. Coast and Geodetic Survey.*

There are two sets of lines on the isogonic chart, Fig. 25, which may be distinguished in two ways: (1) the isoporic lines are much smoother than the isogonic lines; (2) the isoporic lines are numbered in minutes and the isogonic lines in degrees.

The isogonic lines or lines of equal declination (also called "lines of equal variation of the compass") are drawn for January 1950. East of the agonic line, the lines are solid, signifying that the north end of the compass needle points west of true north; west of the agonic line they are dashed, and the compass points east of true north. The lines are drawn to show a smoothed distribution; in the more disturbed regions, the sinuosities of the lines must be regarded as an indication of irregularity rather than as a close representation of the declination.

Magnetic declination is subject to gradual change, the rate of which depends upon time and place. The annual rate of change prevailing from about 1944 to 1950 may be estimated from the isoporic lines. These lines are solid in regions where the prevailing declination was increasing, and dashed in regions in which the declination was decreasing. Note that, when an isoporic line crosses the agonic line, its sign changes.

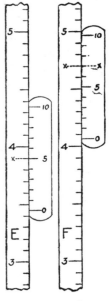

Vernier

Accurate readings on scales will fall somewhere between rather than on the subdivision marks on a scale. The vernier is a supplementary scale designed to aid in evaluating these fractional overages.

It is an adjacent scale against which slides the main scale as illustrated in the figure at the right. The zero of the vernier scale becomes the point from which the reading on the main scale is taken. The divisions of the vernier are a little smaller than those on the main scale. Thus 10 subdivisions on the vernier scale equal 9 subdivisions on the main scale.

From Tracy, Surveying: Theory and Practice, by permission of the author and John Wiley & Sons.

The refinement is given by reading to the nearest subdivision on the main scale opposite the zero on the vernier and looking along the scale until the point is reached where the subdivisions of the vernier scale and the main scale appear coincident. For instance, in the two scales illustrated, if the major subdivisions on the main scales are tenths of a foot, the reading of the scale marked *E* would be 0.345 ft. The reading of the scale marked *F* would be 0.407 ft.

INSTRUMENTS AND THEIR ADJUSTMENTS*

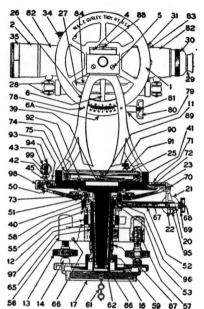

Cross section of Gurley transit.

Parts of Gurley Precise Transits

1. Vertical circle guard.
2. Dust shield, protecting objective slide.
3. Detachable sunshade. (Not illus.)
4. Cap screws to standard.
5. Screws—guard to standard.
6. Vertical circle.
6A. Vertical circle vernier.
7. Side plate level.
9. North (or transverse) plate level.
11. Compass needle.
12. Lower (or leveling head) clamp.
13. Leveling screw.
14. Leveling screw cup.
15. Lower tangent screw. (Not illus.)
16. Shifting center.
17. Bottom plate.
18. Lower clamp screw. (Not illus.)
20. Upper (or limb) clamp screw.
21. Upper (or limb) tangent screw.
22. Upper (or plate) tangent hanger.
23. "A" vernier.
24. Needle lifter.

25. Compass glass cover in metal bezel ring.
26. One piece truss standard.
27. Telescope level.
28. Adjusting nuts for telescope level.
29. Eyepiece cap.
30. Knurled ring for eyepiece focusing.
31. Capstan screw for adjusting cross-wires.
32. Clamp screw for telescope axle.
33. Objective slide adjusting screw.
34. Objective focusing pinion.
35. Objective cap.
36. Side (or longitudinal) level vial.
39. Center pin.
40. Limb centering screws.
41. Screw—plate to spindle.
42. Capstan nut—north (or transverse) vial.
44. Spring guard to north vial.
45. Plate level post.
46. Top plate.
47. Screw—plate to standard.
49. Index pointer for magnetic declination.
50. Limb.
51. Socket.
52. Limb clamp.
53. Screw—clamp sleeve to socket.
54. Clamp sleeve.
55. Clamp collar.
56. Spider, or four-arm piece.
57. Leveling screw nut.
58. Spindle.
59. Half-ball.
61. Jack or plummet chain.
62. Bottom cap.
64. Washer—end of spindle.
65. Shell.
66. Keeper screw.
67. Limb clamp plunger.
68. Locking screw—head to stem of clamp screw.
69. Clamp screw head.
70. Screw—tangent hanger to plate.
71. Vernier glass.
72. Screw—vernier to plate.
73. Screw—limb to socket.
74. Needle circle.
75. Bezel ring.
78. Screw—v.c. vernier to standard.
79. Axis tangent screw stem.
80. Head, axis tangent screw.
81. Locking screw—head to stem of axis tangent screw.
82. Telescope.
83. Collet, for cross-wire adjusting.
84. Telescope level vial.
86. Nut—end of spindle.
87. Half-ball set screw.
88. Capstan adjusting screw in standard cap.
90. Needle lifter screw.
91. Needle lifter housing.
92. Screw—compass to plate.
93. Screw—cover ring to standard base.
94. Nut—top of plate level post.
95. Take-up screw to limb tangent
96. Gib—leveling head clamp.
97. Spacer ring.
98. Cover ring.
99. Plate level adjusting spring.

*From Surveying Instrument Manual, W. & L. E. Gurley, Troy, N. Y.

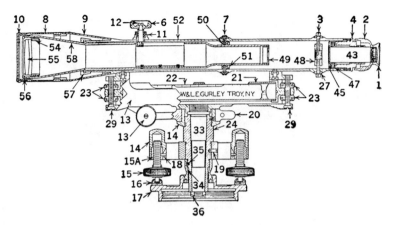

Cross section of Gurley dumpy level.

Parts for Gurley Dumpy Levels

1. Eyepiece cap.
2. Eyepiece focusing ring.
3. Capstan screw for adjusting cross wires.
4. Eyepiece body.
6. Objective focusing pinion.
7. Objective slide centering screw.
8. Dust shield.
9. Main tube head.
10. Objective cap.
11. Objective pinion body.
12. Objective pinion screw.
13. Bar.
14. Leveling head.
15. Leveling screw.
15A. Leveling screw bushing.
16. Leveling screw cup.
17. Bottom plate.
18. Leveling screw keeper screw.
19. Shell set screw.
20. Leveling head clamp.
21. Telescope level.

22. Telescope level vial.
23. Capstan adjusting nuts for telescope level vial.
24. Shell or outer bearing.
27. Collet for cross-wire adjusting screws.
29. Post for adjusting telescope level.
33. Spindle.
34. Half ball.
35. Screw for half ball.
36. Nut, end of spindle.
45. Eyepiece centering ring.
47. Eyepiece centering screw.
48. Cross-wire reticule.
49. Diaphragm in slide.
50. Objective slide centering ring.
51. Babbit, slide centering ring.
52. Main tube.
54. Inner ring, objective setting.
55. Objective lens.
56. Outer ring, objective setting.
57. Babbit, for objective end.
58. Objective slide.

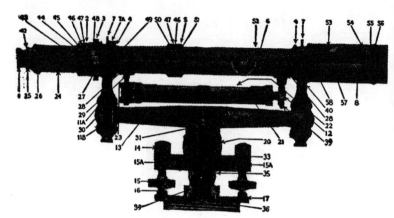

Quarter section of Gurley wye level.

Parts for Gurley Wye Levels

1. Eyepiece cap.
2. Cover ring, covering eyepiece centering screws.
3. Capstan screw, for adjusting cross wires.
4. Wye rings.
5. Cover ring, covering objective slide adjusting screws.
6. Objective focusing pinion.
7. Wye pin.
7A. Wye clip stop pin.
8. Dust shield.
9. Sunshade.*
11A. Wye capstan nuts (upper).
11B. Wye capstan nuts (lower).
12. Level lateral adjusting screw.
13. Wye bar.
14. Leveling head.
15. Leveling screw.
16. Leveling screw cup.
17. Bottom plate.
20. Leveling head clamp.
21. Telescope level complete.
22. Telescope level vial.
23. Vertical adjusting capstan nuts for telescope level.
24. Eyepiece focusing pinion.
25. Sleeve for eyepiece.
26. Eye end ring.

27. Collet, for cross-wire centering screws.
28. Screws for telescope level hanger and post.
29. Telescope level post.
30. Spline.
31. Spindle head.
33. Spindle.
34. Half ball.
35. Screw for half ball.
39. Hanger for telescope level.
40. Wye complete.
42. Babbit ring, in sleeve for eyepiece.
43. Eyepiece.
44. Babbit, in eyepiece centering ring.
45. Eyepiece centering ring.
46. Collet, for eyepiece centering screw.
47. Eyepiece centering screw.
48. Cross-wire reticule.
49. Diaphragm in slide.
50. Slide centering ring.
51. Babbit, slide centering ring.
52. Main tube.
53. Binding ring.
54. Inner ring, objective setting.
55. Objective lens.
56. Outer ring, objective setting.
57. Babbit, for objective end.
58. Objective slide.
59. Objective cap.*

* Not illustrated.

Hints on Adjustments

Before proceeding with any adjustment, read the following suggestions carefully.

Making the Adjustments. Do not attempt to perfect each adjustment the first time as succeeding adjustments may disturb those already made. It is better to keep repeating the entire series until a final check shows each adjustment to be perfect.

Inspection of Instrument. Before adjusting any instrument, clean it thoroughly. Dirt in bearings will not permit a true adjustment. If adjusting screws or nuts are dirty they will not hold adjustment very long. Damaged or worn screws should be replaced by new factory parts as soon as possible. Damaged or worn bearings or damaged structural parts should be repaired and refitted at the factory. Clamps, tangent screws, and tangent springs should be clean and the clamp arm should be examined to make sure there is no indentation where the tangent screw presses. Be sure that the instrument is correctly assembled and that the holding screws are set up solidly but not overstrained. The telescope should be clean, the lenses showing objects sharply and without astigmatism. Be sure that the object lens is tight in its setting and that the setting is screwed tightly in its tube. All axis bearing caps should be screwed up to the proper tension. The proper fit of the telescope axle and the elimination of "walk" is very important. Check the level vials to see that they are firm in their cases. Examine the shoes on the tripod to make sure they are tight.

Select a Suitable Location. Established offices should provide a substantial pier or wall bracket wherewith to support the instrument when adjusting. Targets and scales should be set at convenient distances and elevations. In a limited space, particularly indoors, telescopes focused at infinity should be set up for use as collimators. On construction work an adjusting site should be selected, targets erected and a stake driven to define the instrument position if a tripod and not a permanent support is used. In selecting such sites, avoid places where the line of sight would pass over a railroad track or paved highway, near a heated building, or through successive areas of light and shadow. Protect the instrument from wind and direct rays of the sun, particularly when they strike only one side of the instrument at a time.

Setting up the Instrument. Select a spot where the ground is firm and dry so that moving around the instrument will not disturb it. If the instrument is set on a floor of concrete, brick or stone, make sure that there are no loose sections. Chip holes in a smooth floor to prevent the tripod points from slipping. After screwing the instrument to the tripod, loosen the tripod bolts, then tighten them, in order to remove all residual torque in the tripod head. This helps hold the transit on line. Tighten the leveling screws firmly, but do not force them.

Transits

The adjustments of transits are as follows:

1. Parallax.
2. Rectify cross wires.
3. Collimation at distant focus.
4. Collimation at minimum focus.
5. Telescope axis.
6. Telescope level.
7. Plate levels.
8. Vertical circle vernier.
9. Center eyepiece.
10. Balance compass needle.
11. Straighten needle
12. Center pivot.

Description of Transit

The transit, as generally constructed today, serves to measure angles in azimuth and in altitude. It, therefore, consists of two divided circles or limbs, one of which rotates about a vertical axis and the other about a horizontal axis. Each graduated surface is made perpendicular to its axis of rotation. The pointer of the instrument is a telescope, supported by standards and plate, the plate carrying the indices or verniers. The spindle, carrying the plate and standards, and the socket, carrying the horizontal limb, constitute the "centers" which rotate about each other and within the bearing of the leveling head.

The "centers" or vertical axis is made plumb by two spirit levels mounted on the plate. These levels are adjustable, and they can be readily checked by reversal about the centers.

The telescope is mounted with an axle which rides in bearings on top of the standards. For the axle to form a horizontal axis it must be at right angles to the vertical axis, and adjustment is provided for raising or lowering one end of the axle.

The pointer of the telescope is an optical line of sight passing through the optical center of the objective lens and the intersection of the cross wires. This is commonly called the line of collimation. The cross-wire ring is made adjustable so that the line of collimation can be adjusted at right angles to the horizontal axis or telescope axle.

In order to provide a datum for altitude angles, a spirit level is attached to the telescope so that its axis can be adjusted parallel to the line of collimation.

A clear understanding of the relationship between the various axes of a transit is helpful in performing adjustments. Those outlined can be performed by the instrument man; detailed instructions are given on succeeding pages. Errors of eccentricity should be corrected at the factory. Errors of parallax are due to improper manipulation.

1. Parallax. *Parallax is eliminated by correct focusing of the objective lens on the cross wires.*

Owing to differences in eyesight among individual users, it is necessary also to focus the eyepiece on the cross wires. Strictly speaking, this is not an adjustment but rather a manipulation that should be performed each time an accurate pointing is desired. Since incorrect focusing will affect other adjustments involving the use of the telescope, it is listed herein as the first adjustment, and it is important that every detail be followed carefully.

(*a*) Sight through telescope and make preliminary focus of eyepiece on cross wires. Turn knurled ring at eye end of telescope, until wires appear black and sharp. (On some transits turn eyepiece cap or possibly an eyepiece pinion on side of telescope.)

Eye should be relaxed and time of setting should be brief, otherwise the eye may accommodate itself to the telescope rather than the telescope become adjusted to the eye. If both eyes can be left open, a better focus will be obtained.

(*b*) Focus the objective lens on a clearly defined, well-lighted target about 300 ft. away. Turn the objective focusing pinion slowly backward and forward of the position of focus, at the same time wagging the head. Observe for apparent lateral movement between target image and cross wires. Stop focusing at the point where no lateral displacement appears. Disregard sharpness of image and of cross wires. It is this objective focusing which is important in the elimination of parallax.

(*c*) If necessary to sharpen the image, refocus the eyepiece slightly. It will be found that the cross wires also will be more distinct.

(*d*) Further focusing of the eyepiece will not be necessary unless the eye tires or a different observer uses the instrument, in which event paragraphs *b* and *c* should be repeated.

(*e*) On surveys of a high order, paragraph *b* should be followed on all pointings if the observer wishes surely to eliminate parallax error due to focusing.

It may be pointed out that a young man has more trouble than an old man in getting an eyepiece properly focused. This is due to the greater "accommodation" of the younger eye. The above procedure tends to produce a relaxed and normal condition of the eye when setting the final focus of the eyepiece. Furthermore, greater difficulty is experienced with low magnification and with the simple eyepiece of the inverting telescope.

2. Rectify Cross Wires. *To make the vertical cross wire perpendicular to the telescope axis.*

(*a*) Sight through telescope and set one end of the vertical cross wire on a sharply defined point *A*, Fig. 26.

(*b*) Elevate or depress telescope so that vertical wire traces over point. If wire coincides with point throughout its length, its position is correct.

(*c*) If not, slightly loosen all four capstan screws, located on eyepiece end of telescope.

Fig. 26.

(d) Move cross-wire ring around, in proper direction, until test shows that vertical wire exactly traces point. Hold screw driver against each of the collets and tap lightly against it.

(e) Tighten capstan screws and check.

3. Collimation at Distant Focus. *To make the collimation plane of the vertical cross wire perpendicular to the telescope axis.*

(a) Set up and sight vertical wire on a sharply defined point A (see Fig. 27), 200 or 300 ft. away.

(b) Transit the telescope and set a point B at approximately the same elevation and distance as A.

(c) Leave the telescope reversed, rotate the transit plate a half turn, and again sight on A.

(d) Again transit the telescope (bring it to its normal position), and set point C.

(e) Mark a new point E, one-quarter the distance from C to B.

(f) By turning the horizontal capstan screws shift the vertical cross wire until it is set on point E.

(g) Again set on A and repeat until instrument will make both points, B and C, coincide at D.

(h) Check rectification of vertical wire (refer to section 2).

4. Collimation at Minimum Focus. *In most Gurley transits the objective slide rear bearing is adjustable, so that the slide can be made to move parallel to the line of collimation and make it accurate for sighting at all distances. This adjustment is carefully made in the factory and, barring accident to the transit, should require no changing. With Gurley transits having inner-slide focusing any correction necessary can be made in the field; others should be returned to their makers. With internal focusing telescopes this construction is not permitted.*

(a) Set up and sight vertical wire on a sharply defined point, 200 or 300 ft. away.

(b) Place a horizontal scale or rod about 6 ft. in front of telescope (not nearer than point of minimum clear focus), and so that it appears just under the horizontal cross wire in the field of view, without moving the telescope.

(c) Focus on scale and read vertical wire intersection.

(d) Turn transit plate a half turn, transit telescope, and again set vertical wire on distant point.

(e) Without moving telescope, focus on nearby scale and read vertical wire intersection.

(*f*) If second reading (*e*) coincides with first reading (*c*), the objective slide is in adjustment with the vertical wire.

(*g*) Turn nearby scale or rod to vertical position in field of view and repeat readings using horizontal wire intersection. If two readings coincide, the objective slide is parallel to the horizontal wire.

(*h*) If not, correct for half the error by moving the rear bearing ring of the objective slide up or down or to the right or left as required. Turn slotted screws near or in telescope axis, using screw driver. Turning screw clockwise draws ring towards screw. Loosen opposite screw first. With an erecting telescope, actual movement should be opposite to apparent movement. With many telescopes, screws are on a 45° angle with respect to the cross wires; hence they are to be turned in pairs in order to move the bearing ring as required.

(*i*) Repeat sections 3 and 4 until the conditions of both are satisfied.

F𝚒g. 27.

5. Telescope Axis. *To make the telescope axis perpendicular to the vertical axis or spindle.*

(*a*) Set up transit.

(*b*) Sight on a high point *A* (see Fig. 28).

(*c*) Depress telescope and set point *B* on ground, in front of instrument.

(*d*) Rotate instrument 180° and transit telescope.

(*e*) With telescope in reversed position, again sight on point *B*.

(*f*) Elevate telescope and note point *C*.

(*g*) Note a new point *D* halfway between *B* and *C*.

(*h*) Raise or lower the right end of the telescope axle until the vertical cross wire intersects the halfway point *D*, when elevating telescope from point *B*.

To raise or lower the telescope axle turn the right-hand threaded capstan headed screw which is to be found under the standard cap on the right-hand side. Turn clockwise to raise, counterclockwise to lower.

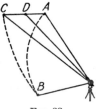

F𝚒g. 28.

Before raising:

On old-model Gurley Transits: Loosen cap screws.

On late-model Gurley Transits: Loosen capstan screw on top of standard.

After adjusting:

On old-model Gurley Transits: Tighten the two cap screws equally until there is sufficient friction on the axle bearing to keep the telescope end from dropping under its own weight. On some models, laminated shims have been placed under the standard cap. In such cases the cap screws should be set up solidly. If the telescope transits too freely, remove laminations from the shims until the proper braking action is arrived at. Check and adjust the cap screws on the left-hand standard so that these provide equal braking power on both ends of telescope axle.

On late-model Gurley Transits: Tighten the two capstan screws on top of standards. Adjust both screws equally until there is sufficient braking action on the axle to keep the telescope end from dropping. Check and adjust the capstan screws on the left-hand standard so that they provide equal braking power on both ends of telescope axle.

(*i*) Check and repeat until transit will make points *A* and *C* coincide.

6. Telescope Level. *To make the axis of the bubble parallel to the line of sight when the latter is horizontal.*

The "Four Peg" Method

For the "Two Peg" method, see p. 360

(*a*) Drive four stakes, *A*, *B*, *C*, and *D*, in line and exactly equidistant, from 50 to 100 ft. apart (see Fig. 29).

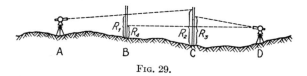

Fig. 29.

(*b*) Set up the transit at *A*.

(*c*) Bring the bubble to the center of the telescope level.

(*d*) Read the elevation of the line of sight on a rod held at both *B* and *C*, calling the first reading R_1 and the second R_2.

(*e*) Set up the transit at *D*.

(*f*) With the bubble in the center of the telescope level, read the rod on *C*, calling it R_3.

(*g*) Add R_1 to R_3, subtract R_2, and set target on rod to this result.

$$R_4 = (R_1 + R_3 - R_2)$$

(*h*) Hold rod on *B*.

(*i*) By means of the axis tangent motion, incline the telescope until the horizontal wire intersects the target.

(*j*) Raise or lower one end of the bubble tube, by turning the capstan nuts, until the bubble returns to the center.

Reversion Vial: A procedure simpler than the peg method can be employed if the telescope level vial is of the reversion type.

(*a*) Set up transit, sight on level rod about 100 ft. distant, and center bubble.

(*b*) Read level rod (middle horizontal wire).

(*c*) Rotate instrument 180° in azimuth, transit telescope, again sight on rod, and center bubble.

(*d*) Read level rod.

(*e*) Average readings *b* and *d*. Set horizontal wire to average reading on rod. Center bubble by capstan adjusting nuts.

7. Plate Levels. *To make the bubble tube axes perpendicular to the vertical axis or spindle.*

(*a*) Set up transit on tripod.

(*b*) Rotate transit plate so that each bubble is in line with a pair of opposite leveling screws.

(*c*) Bring plate level bubbles to the center in both tubes.

(*d*) Turn the plate through 180° in azimuth.

(*e*) Note the amount that the bubbles move from the center.

(*f*) Raise or lower one end of each bubble tube as required to bring the bubbles back one half the amount they moved off.

To raise or lower one end of the bubble tube: On transits having capstan nuts above and below level tube, use adjusting pin to raise or lower both nuts as required. Do not force together so as to spring bubble tube.

On transits having a slotted screw at top of adjusting post, use adjusting pin to raise or lower only the capstan nut, underneath the tube. Coiled spring inside tube supplies proper tension. Adjust end of tube which will keep slotted screw about flush with top of tube.

(*g*) Level up and repeat the above until both bubbles remain in the center when rotating them 180°. *Check and correct the bubbles alternately.*

8. Vertical Circle Vernier. *To make the vertical circle (or arc) read zero when the line of collimation is horizontal.*

(*a*) Level up transit carefully, using telescope level.

(*b*) Center bubble of telescope level, using axis tangent motion. Check bubble adjustment, section 6.

(*c*) Inspect vernier and vertical circle to see if zeros of each coincide.

(*d*) If not, slightly loosen screws which hold vernier to standard.

(*e*) Shift vernier until zeros coincide.

(*f*) Tighten vernier screws and check.

Two-Vernier Vertical Circle. To make the vertical circle read zero when the line of collimation is horizontal.

(a) Level up transit carefully, using telescope level.

(b) Center bubble of telescope level, using axis tangent motion.

(c) Turn capstan headed screw until zeros of one vernier and vertical circle coincide.

To make zeros of verniers read 180° apart.

(a) Make line of collimation horizontal and also one vernier read zero as described above.

(b) If opposite vernier does not read zero, slightly loosen the screws which hold that vernier to the vernier frame.

(c) Shift vernier until zeros coincide.

(d) Adjust spacing between vernier and circle until end graduations on vernier match with limb.

(e) Tighten vernier screws and check.

Beaman Stadia Arc Indices. To make indices read zero when vernier reads zero.

(a) Set vernier to read zero on limb.

(b) If indices H and V do not read zero, slightly loosen index screws.

(c) Shift indices until they both read zero.

(d) Tighten index screws.

9. Center Eyepiece. *To make the cross wires appear in the center of the field of view. This adjustment is not an essential to accuracy but is of convenience to the observer.*

(a) After the cross wires have been adjusted, observe whether they appear in the center of the field.

(b) If not, unscrew the entire eyepiece from the telescope, turning raised rim ahead of knurled ring.

(c) Move the eyepiece slide in proper direction (opposite to apparent direction) by means of opposing flat headed screws in eyepiece. Estimate the amount of movement necessary.

(d) Replace the eyepiece in telescope and, if necessary, repeat until the eyepiece is properly centered.

10. Balance Compass Needle. *The compass needle is balanced horizontally, as near as possible, for the locality to which it is sent. The metal spring or bright coiled wire on the south end of the needle slides along the needle to enable the instrument man to do exact balancing in the field. The needle should be tested for balance when the instrument is moved from one locality to another. Balancing at the office, particularly in a large building, will probably not give satisfactory results.*

(a) Level up the instrument.

(b) Release the needle on its pivot.

(c) Remove the compass glass by pressing the palm of the hand flat on the glass and turning counterclockwise.

Some transits have a set screw in the bezel ring, which should be removed before turning ring. This is located in either the NW or SE quadrants. If glass is tight, tap around bezel ring with handle of screw driver to loosen threads. The compass glass cannot be removed from between the standards on some Gurley transits without first detaching the vertical axis tangent bar which is held to the standard by two screws. However, it is unnecessary to remove the compass glass entirely when making adjustments.

(d) Note the dip of the needle, raise one side of the compass glass, and carefully remove the needle. Slide the counterbalance along the needle toward the high end.

(e) Lower the needle on its pivot point as gently as possible.

(f) Repeat until the needle balances.

(g) Replace the compass glass, taking care not to cross the threads. Finish turning with index pointer at N position. Replace locating set screw.

(h) Raise the needle from its pivot until ready to use.

11. Straighten Needle. *To make both ends of the needle read 180° apart in one position. This makes both ends and the center of the needle lie in the same vertical plane.*

(a) Set up compass, lower needle gently on the center pin, and remove the cover glass.

(b) With a small splinter of wood, bring the north end of the needle exactly opposite the north zero mark of the circle.

(c) Read the south end of the needle.

(d) Rotate the needle a half turn and bring the south end exactly opposite the north zero.

(e) Read the north end of the needle.

(f) If the two readings agree (paragraphs c and e) the needle is straight.

(g) If not, correct for half the error by bending the needle.

(h) Repeat the test until the needle is straight.

12. Center Pivot. *To make both ends of the needle read 180° apart in all positions. This brings the pivot point exactly in the center of the compass circle.*

(a) After straightening needle, bring north end of needle exactly opposite the north zero mark of the circle.

(b) Note whether south end of needle reads zero.

(c) If not, correct for the whole error by bending the center pin in a direction at right angles to the needle. *Use wrench, carried in spare parts kit, to bend center pin.*

(d) Rotate the needle a quarter turn, bring the north end opposite a 90° mark, and note whether the south end of the needle reads 90°.

(e) If not, correct for the whole error by bending the center pin in a direction at right angles to the needle.

(f) Repeat the above, reading first at the zero and then at the 90° marks, until both ends of the needle read alike in both positions.

Adjustments of Wye Levels

The adjustments of Gurley wye levels are as follows:

1. Parallax.
2. Rectify cross wires.
3. Collimation at distant focus.
4. Collimation at minimum focus.
5. Telescope level vial.
6. Wyes.
7. Center eyepiece.

Adjustments of Dumpy Levels

The adjustments of Gurley dumpy levels are as follows:

1. Parallax.
2. Telescope level vial.
3. Rectify cross wires.
4. Collimation at distant focus.
5. Collimation at minimum focus.
6. Center eyepiece.

A level is an instrument used to determine the position of all points in a horizontal plane. It consists of a collimated line of sight adjusted parallel to the axis of a spirit bubble. This fundamental description should be kept in mind when adjusting and using a level of any type.

The type of level is determined from the structural arrangement of the parts necessary to adjust the axis of the bubble parallel to the line of sight and the convenience of keeping the bubble centered when taking a reading.

With the wye level, the telescope is provided with two accurately machined bearing rings, truly circular and of equal diameter, separated by about half the length of the telescope. These rest in wye bearings which are adjustable in the wye bar, which is permanently fixed at right angles to the vertical spindle. Two level posts attached to the telescope (usually underneath) carry the level vial, the position being fixed by adjusting nuts, usually at both ends.

With the dumpy level, the telescope, bar, and spindle are assembled as one unit, the workmanship being such that the axis of the telescope is closely perpendicular to the vertical axis of rotation or spindle. Level posts may be attached either to the bar or to the telescope, these carrying the level vial with adjusting nuts at both ends.

This difference in construction between the wye and dumpy level determines a difference in adjustment procedure. Thus, with the wye level, the collimated line of sight is made concentric with the wye rings by rotating the telescope in the wyes and adjusting the reticule carrying the cross wires. By reversing the telescope rings end for end in the wye bearings, and by adjusting the level vial in the level posts until the bubble holds its central position in both positions of the telescope, the bubble axis is made parallel with the wye rings and thereby parallel with the collimated line of sight. As long as this parallelism holds, it is possible to do accurate leveling with a wye level, provided the level bubble is made central by the leveling screws each time a reading is taken. For convenience in keeping the bubble centered when pointing the telescope in a new direction, the wye adjustment is provided, which, by reversing the telescope about its spindle and by adjusting the wye nuts, makes the bubble axis, also the collimated line of sight, perpendicular to the spindle, or axis of rotation. When making the latter adjustment, the telescope slide should be moved by the focusing screw until the objective end of the telescope balances the eyepiece end. This position of the slide should be noted and the slide brought back to it when subsequently leveling up the instrument. Any movement of the slide from this position changes the balance of the instrument and may cause the bubble to run. This condition does not indicate a change in adjustment, since nothing has been done to change the parallelism between the bubble axis and the collimated line of sight. Therefore such a run of the bubble should be corrected by the leveling screws.

In adjusting the dumpy level, the construction necessitates a different procedure. The level bubble axis is first made perpendicular to the spindle or axis of rotation by reversing the telescope end for end about the spindle, centering the bubble by the level post adjusting nut. The collimated line of sight is then brought parallel to the bubble axis by the peg method of adjustment, the details of which are given on p. 360. For careful adjustment the objective slide should be at the position of balance, and any subsequent run of the bubble should be compensated for by the leveling screws, as explained under the wye level paragraph above.

When using a level, the adjustment or parallelism between bubble axis and collimated line of sight is important but it is equally important to make sure that the bubble is centered each time a reading is taken. To assist in this purpose, various devices from a simple mirror to a complicated prism system are used to enable the observer to see the position of the bubble at the time he reads on the rod.

The tilting type of level has been devised to assist the observer in keeping the bubble centered without recourse to the leveling screws. In addition to the change in balance caused by focusing on rods at different distances, there are other factors which cause a bubble to run without disturbing the fundamental parallelism between bubble axis and line of sight, especially so if a sensitive bubble is used.

The tilting level (used for precise leveling) has a double bar, one part attached parallel to the telescope, the other part at right angles to the spindle. The two bars are arranged to pivot one on the other, being separated by a slow-motion screw with opposing spring. A circular or bull's-eye level on the bar or leveling head serves to plumb the spindle. Final leveling with each reading is done by centering the bubble by the slow-motion screw. Such levels are generally provided with a reflecting device so that both bubble and rod image are visible at the same time.

Tilting levels may be either of the dumpy or of the wye type. In the dumpy type, the parallelism between the bubble axis is established by the peg method of adjustment. In the wye type, the telescope is made with wye rings and with a reversion type of level attached to the side. The advantage of the wye or reversible type of tilting level is the ease of adjusting the line of collimation and the level bubble.

The relative advantages of the wye and dumpy levels boil down to a matter of individual preference. The dumpy level with fewer parts is supposed to remain in adjustment over a longer period of time. However, its adjustment is dependent upon a well-fitted spindle and socket.

The advantage claimed for the wye level is that the adjustments can be checked readily by one person (the dumpy level requires the assistance of a rodman in making the peg adjustment). The principal objection is that the adjustments are dependent upon the wye bearing rings being truly circular and equal in diameter. Since the rings are exposed to wear and to possible damage, some engineers feel that they cannot be sure of the adjustment unless the peg method is used anyway.

For construction engineering the compact solidarity of either the wye or the dumpy level gives these types the preference. However, for accuracy and speed on long lines of differential levels the tilting type is superior.

1. Parallax. *See parallax adjustment for transit, p. 347.*

2. Rectify Cross Wires. *To make the horizontal cross wire perpendicular to the vertical axis or spindle. The vertical wire is set perpendicular to the horizontal wire by the maker.*

(*a*) Set up a level on tripod. Set one end of horizontal wire on a sharply defined point *A*, Fig. 30.

(*b*) Turn level slowly about its spindle, so that horizontal wire traces over the point. Wire should coincide with point throughout its length.

(*c*) If point appears to trace dotted line *AB*, Fig. 30, slightly release pressure on capstan screws. Turn all four capstan screws only slightly and by equal amounts.

(*d*) Gently tap capstan screws in direction to close angle between horizontal wire and dotted line *AB*, Fig. 30. Rotate cross-wire ring (test, paragraph *b* above) until horizontal wire exactly traces point from *A'* to *B'*, Fig. 30.

(*e*) Tighten capstan screws (all four equally), and check.

3. Collimation at Distant Focus. *To make the line of sight (collimation) pass through the axis of the wye rings.*

(*a*) Set up level on tripod, remove wye pins from clips, and raise clips so that telescope is free to rotate.

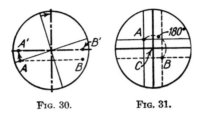

Fɪɢ. 30. Fɪɢ. 31.

(*b*) Set intersection of cross wires on a well-defined point (*A*, Fig. 31), about 300 ft. distant.

(*c*) Carefully rotate the telescope halfway around in its wyes, and note whether the intersection of the cross wires still covers the point.

(*d*) If not, move the telescope by leveling and tangent screws until the error seems to be one-half corrected.

(*e*) Move the cross-wire ring, using each pair of opposite capstan screws successively, until the error is entirely corrected and the cross-wire intersection now covers the point (*C*, Fig. 31).

(*f*) Repeat the rectification (2) and collimation (3) of the cross wires until both adjustments are correct.

4. Collimation at Minimum Focus. *To make the objective slide move parallel to the line of collimation when racked in or out for focusing on distant or near targets.*

This adjustment may be checked on any telescope but can be corrected only on Gurley inner-slide focusing telescopes. It is not on internal focusing telescopes or on the external focusing telescopes of other makes. It is primarily a factory adjustment and, barring accident, should need no correction in the field.

(*a*) Set up level on tripod, remove wye pins from clips, and raise clips so that telescope is free to rotate.

(*b*) Check adjustment of the line of collimation (3) for a remote target.

(*c*) Unscrew the cover ring in center of telescope, exposing the flat-headed screws for adjusting the rear bearing of the objective slide.

(*d*) Set intersection of cross wires on a well-defined point about 15 ft. distant.

(*e*) Carefully rotate telescope halfway round in its wyes, and note whether the intersection of the cross wires still covers the point.

(*f*) If not, move the telescope by leveling and tangent screws until the error seems to be one-half corrected.

(*g*) Correct the remainder of the error by turning the flat-headed screws with a screw driver until the cross wires intersect on the point. Adjust first one pair of screws and then the other. Loosen one screw and tighten the other.

(*h*) Repeat sections 3 and 4 until the conditions of both are satisfied.

(*i*) Replace cover ring.

5. Telescope Level Vial. *To make the axis of the bubble parallel to and in the same vertical plane with the axis of the wye rings. As long as this adjustment and section 3 are correct, accurate leveling can be done with the instrument.*

(*a*) Hold level sideways with spindle horizontal, and turn focusing screw until level balances. Then set up on tripod, clamp telescope over two diagonally opposite leveling screws.

(*b*) Remove wye pins and raise wye clips.

(*c*) Bring bubble to center of tube (see Fig. 32).

(*d*) Lift telescope out of wyes, turn end for end, and replace in wyes. Note whether bubble remains in center of tube (see Fig. 33).

(*e*) If not, bring bubble halfway back to center by the leveling screws.

(*f*) Correct balance of error by turning the capstan nuts at eyepiece end of bubble tube until bubble returns to center (see Fig. 34).

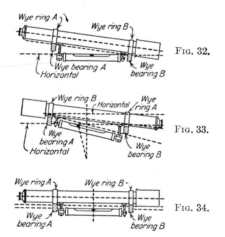

Fig. 32.

Fig. 33.

Fig. 34.

(*g*) Rotate telescope in its wyes, about 30° either side of the vertical, and note whether bubble remains in center of tube.

(*h*) If not, bring bubble all the way back to center by turning the lateral capstan screws on each side of the bubble tube post at the objective end of the level.

(*i*) Repeat the vertical adjustment, as given under section 5, paragraphs *c*, *d*, *e*, and *f* above.

(*j*) Check alternately until both the lateral adjustment and the vertical adjustment of the vial are correct.

Note: Bubble will run if balance is changed, by running objective slide in or out. This does not indicate adjustment is out. See p. 355.

6. Wyes. *To make the axis of the wyes perpendicular to the vertical axis or spindle.*

This adjustment is made as a convenience, rather than as a necessity. Accurate leveling can be done if the bubble is in adjustment, and is centered by the leveling screws before each rod reading.

(a) Set up level, rotate about spindle until telescope is over two diagonally opposite leveling screws, and bring bubble to the center of tube (see Fig. 35). Check telescope bubble adjustment, section 5, very carefully. Telescope slide must be in position of balance.

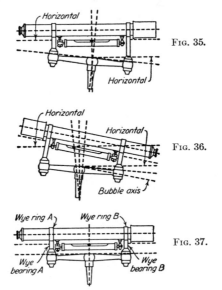

Fig. 35.

Fig. 36.

Fig. 37.

(b) Rotate level about spindle 180°, and note whether bubble remains in center of tube (see Fig. 36).

(c) If not, bring bubble halfway back to the center by the leveling screws. Raise or lower one end of the wye bar, until the bubble returns to the center, by turning a pair of capstan nuts at either end of the wye bar.

(d) Repeat until the bubble remains in center of tube when rotated about spindle (see Fig. 37).

7. Center Eyepiece. *To make cross wires appear in center of field.*

This is not essential to the accuracy of the work, but it is a convenience to the observer to have the cross wires appear in the center of the field.

(a) Set up level, and observe whether cross wires appear in center of field.

(b) If not, unscrew cover ring between cross wires and eye end of telescope.

(c) Turn the flat-headed screws with a screw driver until the cross wires appear in the center of the field.

Adjust first one pair of screws, and then the other. Loosen one screw and tighten the opposite one. Correct in a direction opposite to the apparent error.

(d) Replace cover ring.

Adjustments of Dumpy Levels

1. Parallax. *See parallax adjustment for transit, p. 347.*

2. Telescope Level Vial. *To make the axis of the bubble perpendicular to the vertical axis or spindle.*

(a) Set up level on tripod, rotate about spindle until telescope is over two diagonally opposite leveling screws, and bring bubble to center of tube.

(b) Rotate level about spindle 180°, and note whether bubble remains in center of tube.

(c) If not, bring the bubble halfway back to the center by the leveling screws.

(d) Correct balance of error by turning capstan nuts at either end of bubble tube, until bubble returns to center.

(e) Alternate over both pairs of leveling screws until the bubble remains in center of tube when rotated about spindle.

3. Rectify Cross Wires. *To make the horizontal cross wire perpendicular to the vertical axis or spindle. Vertical wires are set by the maker at right angles to the horizontal wire.*

(a) Set up level on tripod, and set one end of horizontal cross wire on a sharply defined point (A, Fig. 30).

(b) Turn level slowly about its spindle, so that horizontal wire traces over the point. If wire coincides with point throughout its length, its position is correct.

(c) If not, slightly loosen all four capstan screws located on eyepiece end of telescope.

(d) Move cross-wire ring around, in proper direction, until test shows that horizontal wire exactly traces point (A' B', Fig. 30).

(e) Tighten capstan screws and check.

4. Collimation at Distant Focus. *To make the line of sight parallel to the axis of the bubble.*

The "Two Peg" Method

For the "Four Peg" method, see p. 350.

(a) Set up level at some convenient point A, Fig. 38, holding rod at C, distant at least 100 ft. With instrument carefully leveled and bubble in center of telescope level, read rod on C, calling the reading R_c.

(b) Locate point B directly behind instrument and so that distance AB equals AC.

(c) Point telescope toward B, bring bubble to center of telescope tube, and take rod reading R_b.

(d) Set up level beside point B, so that eyepiece of telescope is directly over point. Level up carefully, bringing bubble to center of telescope tube.

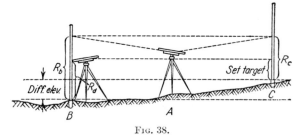

Fig. 38.

(e) Point eyepiece of telescope toward rod at B, and read through objective end of telescope, calling this reading R_d. If more convenient, measure along the outside center line of telescope.

(f) Add to R_d the difference between the first readings ($R_c - R_b$).

(g) Set rod target to this result, and hold the rod on point C.

(h) Move the cross-wire ring up or down until the horizontal wire cuts the target, by turning the vertical pair of opposite capstan screws.

(i) Check by again reading rod on B, computing rod reading for C, and observing whether horizontal wire cuts the target.

5. Collimation at Minimum Focus. *To make the objective slide move parallel to the line of collimation when racked in or out for focusing on distant or near targets.*

This adjustment may be checked on any telescope but can be corrected only on Gurley inner-slide focusing telescopes. It is not on internal focusing telescopes or on the external focusing telescopes of other makes. It is primarily a factory adjustment and, barring accident, should need no correction in the field.

(a) After doing section 4, set up level about 15 ft. from B (Fig. 38) toward C, which is the same distance away.

(b) On old-model Gurley dumpy levels unscrew the cover ring in center of telescope, exposing the flat-headed screws for adjusting the rear bearing of the objective slide.

(c) Level carefully and read rod C.

(d) Rotate level and focus on rod B. Moving objective slide out will probably cause the bubble to run, owing to the change in balance. Bring the bubble to the center by turning the leveling screws.

(e) Set target on rod B to proper reading to give true difference in elevation ($R_c - R_b$) as determined in section 4. Cross wires should bisect target at this setting.

(*f*) If not, turn the flat-headed screws, moving the rear bearing up or down, until the horizontal wire cuts the target.

(*g*) Check sections 4 and 5 alternately, until both are correct.

6. Center Eyepiece. *To make the cross wires appear in the center of the field of view. This adjustment is not an essential to accuracy but is of convenience to the observer.*

(*a*) After the cross wires have been adjusted, observe whether they appear in the center of the field.

(*b*) If not, unscrew the entire eyepiece from the telescope, turning raised rim ahead of knurled ring.

(*c*) Move the eyepiece slide in proper direction (opposite to apparent direction) by means of opposing flat-headed screws in eyepiece. Estimate the amount of movement necessary.

(*d*) Replace the eyepiece in telescope, and, if necessary, repeat until the eyepiece is properly centered.

Taping

Changes in Temperature

Correction in feet $= C \times L(T - T_s)$.

$C = 0.0000065$ for steel tape.

$C = 0.00000056$ for Invar tape.

$L =$ length of tape in feet.

$T =$ temperature in degrees Fahrenheit at which tape is used.

$T_s =$ temperature at which tape was standardized (62° F. or 68° F.).

Variation in Tension

Correction in feet $= \dfrac{(P - P_s)L}{AE}$.

$P =$ tension applied.

$P_s =$ standard tension (10 to 15 lb.).

$L =$ length of tape in feet.

$A =$ cross section area of tape in square inches (light steel tape = 0.0025 ±; heavy steel tape = 0.01±).

$E =$ modulus of elasticity in pounds per square inch (30,000,000 for steel tapes).

Sag

Correction in feet between points of support $= \dfrac{W^2 L}{24P^2}$.

$W =$ weight of tape in pounds between supports (a light tape = 1.0± lb. per 100 ft.; a heavy tape = 3.0± per 100 ft.).

$L =$ length in feet between supports.

$P =$ tension used in pounds.

MAPPING

PLOTTING TRAVERSES

1. Plotting by Protractor

Procedure. Fix position of first line, and lay off its length AB by

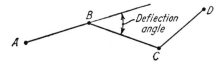

scaling. Orient the protractor at the forward point B; lay off the deflection angle to the succeeding line, and draw a light line of indefinite length. Scale off the given distance BC to the next traverse point C, etc.

Hints and Precautions. Orient the position of the first line so that the succeeding lines will not run off the paper. Carefully check the deflection angles as to their direction right or left. Calculated bearings should check reasonably with observed magnetic bearings. When azimuths or calculated bearings are used, a meridian line may be drawn through each station and the direction of the succeeding line laid off from the meridian.

2. Plotting by Tangents

Procedure. Fix position of first line, and lay off its length AB by

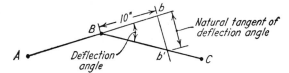

scaling. Prolong the line AB some convenient distance, to form a base line Bb. Erect a perpendicular bb' of sufficient length. Scale off the distance bb' equal to the length of the base line Bb multiplied by the natural tangent of the deflection angle. Draw a line from B through b' to define the direction of BC, etc.

Hints and Precautions. Time and accuracy can be gained by laying off the base line Bb 10 in. in length and scaling off the natural tangent along the perpendicular with an engineer's scale. Because the 50 scale has more graduations than the 10 scale, it is customary to scale off one-half the natural tangent with the 50 scale. Scale all distances and erect all perpendiculars carefully. Where the deflection angle is greater than 90° the perpendicular is erected by measuring the base line back on the course from the last point and scaling off the tangent for 180°—the deflection angle. When the deflection angle is greater than 45°, erect a perpendicular from the last point set, scale off a 10-in. base line, and erect a line parallel to the last course, along which scale off the cotangent of the deflection angle. Check all plotted angles with a protractor. For increased accuracy the base lines may be made 20 in. and the tangents scaled direct with the 50 scale. For checking the erected perpendiculars the diagonal distance on the hypotenuse of the 10-in. sides should scale 14.14 in.

3. Plotting by Chords

Procedure. Proceed the same as in plotting by tangents except that, instead of erecting a perpendicular at the end of the 10-in. base line, describe an arc of 10-in. radius. Scale the chord distance *bb'*. Draw a

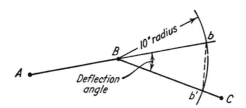

line through *Bb'*, and plot the distance *BC*. The length of the chord *bb'* is equal to $20 \cdot \sin \frac{1}{2}$, the deflection angle.

Hints and Precautions. In swinging the 10-in. arc use a beam compass or improvise one by inserting a needle point and a pencil point exactly 10 in. apart in a thin strip of wood. If a table of chords is available no computations are necessary. Check the plotted angles with a protractor.

4. Plotting by Rectangular Coordinates—Latitudes and Departures

Procedure. (1) Transpose the survey data to a computation book as shown in the sample form on p. 333. (2) Compute the latitudes and departures of the courses, and, if a closed traverse, balance the survey. Assume one of the traverse points as the origin of coordinates, calculate total latitudes and departures, and check the computations. (3) Determine the size of the enclosing rectangle, the four sides of which pass through the eastern, western, northern, and southern points of the traverse. (4) Plot the enclosing rectangle to required scale on drawing paper, estimating its position on the sheet by means of a small-scale sketch. Place the traverse symmetrical with the sheet (the sides of the rectangle may or may not be parallel to the edges of the paper). (5) Test the accuracy of the plotting by scaling the length of diagonals. Plot the reference meridian, and plot and check the reference parallel. (6) Construct coordinate lines (other meridians and parallels) so that the area will be divided in squares with sides less than the length of the scale to be used. Number each of these lines with its distance from the reference meridian or parallel. (7) Locate each traverse point by plotting its latitude and departure. (8) Check the length of the traverse lines connecting the points by scaling, and check the angles with the protractor.

Hints and Precautions. Accurately construct the meridians and parallels. After the enclosing triangle has been constructed and adjusted by trial the other lines should be plotted entirely by scaling. Do not use a T-square and triangles in the usual way but use straightedges only. The best way to lay out the rectangle and coordinates is with a beam compass and steel straightedge, checking all rectangles by diagonals. If the southwest corner of the enclosing rectangle is taken as the origin of coordinates, all the total latitudes and departures will be positive.

Practical Applications. Plotting by coordinates is the best method for plotting most traverses. When the area of a closed traverse is to be computed the latitudes and departures are necessary. The size and shape of the drawing can be determined before plotting. Errors of plotting are not cumulative. The method of checking is simple, and in closed traverses the survey is balanced before plotting.

PLOTTING TOPOGRAPHY

1. Stadia Topography by Protractor

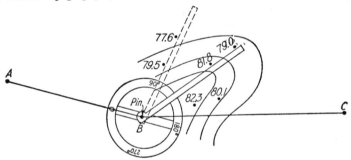

Procedure. First lay out the traverse from which topography was taken. To facilitate plotting use a full circle protractor and a scale that can be pinned at the center. Orient the zero of the protractor on the line to the point on which the transit was sighted in the field. Move the scale to the horizontal angle desired and lay off the horizontal distance.

Hints and Precautions. One way of marking points as they are plotted is to note the elevation; another is to note the number of the point. Points which are to be connected should be connected before beginning a new station, i.e., points along a road, corners of a building, etc. When each traverse point occupied requires the plotting of a considerable number of points, speed and accuracy will be attained by two persons working as a team, one reading the notes and the other plotting the points.

MAPPING SYMBOLS *

FENCES AND WALLS

In General	(State type)
Woven Wire	—o——o——o—
Barbed Wire	—×——×——×—
Board Fence	— — — — —
Picket Fence	—//——//——//—
Rail Fence	⟩⟨⟩⟨⟩⟨
Stone Wall	○◦○◦○○◦○◦○○○
Retaining Wall	⌄⌄⌄⌄⌄
Hedge	(hedge symbol)

STRUCTURES

Buildings (Large Scale)	▨
Buildings (Small scale)	■ ■ ■
Barn or Garage	⊠
Bridge	
Dam	
Tunnel	→—=====—←

ROADS AND RAILROADS

Path of trail	— — — — —
Secondary Road	=======
Improved Road	═══════
Single Track R.R.	++++++++++++
Double Track R.R.	++++++++++++

BOUNDARY LINES

In General	(State type)
Property Line	— ·· — ·· —
Street Line	——— ·· ———
Curb Line	———————
Easement Line	—— · —— · ——
National, State	——— — — ———
County	—— · — · ——
City or Town	— ·· — · — ·· —

SURVEY SYMBOLS

Transit Station	⊙
Stadia Station	⊡
Triangulation Station	△
Bench Mark	$\overset{B.M.}{\underset{481}{\times}}$ or 481 × B.M.

MISCELLANEOUS

Stone Bound	⁂
Monument	⊡
Tree (State size and species)	◉
Edge of Woods	(woods symbol)
Ledge	(ledge symbol)
Stream	(stream symbol)
Wire Line	T T T T
Power Line	—●——●——●—

* From Tracy, *Surveying Theory and Practice,* John Wiley & Sons.

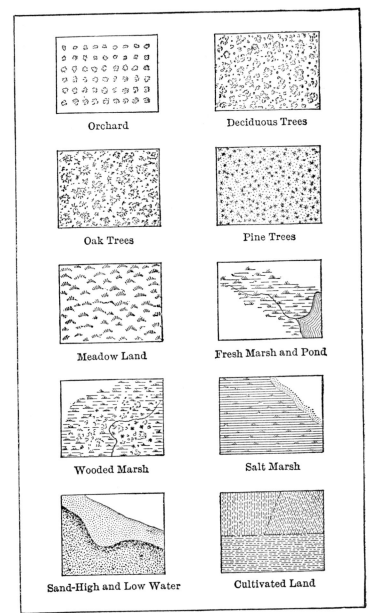

Orchard

Deciduous Trees

Oak Trees

Pine Trees

Meadow Land

Fresh Marsh and Pond

Wooded Marsh

Salt Marsh

Sand-High and Low Water

Cultivated Land

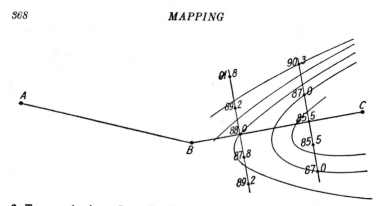

2. Topography from Cross Sections

Procedure. Indicate the line of cross sections by drawing a light line on the map. Scale off the distance right or left from the base line and mark the elevation.

Hints and Precautions. Orient the base line so that right on the map corresponds to right in the notes.

LAND MEASURE *

A rod is 16½ feet.
A chain is 66 feet or 4 rods.
A mile is 320 rods, 80 chains, or 5280 feet.
A square rod is 272¼ square feet.
An acre contains 43,560 square feet.
An acre contains 160 square rods
An acre is about 208¾ feet square.

From Water Works & Sewerage, Vol. 91, No. 6, June 1944.

TRIGONOMETRIC FORMULAS *

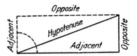

Functions of Angle	Opposite	Adjacent	Hyp
sin = Op ÷ Hyp	Hyp × sin		Op ÷ sin
cos = Ad ÷ Hyp		Hyp × cos	Ad ÷ cos
tan = Op ÷ Ad	Ad × tan	Op ÷ tan	
cot = Ad ÷ Op	Ad ÷ cot	Op × cot	
sec = Hyp ÷ Ad		Hyp ÷ sec	Ad × sec
cosec = Hyp ÷ Op	Hyp ÷ cosec		Op × cosec

Data by American Bridge Co., from Manual of Structural Design by Singleton.

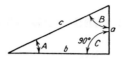

Given	To Find	Formula
ab	A	$\tan = a \div b \quad \cot = b \div a$
ab	B	$\cot = a \div b \quad \tan = b \div a$
ac	A	$\sin = a \div c \quad \operatorname{cosec} = c \div a$
ac	B	$\cos = a \div c \quad \sec = c \div a$
bc	A	$\sec = c \div b \quad \cos = b \div c$
bc	B	$\operatorname{cosec} = c \div b \quad \sin = b \div c$
Aa	b	$a \cot A \quad a \div \tan A$
Aa	c	$a \operatorname{cosec} A \quad a \div \sin A$
Ab	a	$b \tan A \quad b \div \cot A$
Ab	c	$b \sec A \quad b \div \cos A$
Ac	a	$c \sin A \quad c \div \operatorname{cosec} A$
Ac	b	$c \cos A \quad c \div \sec A$

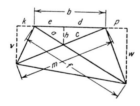

Given	To Find	Formula
bpw	f	$\sqrt{(b+p)^2 + w^2}$
bkv	m	$\sqrt{(b+k)^2 + v^2}$
bkp	d	$bw(b+k) \div [v(b+p) + w(b+k)]$
vw	e	$bv(b+p) \div [v(b+p) + w(b+k)]$
$\left.\begin{matrix} bfk \\ pvw \end{matrix}\right\}$	a	$fbv \div [v(b+p) + w(b+k)]$
$\left.\begin{matrix} bkm \\ pvw \end{matrix}\right\}$	c	$bmw \div [v(b+p) + w(b+k)]$
$bkpvw$	h	$bvw \div [v(b+p) + w(b+k)]$
afw	h	$aw \div f$
cmv	h	$cv \div m$

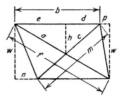

Given	To Find	Formula
bpw	f	$\sqrt{(b + p)^2 + w^2}$
bnw	m	$\sqrt{(b - n)^2 + w^2}$
bnp	d	$b(b - n) \div (2b + p - n)$
bnp	e	$b(b + p) \div (2b + p - n)$
$bfnp$	a	$bf \div (2b + p - n)$
$bmnp$	c	$bm \div (2b + p - n)$
$bnpw$	h	$bw \div (2b + p - n)$
afw	h	$aw \div f$
cmw	h	$cw \div m$

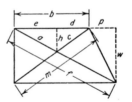

Given	To Find	Formula
bpw	f	$\sqrt{(b + p)^2 + w^2}$
bw	m	$\sqrt{b^2 + w^2}$
bp	d	$b^2 \div (2b + p)$
bp	e	$b(b + p) \div (2b + p)$
bfp	a	$bf \div (2b + p)$
bmp	c	$bm \div (2b + p)$
bpw	h	$bw \div (2b + p)$
afw	h	$aw \div f$
cmw	h	$cw \div m$

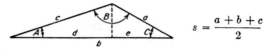

$$s = \frac{a + b + c}{2}$$

Given	To Find	Formula
ABa	b	$a \sin B \div \sin A$
ABa	c	$a \sin (A + B) \div \sin A$
ABb	a	$b \sin A \div \sin B$
ABb	c	$b \sin (A + B) \div \sin B$
ABc	a	$c \sin A \div \sin (A + B)$
ABc	b	$c \sin B \div \sin (A + B)$
ACa	b	$a \sin (A + C) \div \sin A$
ACa	c	$a \sin C \div \sin A$
ACb	a	$b \sin A \div \sin (A + C)$
ACb	c	$b \sin C \div \sin (A + C)$
ACc	a	$c \sin A \div \sin C$
ACc	b	$c \sin (A + C) \div \sin C$
BCa	b	$a \sin B \div \sin (B + C)$
BCa	c	$a \sin C \div \sin (B + C)$
BCb	a	$b \sin (B + C) \div \sin B$
BCb	c	$b \sin C \div \sin B$
BCc	a	$c \sin (B + C) \div \sin C$
BCc	b	$c \sin B \div \sin C$
abc	S	$(a + b + c) \div 2$
$abcs$	A	$\sin \tfrac{1}{2}A = \sqrt{(s - b)(s - c) \div bc}$
$abcs$	A	$\cos \tfrac{1}{2}A = \sqrt{s(s - a) \div bc}$
$abcs$	A	$\tan \tfrac{1}{2}A = \sqrt{(s - b)(s - c) \div s(s - a)}$
$abcs$	B	$\sin \tfrac{1}{2}B = \sqrt{(s - a)(s - c) \div ac}$
$abcs$	B	$\cos \tfrac{1}{2}B = \sqrt{s(s - b) \div ac}$
$abcs$	B	$\tan \tfrac{1}{2}B = \sqrt{(s - a)(s - c) \div s(s - b)}$
$abcs$	C	$\sin \tfrac{1}{2}C = \sqrt{(s - a)(s - b) \div ab}$
$abcs$	C	$\cos \tfrac{1}{2}C = \sqrt{s(s - c) \div ab}$
$abcs$	C	$\tan \tfrac{1}{2}C = \sqrt{(s - a)(s - b) \div s(s - c)}$
$abcs$	d	$(b^2 + c^2 - a^2) \div 2b$
$abcs$	e	$(a^2 + b^2 - c^2) \div 2b$
Aab	B	$\sin = b \sin A \div a$
Aac	C	$\sin = c \sin A \div a$
Bab	A	$\sin = a \sin B \div b$
Bbc	C	$\sin = c \sin B \div b$
Cac	A	$\sin = a \sin C \div c$
Cbc	B	$\sin = b \sin C \div c$
Abc	$\tfrac{1}{2}(B + C)$	$90° - \tfrac{1}{2}A$

Given	To Find	Formula
Abc	$\frac{1}{2}(B-C)$	$\tan = [(b-c)\tan(90° - \frac{1}{2}A)] \div (b+c)$
Abc	B	$\frac{1}{2}(B+C) + \frac{1}{2}(B-C)$
Abc	C	$\frac{1}{2}(B+C) - \frac{1}{2}(B-C)$
Abc	a	$\sqrt{b^2 + c^2 - 2bc\cos A}$
Bac	$\frac{1}{2}(A+C)$	$90° - \frac{1}{2}B$
Bac	$\frac{1}{2}(A-C)$	$\tan = [(a-c)\tan(90° - \frac{1}{2}B)] \div (a+c)$
Bac	A	$\frac{1}{2}(A+C) + \frac{1}{2}(A-C)$
Bac	C	$\frac{1}{2}(A+C) - \frac{1}{2}(A-C)$
Bac	b	$\sqrt{a^2 + c^2 - 2ac\cos B}$
Cab	$\frac{1}{2}(A+B)$	$90° - \frac{1}{2}C$
Cab	$\frac{1}{2}(A-B)$	$\tan = [(a-b)\tan(90° - \frac{1}{2}C)] \div (a+b)$
Cab	A	$\frac{1}{2}(A+B) + \frac{1}{2}(A-B)$
Cab	B	$\frac{1}{2}(A+B) - \frac{1}{2}(A-B)$
Cab	c	$\sqrt{a^2 + b^2 - 2ab\cos C}$

Given	To Find	Formula
drB	b	$d\sin^2 B$
drB	f	$r\sin 2B$
drB	e	$d\sin B$
drb	Ang B	$\sin B = \sqrt{b \div d}$
drb	f	$\sqrt{b(d-b)}$
drb	e	$\sqrt{db}$
dre	Ang B	$\sin B = e \div d$
dre	b	$e^2 \div d$
dre	f	$e\sqrt{d^2 - e^2} \div d$
bB	r	$\frac{1}{2}b \div \sin^2 B$
eB	r	$\frac{1}{2}e \div \sin B$
bf	Ang B	$\tan B = b \div f$

Given	To Find	Formula
bf	r	$(f^2 + b^2) \div 2b$
fe	Ang B	$\sin B = \sqrt{e^2 - f^2} \div e$
fe	r	$\tfrac{1}{2}e^2 \div \sqrt{e^2 - f^2}$
be	Ang B	$\sin B = b \div e$
be	r	$\tfrac{1}{2}e^2 \div b$
rxy	Ang B	$\cos 2B = (\sqrt{r^2 - x^2} - y) \div r$
rxy	b	$r + y - \sqrt{r^2 - x^2}$
brx	y	$b + \sqrt{r^2 - x^2} - r$
bry	x	$\sqrt{r^2 - (r + y - b)^2}$
bxy	r	$[x^2 + (b - y)^2] \div (2b - 2y)$
r	Circ	$6.2832r$
rD	Arc a	$.0174533rD°$
rD	Arc a	$.0002909\ rD'$
rD	Arc a	$.00000485\ rD''$
r	Area	Circle $= 3.1416\ r^2$
d	Area	Circle $= 0.7854\ d^2$
c	Area	Circle $= 0.0796\ c^2$
ar	Area	Sector $= 0.5\ ar$
$arfh$	Area	Segment $= 0.5\ ar - fh$

TABLE 20. NATURAL TRIGONOMETRIC FUNCTIONS *

Deg.	Min.	Sine	Covers	Cosec	Tan	Cotan	Secant	Versin	Cosine		
0	0	0.00000	1.00000	Infinite	0.00000	Infinite	1.0000	0.00000	1.00000	90	0
	15	.00436	.99564	229.18	.00436	229.18	1.0000	.00001	.99999		45
	30	.00873	.99127	114.59	.00873	114.59	1.0000	.00004	.99996		30
	45	.01309	.98691	76.397	.01309	76.390	1.0001	.00009	.99991		15
1	0	.01745	.98255	57.299	.01745	57.290	1.0001	.00015	.99985	89	0
	15	.02181	.97819	45.840	.02182	45.829	1.0002	.00024	.99976		45
	30	.02618	.97382	38.202	.02618	38.188	1.0003	.00034	.99966		30
	45	.03054	.96946	32.746	.03055	32.730	1.0005	.00047	.99953		15
2	0	.03490	.96510	28.654	.03492	28.636	1.0006	.00061	.99939	88	0
	15	.03926	.96074	25.471	.03929	25.452	1.0008	.00077	.99923		45
	30	.04362	.95638	22.926	.04366	22.904	1.0009	.00095	.99905		30
	45	.04798	.95202	20.843	.04803	20.819	1.0011	.00115	.99885		15
3	0	.05234	.94766	19.107	.05241	19.081	1.0014	.00137	.99863	87	0
	15	.05669	.94331	17.639	.05678	17.611	1.0016	.00161	.99839		45
	30	.06105	.93895	16.380	.06116	16.350	1.0019	.00187	.99813		30
	45	.06540	.93460	15.290	.06554	15.257	1.0021	.00214	.99786		15
4	0	.06976	.93024	14.336	.06993	14.301	1.0024	.00244	.99756	86	0
	15	.07411	.92589	13.494	.07431	13.457	1.0028	.00275	.99725		45
	30	.07846	.92154	12.745	.07870	12.706	1.0031	.00308	.99692		30
	45	.08281	.91719	12.076	.08309	12.035	1.0034	.00343	.99656		15
5	0	.08716	.91284	11.474	.08749	11.430	1.0038	.00381	.99619	85	0
	15	.09150	.90850	10.929	.09189	10.883	1.0042	.00420	.99580		45
	30	.09585	.90415	10.433	.09629	10.385	1.0046	.00466	.99540		30
	45	.10019	.89981	9.9812	.10069	9.9310	1.0051	.00503	.99497		15
6	0	.10453	.89547	9.5668	.10510	9.5144	1.0055	.00548	.99452	84	0
	15	.10887	.89113	9.1855	.10952	9.1309	1.0060	.00594	.99406		45
	30	.11320	.88680	8.8337	.11393	8.7769	1.0065	.00643	.99357		30
	45	.11754	.88246	8.5079	.11836	8.4490	1.0070	.00693	.99307		15
7	0	.12187	.87813	8.2055	.12278	8.1443	1.0075	.00745	.99255	83	0
	15	.12620	.87380	7.9240	.12722	7.8606	1.0081	.00800	.99200		45
	30	.13053	.86947	7.6613	.13165	7.5958	1.0086	.00856	.99144		30
	45	.13485	.86515	7.4156	.13609	7.3479	1.0092	.00913	.99086		15
8	0	.13917	.86083	7.1853	.14054	7.1154	1.0098	.00973	.99027	82	0
	15	.14349	.85651	6.9690	.14499	6.8969	1.0105	.01035	.98965		45
	30	.14781	.85219	6.7655	.14945	6.6912	1.0111	.01098	.98902		30
	45	.15212	.84788	6.5736	.15391	6.4971	1.0118	.01164	.98836		15
9	0	.15643	.84357	6.3924	.15838	6.3138	1.0125	.01231	.98769	81	0
	15	.16074	.83926	6.2211	.16286	6.1402	1.0132	.01300	.98700		45
	30	.16505	.83495	6.0589	.16734	5.9758	1.0139	.01371	.98629		30
	45	.16935	.83065	5.9049	.17183	5.8197	1.0147	.01444	.98556		15
10	0	.17365	.82635	5.7588	.17633	5.6713	1.0154	.01519	.98481	80	0
	15	.17794	.82206	5.6198	.18083	5.5301	1.0162	.01596	.98404		45
	30	.18224	.81776	5.4874	.18534	5.3955	1.0170	.01675	.98325		30
	45	.18652	.81348	5.3612	.18986	5.2672	1.0179	.01755	.98245		15
11	0	.19081	.80919	5.2408	.19438	5.1446	1.0187	.01837	.98163	79	0
	15	.19509	.80491	5.1258	.19891	5.0273	1.0196	.01921	.98079		45
	30	.19937	.80063	5.0158	.20345	4.9152	1.0205	.02008	.97992		30
	45	.20364	.79636	4.9106	.20800	4.8077	1.0214	.02095	.97905		15
12	0	.20791	.79209	4.8097	.21256	4.7046	1.0223	.02185	.97815	78	0
	15	.21218	.78782	4.7130	.21712	4.6057	1.0233	.02277	.97723		45
	30	.21644	.78356	4.6202	.22169	4.5107	1.0243	.02370	.97630		30
	45	.22070	.77930	4.5311	.22628	4.4194	1.0253	.02466	.97534		15
13	0	.22495	.77505	4.4454	.23087	4.3315	1.0263	.02563	.97437	77	0
	15	.22920	.77080	4.3630	.23547	4.2468	1.0273	.02662	.97338		45
	30	.23345	.76655	4.2837	.24008	4.1653	1.0284	.02763	.97237		30
	45	.23769	.76231	4.2072	.24470	4.0867	1.0295	.02866	.97134		15
14	0	.24192	.75808	4.1336	.24933	4.0108	1.0306	.02970	.97030	76	0
	15	.24615	.75385	4.0625	.25397	3.9375	1.0317	.03077	.96923		45
	30	.25038	.74962	3.9939	.25862	3.8667	1.0329	.03185	.96815		30
	45	.25460	.74540	3.9277	.26328	3.7983	1.0341	.03295	.96705		15
15	0	.25882	.74118	3.8637	.26795	3.7320	1.0353	.03407	.96593	75	0
		Cosine	Versin	Secant	Cotan	Tan	Cosec	Covers	Sine	Deg.	Min.

From 75° to 90° read from bottom of table upwards.

* From Peele, *Mining Engineers' Handbook*, John Wiley & Sons.

TABLE 20. NATURAL TRIGONOMETRIC FUNCTIONS—*Continued*

Deg.	Min.	Sine	Covers	Cosec	Tan	Cotan	Secant	Versin	Cosine		
15	0	0.25882	0.74118	3.8637	0.26795	3.7320	1.0353	0.03407	0.96593	75	0
	15	.26303	.73697	3.8018	.27263	3.6680	1.0365	.03521	.96479		45
	30	.26724	.73276	3.7420	.27732	3.6059	1.0377	.03637	.96363		30
	45	.27144	.72856	3.6840	.28203	3.5457	1.0390	.03754	.96246		15
16	0	.27564	.72436	3.6280	.28674	3.4874	1.0403	.03874	.96126	74	0
	15	.27983	.72017	3.5736	.29147	3.4308	1.0416	.03995	.96005		45
	30	.28402	.71598	3.5209	.29621	3.3759	1.0429	.04118	.95882		30
	45	.28820	.71180	3.4699	.30096	3.3226	1.0443	.04243	.95757		15
17	0	.29237	.70763	3.4203	.30573	3.2709	1.0457	.04370	.95630	73	0
	15	.29654	.70346	3.3722	.31051	3.2205	1.0471	.04498	.95502		45
	30	.30070	.69929	3.3255	.31530	3.1716	1.0485	.04628	.95372		30
	45	.30486	.69514	3.2801	.32010	3.1240	1.0500	.04760	.95240		15
18	0	.30902	.69098	3.2361	.32492	3.0777	1.0515	.04894	.95106	72	0
	15	.31316	.68684	3.1932	.32975	3.0326	1.0530	.05030	.94970		45
	30	.31730	.68270	3.1515	.33459	2.9887	1.0545	.05168	.94832		30
	45	.32144	.67856	3.1110	.33945	2.9459	1.0560	.05307	.94693		15
19	0	.32557	.67443	3.0715	.34433	2.9042	1.0576	.05448	.94552	71	0
	15	.32969	.67031	3.0331	.34921	2.8636	1.0592	.05591	.94409		45
	30	.33381	.66619	2.9957	.35412	2.8239	1.0608	.05736	.94264		30
	45	.33792	.66208	2.9593	.35904	2.7852	1.0625	.05882	.94118		15
20	0	.34202	.65798	2.9238	.36397	2.7475	1.0642	.06031	.93969	70	0
	15	.34612	.65388	2.8892	.36892	2.7106	1.0659	.06181	.93819		45
	30	.35021	.64979	2.8554	.37388	2.6746	1.0676	.06333	.93667		30
	45	.35429	.64571	2.8225	.37887	2.6395	1.0694	.06486	.93514		15
21	0	.35837	.64163	2.7904	.38386	2.6051	1.0711	.06642	.93358	69	0
	15	.36244	.63756	2.7591	.38888	2.5715	1.0729	.06799	.93201		45
	30	.36650	.63350	2.7285	.39391	2.5386	1.0748	.06958	.93042		30
	45	.37056	.62944	2.6986	.39896	2.5065	1.0766	.07119	.92881		15
22	0	.37461	.62539	2.6695	.40403	2.4751	1.0785	.07282	.92718	68	0
	15	.37865	.62135	2.6410	.40911	2.4443	1.0804	.07446	.92554		45
	30	.38268	.61732	2.6131	.41421	2.4142	1.0824	.07612	.92388		30
	45	.38671	.61329	2.5859	.41933	2.3847	1.0844	.07780	.92220		15
23	0	.39073	.60927	2.5593	.42447	2.3559	1.0864	.07950	.92050	67	0
	15	.39474	.60526	2.5333	.42963	2.3276	1.0884	.08121	.91879		45
	30	.39875	.60125	2.5078	.43481	2.2998	1.0904	.08294	.91706		30
	45	.40275	.59725	2.4829	.44001	2.2727	1.0925	.08469	.91531		15
24	0	.40674	.59326	2.4586	.44523	2.2460	1.0946	.08645	.91355	66	0
	15	.41072	.58928	2.4348	.45047	2.2199	1.0968	.08824	.91176		45
	30	.41469	.58531	2.4114	.45573	2.1943	1.0989	.09004	.90996		30
	45	.41866	.58134	2.3886	.46101	2.1692	1.1011	.09186	.90814		15
25	0	.42262	.57738	2.3662	.46631	2.1445	1.1034	.09369	.90631	65	0
	15	.42657	.57343	2.3443	.47163	2.1203	1.1056	.09554	.90446		45
	30	.43051	.56949	2.3228	.47697	2.0965	1.1079	.09741	.90259		30
	45	.43445	.56555	2.3018	.48234	2.0732	1.1102	.09930	.90070		15
26	0	.43837	.56163	2.2812	.48773	2.0503	1.1126	.10121	.89879	64	0
	15	.44229	.55771	2.2610	.49314	2.0278	1.1150	.10313	.89687		45
	30	.44620	.55380	2.2412	.49858	2.0057	1.1174	.10507	.89493		30
	45	.45010	.54990	2.2217	.50404	1.9840	1.1198	.10702	.89298		15
27	0	.45399	.54601	2.2027	.50952	1.9626	1.1223	.10899	.89101	63	0
	15	.45787	.54213	2.1840	.51503	1.9416	1.1248	.11098	.88902		45
	30	.46175	.53825	2.1657	.52057	1.9210	1.1274	.11299	.88701		30
	45	.46561	.53439	2.1477	.52612	1.9007	1.1300	.11501	.88499		15
28	0	.46947	.53053	2.1300	.53171	1.8807	1.1326	.11705	.88295	62	0
	15	.47332	.52668	2.1127	.53732	1.8611	1.1352	.11911	.88089		45
	30	.47716	.52284	2.0957	.54295	1.8418	1.1379	.12118	.87882		30
	45	.48099	.51901	2.0790	.54862	1.8228	1.1406	.12327	.87673		15
29	0	.48481	.51519	2.0627	.55431	1.8040	1.1433	.12538	.87462	61	0
	15	.48862	.51138	2.0466	.56003	1.7856	1.1461	.12750	.87250		45
	30	.49242	.50758	2.0308	.56577	1.7675	1.1490	.12964	.87036		30
	45	.49622	.50378	2.0152	.57155	1.7496	1.1518	.13180	.86820		15
30	0	.50000	.50000	2.0000	.57735	1.7320	1.1547	.13397	.86603	60	0
		Cosine	Versin	Secant	Cotan	Tan	Cosec	Covers	Sine	Deg.	Min.

From 60° to 75° read from bottom of table upwards.

TABLE 20. NATURAL TRIGONOMETRIC FUNCTIONS—*Concluded*

Deg.	Min.	Sine	Covers	Cosec	Tan	Cotan	Secant	Versin	Cosine		
30	0	0.50000	0.50000	2.0000	0.57735	1.7320	1.1547	0.13397	0.86603	60	0
	15	.50377	.49623	1.9850	.58318	1.7147	1.1576	.13616	.86384		45
	30	.50754	.49246	1.9703	.58904	1.6977	1.1606	.13837	.86163		30
	45	.51129	.48871	1.9558	.59494	1.6808	1.1636	.14059	.85941		15
31	0	.51504	.48496	1.9416	.60086	1.6643	1.1666	.14283	.85717	59	0
	15	.51877	.48123	1.9276	.60681	1.6479	1.1697	.14509	.85491		45
	30	.52250	.47750	1.9139	.61280	1.6319	1.1728	.14736	.85264		30
	45	.52621	.47379	1.9004	.61882	1.6160	1.1760	.14965	.85035		15
32	0	.52992	.47008	1.8871	.62487	1.6003	1.1792	.15195	.84805	58	0
	15	.53361	.46639	1.8740	.63095	1.5849	1.1824	.15427	.84573		45
	30	.53730	.46270	1.8612	.63707	1.5697	1.1857	.15661	.84339		30
	45	.54097	.45903	1.8485	.64322	1.5547	1.1890	.15896	.84104		15
33	0	.54464	.45536	1.8361	.64941	1.5399	1.1924	.16133	.83867	57	0
	15	.54829	.45171	1.8238	.65563	1.5253	1.1958	.16371	.83629		45
	30	.55194	.44806	1.8118	.66188	1.5108	1.1992	.16611	.83389		30
	45	.55557	.44443	1.7999	.66818	1.4966	1.2027	.16853	.83147		15
34	0	.55919	.44081	1.7883	.67451	1.4826	1.2062	.17096	.82904	56	0
	15	.56280	.43720	1.7768	.68087	1.4687	1.2098	.17341	.82659		45
	30	.56641	.43359	1.7655	.68728	1.4550	1.2134	.17587	.82413		30
	45	.57000	.43000	1.7544	.69372	1.4415	1.2171	.17835	.82165		15
35	0	.57358	.42642	1.7434	.70021	1.4281	1.2208	.18085	.81915	55	0
	15	.57715	.42285	1.7327	.70673	1.4150	1.2245	.18336	.81664		45
	30	.58070	.41930	1.7220	.71329	1.4019	1.2283	.18588	.81412		30
	45	.58425	.41575	1.7116	.71990	1.3891	1.2322	.18843	.81157		15
36	0	.58779	.41221	1.7013	.72654	1.3764	1.2361	.19098	.80902	54	0
	15	.59131	.40869	1.6912	.73323	1.3638	1.2400	.19356	.80644		45
	30	.59482	.40518	1.6812	.73996	1.3514	1.2440	.19614	.80386		30
	45	.59832	.40168	1.6713	.74673	1.3392	1.2480	.19875	.80125		15
37	0	.60181	.39819	1.6616	.75355	1.3270	1.2521	.20136	.79864	53	0
	15	.60529	.39471	1.6521	.76042	1.3151	1.2563	.20400	.79600		45
	30	.60876	.39124	1.6427	.76733	1.3032	1.2605	.20665	.79335		30
	45	.61222	.38778	1.6334	.77428	1.2915	1.2647	.20931	.79069		15
38	0	.61566	.38434	1.6243	.78129	1.2799	1.2690	.21199	.78801	52	0
	15	.61909	.38091	1.6153	.78834	1.2685	1.2734	.21468	.78532		45
	30	.62251	.37749	1.6064	.79543	1.2572	1.2778	.21739	.78261		30
	45	.62592	.37408	1.5976	.80258	1.2460	1.2822	.22012	.77988		15
39	0	.62932	.37068	1.5890	.80978	1.2349	1.2868	.22285	.77715	51	0
	15	.63271	.36729	1.5805	.81703	1.2239	1.2913	.22561	.77439		45
	30	.63608	.36392	1.5721	.82434	1.2131	1.2960	.22838	.77162		30
	45	.63944	.36056	1.5639	.83169	1.2024	1.3007	.23116	.76884		15
40	0	.64279	.35721	1.5557	.83910	1.1918	1.3054	.23396	.76604	50	0
	15	.64612	.35388	1.5477	.84656	1.1812	1.3102	.23677	.76323		45
	30	.64945	.35055	1.5398	.85408	1.1708	1.3151	.23959	.76041		30
	45	.65276	.34724	1.5320	.86165	1.1606	1.3200	.24244	.75756		15
41	0	.65606	.34394	1.5242	.86929	1.1504	1.3250	.24529	.75471	49	0
	15	.65935	.34065	1.5166	.87698	1.1403	1.3301	.24816	.75184		45
	30	.66262	.33738	1.5092	.88472	1.1303	1.3352	.25104	.74896		30
	45	.66588	.33412	1.5018	.89253	1.1204	1.3404	.25394	.74606		15
42	0	.66913	.33087	1.4945	.90040	1.1106	1.3456	.25686	.74314	48	0
	15	.67237	.32763	1.4873	.90834	1.1009	1.3509	.25978	.74022		45
	30	.67559	.32441	1.4802	.91633	1.0913	1.3563	.26272	.73728		30
	45	.67880	.32120	1.4732	.92439	1.0818	1.3618	.26568	.73432		15
43	0	.68200	.31800	1.4663	.93251	1.0724	1.3673	.26865	.73135	47	0
	15	.68518	.31482	1.4595	.94071	1.0630	1.3729	.27163	.72837		45
	30	.68835	.31165	1.4527	.94896	1.0538	1.3786	.27463	.72537		30
	45	.69151	.30849	1.4461	.95729	1.0446	1.3843	.27764	.72236		15
44	0	.69466	.30534	1.4396	.96569	1.0355	1.3902	.28066	.71934	46	0
	15	.69779	.30221	1.4331	.97416	1.0265	1.3961	.28370	.71630		45
	30	.70091	.29909	1.4267	.98270	1.0176	1.4020	.28675	.71325		30
	45	.70401	.29599	1.4204	.99131	1.0088	1.4081	.28981	.71019		15
45	0	.70711	.29289	1.4142	.10000	1.0000	1.4142	.29289	.70711	45	0
		Cosine	Versin	Secant	Cotan	Tan	Cosec	Covers	Sine	Deg.	Min.

From 45° to 60° read from bottom of table upwards.

TABLE 21. LOGARITHMIC TRIGONOMETRIC FUNCTIONS *

Deg.	Sine	Cosec	Versin	Tangent	Cotan	Covers	Secant	Cosine	Deg
0	− ∞	+ ∞	− ∞	− ∞	+ ∞	10.00000	10.00000	10.00000	90
1	8.24186	11.75814	6.18271	8.24192	11.75808	9.99235	10.00007	9.99993	89
2	8.54282	11.45718	6.78474	8.54308	11.45692	9.98457	10.00026	9.99974	88
3	8.71880	11.28120	7.13687	8.71940	11.28060	9.97665	10.00060	9.99940	87
4	8.84358	11.15642	7.38667	8.84464	11.15536	9.96860	10.00106	9.99894	86
5	8.94030	11.05970	7.58039	8.94195	11.05805	9.96040	10.00166	9.99834	85
6	9.01923	10.98077	7.73863	9.02162	10.97838	9.95205	10.00239	9.99761	84
7	9.08589	10.91411	7.87238	9.08914	10.91086	9.94356	10.00325	9.99675	83
8	9.14356	10.85644	7.98820	9.14780	10.85220	9.93492	10.00425	9.99575	82
9	9.19433	10.80567	8.09032	9.19971	10.80029	9.92612	10.00538	9.99462	81
10	9.23967	10.76033	8.18162	9.24632	10.75368	9.91717	10.00665	9.99335	80
11	9.28060	10.71940	8.26418	9.28865	10.71135	9.90805	10.00805	9.99195	79
12	9.31788	10.68212	8.33950	9.32747	10.67253	9.89877	10.00960	9.99040	78
13	9.35209	10.64791	8.40875	9.36336	10.63664	9.88933	10.01128	9.98872	77
14	9.38368	10.61632	8.47282	9.39677	10.60323	9.87971	10.01310	9.98690	76
15	9.41300	10.58700	8.53243	9.42805	10.57195	9.86992	10.01506	9.98494	75
16	9.44034	10.55966	8.58814	9.45750	10.54250	9.85996	10.01716	9.98284	74
17	9.46594	10.53406	8.64043	9.48534	10.51466	9.84981	10.01940	9.98060	73
18	9.48998	10.51002	8.68969	9.51178	10.48822	9.83947	10.02179	9.97821	72
19	9.51264	10.48736	8.73625	9.53697	10.46303	9.82894	10.02433	9.97567	71
20	9.53405	10.46595	8.78037	9.56107	10.43893	9.81821	10.02701	9.97299	70
21	9.55433	10.44567	8.82230	9.58418	10.41582	9.80729	10.02985	9.97015	69
22	9.57358	10.42642	8.86223	9.60641	10.39359	9.79615	10.03283	9.96717	68
23	9.59188	10.40812	8.90034	9.62785	10.37215	9.78481	10.03597	9.96403	67
24	9.60931	10.39069	8.93679	9.64858	10.35142	9.77325	10.03927	9.96073	66
25	9.62595	10.37405	8.97170	9.66867	10.33133	9.76146	10.04272	9.95728	65
26	9.64184	10.35816	9.00521	9.68818	10.31182	9.74945	10.04634	9.95366	64
27	9.65705	10.34295	9.03740	9.70717	10.29283	9.73720	10.05012	9.94988	63
28	9.67161	10.32839	9.06838	9.72567	10.27433	9.72471	10.05407	9.94593	62
29	9.68557	10.31443	9.09823	9.74375	10.25625	9.71197	10.05818	9.94182	61
30	9.69897	10.30103	9.12702	9.76144	10.23856	9.69897	10.06247	9.93753	60
31	9.71184	10.28816	9.15483	9.77877	10.22123	9.68571	10.06693	9.93307	59
32	9.72421	10.27579	9.18171	9.79579	10.20421	9.67217	10.07158	9.92842	58
33	9.73611	10.26389	9.20771	9.81252	10.18748	9.65836	10.07641	9.92359	57
34	9.74756	10.25244	9.23290	9.82899	10.17101	9.64425	10.08143	9.91857	56
35	9.75859	10.24141	9.25731	9.84523	10.15477	9.62984	10.08664	9.91336	55
36	9.76922	10.23078	9.28099	9.86126	10.13874	9.61512	10.09204	9.90796	54
37	9.77946	10.22054	9.30398	9.87711	10.12289	9.60008	10.09765	9.90235	53
38	9.78934	10.21066	9.32631	9.89281	10.10719	9.58471	10.10347	9.89653	52
39	9.79887	10.20113	9.34802	9.90837	10.09163	9.56900	10.10950	9.89050	51
40	9.80807	10.19193	9.36913	9.92381	10.07619	9.55293	10.11575	9.88425	50
41	9.81694	10.18306	9.38968	9.93916	10.06084	9.53648	10.12222	9.87778	49
42	9.82551	10.17449	9.40969	9.95444	10.04556	9.51966	10.12893	9.87107	48
43	9.83378	10.16622	9.42918	9.96966	10.03034	9.50243	10.13587	9.86413	47
44	9.84177	10.15823	9.44818	9.98484	10.01516	9.48479	10.14307	9.85693	46
45	9.84949	10.15052	9.46671	10.00000	10.00000	9.46671	10.15052	9.84949	45
	Cosine	Secant	Covers	Cotan	Tangent	Versin	Cosec	Sine	

From 45° to 90° read from bottom of table upwards.

From Kent, Mechanical Engineers' Handbook, Power Volume, John Wiley & Sons.

TABLE 22. MINUTES INTO DECIMALS OF A DEGREE *

′	0″	10″	15″	20″	30″	40″	45″	50″	′
0	.00000	.00278	.00417	.00556	.00833	.01111	.01250	.01389	0
1	.01667	.01944	.02083	.02222	.02500	.02778	.02917	.03055	1
2	.03333	.03611	.03750	.03889	.04167	.04444	.04583	.04722	2
3	.05000	.05278	.05417	.05556	.05833	.06111	.06250	.06389	3
4	.06667	.06944	.07083	.07222	.07500	.07778	.07917	.08056	4
5	.08333	.08611	.08750	.08889	.09167	.09444	.09583	.09722	5
6	.10000	.10278	.10417	.10556	.10833	.11111	.11250	.11389	6
7	.11667	.11944	.12083	.12222	.12500	.12778	.12917	.13056	7
8	.13333	.13611	.13750	.13889	.14167	.14444	.14583	.14722	8
9	.15000	.15278	.15417	.15556	.15833	.16111	.16250	.16389	9
10	.16667	.16944	.17083	.17222	.17500	.17778	.17917	.18056	10
11	.18333	.18611	.18750	.18889	.19167	.19444	.19583	.19722	11
12	.20000	.20278	.20417	.20556	.20833	.21111	.21250	.21389	12
13	.21667	.21944	.22083	.22222	.22500	.22778	.22917	.23056	13
14	.23333	.23611	.23750	.23889	.24167	.24444	.24583	.24722	14
15	.25000	.25278	.25417	.25556	.25833	.26111	.26250	.26389	15
16	.26667	.26944	.27083	.27222	.27500	.27778	.27917	.28056	16
17	.28333	.28611	.28750	.28889	.29167	.29444	.29583	.29722	17
18	.30000	.30278	.30417	.30556	.30833	.31111	.31250	.31389	18
19	.31667	.31944	.32083	.32222	.32500	.32778	.32917	.33056	19
20	.33333	.33611	.33750	.33889	.34167	.34444	.34583	.34722	20
21	.35000	.35278	.35417	.35556	.35833	.36111	.36250	.36389	21
22	.36667	.36944	.37083	.37222	.37500	.37778	.37917	.38056	22
23	.38333	.38611	.38750	.38889	.39167	.39444	.39583	.39722	23
24	.40000	.40278	.40417	.40556	.40833	.41111	.41250	.41389	24
25	.41667	.41944	.42083	.42222	.42500	.42778	.42917	.43056	25
26	.43333	.43611	.43750	.43889	.44167	.44444	.44583	.44722	26
27	.45000	.45278	.45417	.45556	.45833	.46111	.46250	.46389	27
28	.46667	.46944	.47083	.47222	.47500	.47778	.47917	.48056	28
29	.48333	.48611	.48750	.48889	.49167	.49444	.49583	.49722	29
30	.50000	.50278	.50417	.50556	.50833	.51111	.51250	.51389	30
31	.51667	.51944	.52083	.52222	.52500	.52778	.52917	.53056	31
32	.53333	.53611	.53750	.53889	.54167	.54444	.54583	.54722	32
33	.55000	.55278	.55417	.55556	.55833	.56111	.56250	.56389	33
34	.56667	.56944	.57083	.57222	.57500	.57778	.57917	.58056	34
35	.58333	.58611	.58750	.58889	.59167	.59444	.59583	.59722	35
36	.60000	.60278	.60417	.60556	.60833	.61111	.61250	.61389	36
37	.61667	.61944	.62083	.62222	.62500	.62778	.62917	.63056	37
38	.63333	.63611	.63750	.63889	.64167	.64444	.64583	.64722	38
39	.65000	.65278	.65417	.65556	.65833	.66111	.66250	.66389	39
40	.66667	.66944	.67083	.67222	.67500	.67778	.67917	.68056	40
41	.68333	.68611	.68750	.68889	.69167	.69444	.69583	.69722	41
42	.70000	.70278	.70417	.70556	.70833	.71111	.71250	.71389	42
43	.71667	.71944	.72083	.72222	.72500	.72778	.72917	.73056	43
44	.73333	.73611	.73750	.73889	.74167	.74444	.74583	.74722	44
45	.75000	.75278	.75417	.75556	.75833	.76111	.76250	.76389	45
46	.76667	.76944	.77083	.77222	.77500	.77778	.77917	.78056	46
47	.78333	.78611	.78750	.78889	.79167	.79444	.79583	.79722	47
48	.80000	.80278	.80417	.80556	.80833	.81111	.81250	.81389	48
49	.81667	.81944	.82083	.82222	.82500	.82778	.82917	.83056	49
50	.83333	.83611	.83750	.83889	.84167	.84444	.84583	.84722	50
51	.85000	.85278	.85417	.85556	.85833	.86111	.86250	.86389	51
52	.86667	.86944	.87083	.87222	.87500	.87778	.87917	.88056	52
53	.88333	.88611	.88750	.88889	.89167	.89444	.89583	.89722	53
54	.90000	.90278	.90417	.90556	.90833	.91111	.91250	.91389	54
55	.91667	.91944	.92083	.92222	.92500	.92778	.92917	.93056	55
56	.93333	.93611	.93750	.93889	.94167	.94444	.94583	.94722	56
57	.95000	.95278	.95417	.95556	.95833	.96111	.96250	.96389	57
58	.96667	.96944	.97083	.97222	.97500	.97778	.97917	.98056	58
59	.98333	.98611	.98750	.98889	.99167	.99444	.99583	.99722	59
′	0″	10″	15″	20″	30″	40″	45″	50″	′

* From Ives, Seven Place Natural Trigonometric Functions, John Wiley & Sons.

TABLE 23. LOGARITHMS OF NUMBERS *

n	0	1	2	3	4	5	6	7	8	9
10	00000	00432	00860	01284	01703	02119	02531	02938	03342	03743
11	04139	04532	04922	05308	05690	06070	06446	06819	07188	07555
12	07918	08279	08636	08991	09342	09691	10037	10380	10721	11059
13	11394	11727	12057	12385	12710	13033	13354	13672	13988	14301
14	14613	14922	15229	15534	15836	16137	16435	16732	17026	17319
15	17609	17898	18184	18469	18752	19033	19312	19590	19866	20140
16	20412	20683	20952	21219	21484	21748	22011	22272	22531	22789
17	23045	23300	23553	23805	24055	24304	24551	24797	25042	25285
18	25527	25768	26007	26245	26482	26717	26951	27184	27416	27646
19	27875	28103	28330	28556	28780	29003	29226	29447	29667	29885
20	30103	30320	30535	30750	30963	31175	31387	31597	31806	32015
21	32222	32428	32634	32838	33041	33244	33445	33646	33846	34044
22	34242	34439	34635	34830	35025	35218	35411	35603	35793	35984
23	36173	36361	36549	36736	36922	37107	37291	37475	37658	37840
24	38021	38202	38382	38561	38739	38917	39094	39270	39445	39620
25	39794	39967	40140	40312	40483	40654	40824	40993	41162	41330
26	41497	41664	41830	41996	42160	42325	42488	42651	42813	42975
27	43136	43297	43457	43616	43775	43933	44091	44248	44404	44560
28	44716	44871	45025	45179	45332	45484	45637	45788	45939	46090
29	46240	46389	46538	46687	46835	46982	47129	47276	47422	47567
30	47712	47857	48001	48144	48287	48430	48572	48714	48855	48996
31	49136	49276	49415	49554	49693	49831	49969	50106	50243	50379
32	50515	50651	50786	50920	51055	51188	51322	51455	51587	51720
33	51851	51983	52114	52244	52375	52504	52634	52763	52892	53020
34	53148	53275	53403	53529	53656	53782	53908	54033	54158	54283
35	54407	54531	54654	54777	54900	55023	55145	55267	55388	55509
36	55630	55751	55871	55991	56110	56229	56348	56467	56585	56703
37	56820	56937	57054	57171	57287	57403	57519	57634	57749	57864
38	57978	58092	58206	58320	58433	58546	58659	58771	58883	58995
39	59106	59218	59329	59439	59550	59660	59770	59879	59988	60097
40	60206	60314	60423	60531	60638	60746	60853	60959	61066	61172
41	61278	61384	61490	61595	61700	61805	61909	62014	62118	62221
42	62325	62428	62531	62634	62737	62839	62941	63043	63144	63246
43	63347	63448	63548	63649	63749	63849	63949	64048	64147	64246
44	64345	64444	64542	64640	64738	64836	64933	65031	65128	65225
45	65321	65418	65514	65610	65706	65801	65896	65992	66087	66181
46	66276	66370	66464	66558	66652	66745	66839	66932	67025	67117
47	67210	67302	67394	67486	67578	67669	67761	67852	67943	68034
48	68124	68215	68305	68395	68485	68574	68664	68753	68842	68931
49	69020	69108	69197	69285	69373	69461	69548	69636	69723	69810
50	69897	69984	70070	70157	70243	70329	70415	70501	70586	70672
51	70757	70842	70927	71012	71096	71181	71265	71349	71433	71517
52	71600	71684	71767	71850	71933	72016	72099	72181	72263	72346
53	72428	72509	72591	72673	72754	72835	72916	72997	73078	73159
54	73239	73320	73400	73480	73560	73640	73719	73799	73878	73957
	0	1	2	3	4	5	6	7	8	9

* From *American Civil Engineers' Handbook* by Merriam and Wiggin, John Wiley & Sons.

TABLE 23. LOGARITHMS OF NUMBERS (*Continued*)

n	0	1	2	3	4	5	6	7	8	9
55	74036	74115	74194	74273	74351	74429	74507	74586	74663	74741
56	74819	74896	74974	75051	75128	75205	75282	75358	75435	75511
57	75587	75664	75740	75815	75891	75967	76042	76118	76193	76268
58	76343	76418	76492	76567	76641	76716	76790	76864	76938	77012
59	77085	77159	77232	77305	77379	77452	77525	77597	77670	77743
60	77815	77887	77960	78032	78104	78176	78247	78319	78390	78462
61	78533	78604	78675	78746	78817	78888	78958	79029	79099	79169
62	79239	79309	79379	79449	79518	79588	79657	79727	79796	79865
63	79934	80003	80072	80140	80209	80277	80346	80414	80482	80550
64	80618	80686	80754	80821	80889	80956	81023	81090	81158	81224
65	81291	81358	81425	81491	81558	81624	81690	81757	81823	81889
66	81954	82020	82086	82151	82217	82282	82347	82413	82478	82543
67	82607	82672	82737	82802	82866	82930	82995	83059	83123	83187
68	83251	83315	83378	83442	83506	83569	83632	83696	83759	83822
69	83885	83948	84011	84073	84136	84198	84261	84323	84386	84448
70	84510	84572	84634	84696	84757	84819	84880	84942	85003	85065
71	85126	85187	85248	85309	85370	85431	85491	85552	85612	85673
72	85733	85794	85854	85914	85974	86034	86094	86153	86213	86273
73	86332	86392	86451	86510	86570	86629	86688	86747	86806	86864
74	86923	86982	87040	87099	87157	87216	87274	87332	87390	87448
75	87506	87564	87622	87679	87737	87795	87852	87910	87967	88024
76	88081	88138	88195	88252	88309	88366	88423	88480	88536	88593
77	88649	88705	88762	88818	88874	88930	88986	89042	89098	89154
78	89209	89265	89321	89376	89432	89487	89542	89597	89653	89708
79	89763	89818	89873	89927	89982	90037	90091	90146	90200	90255
80	90309	90363	90417	90472	90526	90580	90634	90687	90741	90795
81	90849	90902	90956	91009	91062	91116	91169	91222	91275	91328
82	91381	91434	91487	91540	91593	91645	91698	91751	91803	91855
83	91908	91960	92012	92065	92117	92169	92221	92273	92324	92376
84	92428	92480	92531	92583	92634	92686	92737	92788	92840	92891
85	92942	92993	93044	93095	93146	93197	93247	93298	93349	93399
86	93450	93500	93551	93601	93651	93702	93752	93802	93852	93902
87	93952	94002	94052	94101	94151	94201	94250	94300	94349	94399
88	94448	94498	94547	94596	94645	94694	94743	94792	94841	94890
89	94939	94988	95036	95085	95134	95182	95231	95279	95328	95376
90	95424	95472	95521	95569	95617	95665	95713	95761	95809	95856
91	95904	95952	95999	96047	96095	96142	96190	96237	96284	96332
92	96379	96426	96473	96520	96567	96614	96661	96708	96755	96802
93	96848	96895	96942	96988	97035	97081	97128	97174	97220	97267
94	97313	97359	97405	97451	97497	97543	97589	97635	97681	97727
95	97772	97818	97864	97909	97955	98000	98046	98091	98137	98182
96	98227	98272	98318	98363	98408	98453	98498	98543	98588	98632
97	98677	98722	98767	98811	98856	98900	98945	98989	99034	99078
98	99123	99167	99211	99255	99300	99344	99388	99432	99476	99520
99	99564	99607	99651	99695	99739	99782	99826	99870	99913	99957
	0	1	2	3	4	5	6	7	8	9

TABLE 24. DECIMAL EQUIVALENTS OF COMMON FRACTIONS *

The given decimals are the parts of inches corresponding to fraction of inches in first column; also, the parts of feet for the fraction of inches in third column.

	0.0052	1/16		0.2552	3 1/16		0.5052	6 1/16		0.7552	9 1/16
	0.0104	1/8		0.2604	3 1/8		0.5104	6 1/8		0.7604	9 1/8
1/64	0.015625	3/16	17/64	0.265625	3 3/16	33/64	0.515625	6 3/16	49/64	0.765625	9 3/16
	0.0208	1/4		0.2708	3 1/4		0.5208	6 1/4		0.7708	9 1/4
	0.0260	5/16		0.2760	3 5/16		0.5260	6 5/16		0.7760	9 5/16
1/32	0.03125	3/8	9/32	0.28125	3 3/8	17/32	0.53125	6 3/8	25/32	0.78125	9 3/8
	0.0364	7/16		0.2865	3 7/16		0.5364	6 7/16		0.7865	9 7/16
	0.0417	1/2		0.2917	3 1/2		0.5417	6 1/2		0.7917	9 1/2
3/64	0.046875	9/16	19/64	0.296875	3 9/16	35/64	0.546875	6 9/16	51/64	0.796875	9 9/16
	0.0521	5/8		0.3021	3 5/8		0.5521	6 5/8		0.8021	9 5/8
	0.0573	11/16		0.3073	3 11/16		0.5573	6 11/16		0.8073	9 11/16
1/16	0.0625	3/4	5/16	0.3125	3 3/4	9/16	0.5625	6 3/4	13/16	0.8125	9 3/4
	0.0677	13/16		0.3177	3 13/16		0.5677	6 13/16		0.8177	9 13/16
	0.0729	7/8		0.3229	3 7/8		0.5729	6 7/8		0.8229	9 7/8
5/64	0.078125	15/16	21/64	0.328125	3 15/16	37/64	0.578125	6 15/16	53/64	0.828125	9 15/16
	0.0833	1		0.3333	4		0.5833	7		0.8333	10
	0.0885	1 1/16		0.3385	4 1/16		0.5885	7 1/16		0.8385	10 1/16
3/32	0.09375	1 1/8	11/32	0.34375	4 1/8	19/32	0.59375	7 1/8	27/32	0.84375	10 1/8
	0.0990	1 3/16		0.3490	4 3/16		0.5990	7 3/16		0.8490	10 3/16
	0.1042	1 1/4		0.3542	4 1/4		0.6042	7 1/4		0.8542	10 1/4
7/64	0.109375	1 5/16	23/64	0.359375	4 5/16	39/64	0.609375	7 5/16	55/64	0.859375	10 5/16
	0.1146	1 3/8		0.3646	4 3/8		0.6146	7 3/8		0.8646	10 3/8
	0.1198	1 7/16		0.3698	4 7/16		0.6198	7 7/16		0.8698	10 7/16
1/8	0.1250	1 1/2	3/8	0.3750	4 1/2	5/8	0.6250	7 1/2	7/8	0.8750	10 1/2
	0.1302	1 9/16		0.3802	4 9/16		0.6302	7 9/16		0.8802	10 9/16
	0.1354	1 5/8		0.3854	4 5/8		0.6354	7 5/8		0.8854	10 5/8
9/64	0.140625	1 11/16	25/64	0.390625	4 11/16	41/64	0.640625	7 11/16	57/64	0.890625	10 11/16
	0.1458	1 3/4		0.3958	4 3/4		0.6458	7 3/4		0.8958	10 3/4
	0.1510	1 13/16		0.4010	4 13/16		0.6510	7 13/16		0.9010	10 13/16
5/32	0.15625	1 7/8	13/32	0.40625	4 7/8	21/32	0.65625	7 7/8	29/32	0.90625	10 7/8
	0.1615	1 15/16		0.4114	4 15/16		0.6615	7 15/16		0.9115	10 15/16
	0.1667	2		0.4167	5		0.6667	8		0.9167	11
11/64	0.171875	2 1/16	27/64	0.421875	5 1/16	43/64	0.671875	8 1/16	59/64	0.921875	11 1/16
	0.1771	2 1/8		0.4271	5 1/8		0.6771	8 1/8		0.9271	11 1/8
	0.1823	2 3/16		0.4323	5 3/16		0.6823	8 3/16		0.9323	11 3/16
3/16	0.1875	2 1/4	7/16	0.4375	5 1/4	11/16	0.6875	8 1/4	15/16	0.9375	11 1/4
	0.1927	2 5/16		0.4427	5 5/16		0.6927	8 5/16		0.9427	11 5/16
	0.1979	2 3/8		0.4479	5 3/8		0.6979	8 3/8		0.9479	11 3/8
13/64	0.203125	2 7/16	29/64	0.453125	5 7/16	45/64	0.703125	8 7/16	61/64	0.953125	11 7/16
	0.2083	2 1/2		0.4583	5 1/2		0.7083	8 1/2		0.9583	11 1/2
	0.2135	2 9/16		0.4635	5 9/16		0.7135	8 9/16		0.9635	11 9/16
7/32	0.21875	2 5/8	15/32	0.46875	5 5/8	23/32	0.71875	8 5/8	31/32	0.96875	11 5/8
	0.2240	2 11/16		0.4740	5 11/16		0.7240	8 11/16		0.9740	11 11/16
	0.2292	2 3/4		0.4792	5 3/4		0.7292	8 3/4		0.9792	11 3/4
15/64	0.234375	2 13/16	31/64	0.484375	5 13/16	47/64	0.734375	8 13/16	63/64	0.984375	11 13/16
	0.2395	2 7/8		0.4896	5 7/8		0.7396	8 7/8		0.9896	11 7/8
	0.2448	2 15/16		0.4948	5 15/16		0.7448	8 15/16		0.9948	11 15/16
1/4	0.2500	3	1/2	0.5000	6	3/4	0.7500	9	1	1.0000	12

From Peele, Mining Engineers' Handbook, John Wiley & Sons.

SURVEYING SIGNALS *

Except for short distances a good system of hand signals between different members of the party makes an efficient means of communication. The number of signals necessary will depend upon the kind of work and the nature of the country. A few of the more common are given below:

"Right" or "Left." The arm is extended in the direction of the desired movement, the right arm being extended for a movement to the right and the left arm for a movement to the left. A long, slow, sweeping motion of the hand indicates a long movement; a short, quick motion indicates a short movement. This signal may be given by the transitman in directing the chainman on line, by the leveler in directing the rodman for a turning point, by the chief of the party to any member, or by one chainman to another chainman.

"All Right." Both arms are extended horizontally and the forearms waved vertically. The signal may be given by any member of any party.

"Plumb the Flag" or "Plumb the Rod." The arm is held vertically and moved in the direction that the flag or rod is to be plumbed. It is given by the transitman or leveler.

"Give a Foresight." The instrumentman holds one arm vertically above his head.

"Establish a Turning Point" or "Set a Hub." The instrumentman holds one arm above his head and waves it in a circle.

"Give Line." The flagman holds the flag horizontally in both hands above his head and brings it down and turns it to a vertical position. If he desires to set a hub, he waves the flag with one end in the ground from side to side.

"Turning Point" or "Bench Mark." In profile leveling the rodman holds the rod horizontally above his head and then brings it down on the point.

"Wave the Rod." The leveler holds one arm above his head and moves it from side to side.

"Pick up the Instrument." Both arms are extended downward and outward, then inward and up, as one would do in grasping the legs of the tripod and shouldering the instrument. It is given by the chief of the party or by the head chainman when the transit is to be moved.

Care should be taken to make the signals so clear that they may be readily understood. Where long sights are taken or where the peculiar color of the background renders hand signals indistinct, colored flags similar to those of railroad trainmen may be used to good advantage. Of course the color should be in contrast with that of the background. Red can be seen very well against snow, and white can be distinguished clearly against the dark green of the forest.

* *From Raymond E. Davis, Manual of Surveying for Field and Office, 1915.*

INDEX